The HTML and CSS Workshop

A New, Interactive Approach to Learning HTML and CSS

Lewis Coulson, Brett Jephson, Rob Larsen, Matt Park, and Marian Zburlea

The HTML and CSS Workshop

Authors: Lewis Coulson, Brett Jephson, Rob Larsen, Matt Park, and Marian Zburlea

Reviewers: Tarunkumar Bagmar, Tiffany Ford, Terry O'Brien, Kurri Sudarshan Reddy, and Adam Rosson

Managing Editors: Anush Kumar Mehalavarunan and Anushree Arun Tendulkar

Acquisitions Editors: Sarah Lawton and Sneha Shinde

Production Editor: Shantanu Zagade

Editorial Board: Shubhopriya Banerjee, Bharat Botle, Ewan Buckingham, Megan Carlisle, Mahesh Dhyani, Manasa Kumar, Alex Mazonowicz, Bridget Neale, Dominic Pereira, Shiny Poojary, Abhishek Rane, Erol Staveley, Ankita Thakur, Nitesh Thakur, and Jonathan Wray

First Published: November 2019

Production Reference: 4110620

ISBN: 978-1-83882-453-2

Published by Packt Publishing Ltd.

Livery Place, 35 Livery Street

Birmingham B3 2PB, UK

Experience the Workshop Online

Thank you for purchasing the print edition of *The HTML and CSS Workshop*. Every physical print copy includes free online access to the premium interactive edition. There are no extra costs or hidden charges.

With the interactive edition you'll unlock:

- **Screencasts and Quizzes.** Supercharge your progress with screencasts of all exercises and activities. Take optional quizzes to help embed your new understanding.

- **Built-In Discussions.** Engage in discussions where you can ask questions, share notes and interact. Tap straight into insight from expert instructors and editorial teams.

- **Skill Verification.** Complete the course online to earn a Packt credential that is easy to share and unique to you. All authenticated on the public Bitcoin blockchain.

- **Download PDF and EPUB.** Download a digital version of the course to read offline. Available as PDF or EPUB, and always DRM-free.

To redeem your free digital copy of *The HTML and CSS Workshop* you'll need to follow these simple steps:

1. Visit us at https://courses.packtpub.com/pages/redeem.

2. Login with your Packt account, or register as a new Packt user.

3. Select your course from the list, making a note of the three page numbers for your product. Your unique redemption code needs to match the order of the pages specified.

4. Open up your print copy and find the codes at the bottom of the pages specified. They'll always be in the same place:

Accessible Images

As the saying goes, a picture is worth a thousand words, and images can add a lot to a web page. Some pictures are decorative, whereas others are an important piece of content that gets your web page's message across with great impact.

Every time we add an image to a web page, there are accessibility considerations we should take into consideration in order to make an accessible image. Not all users may be able to see an image. For example, if a user requires a screen reader or other form of non-visual browser to be able to navigate through a web page, they will require a textual description of an image to be able to garner any meaning from it. It can also be the case that a user does not download images because of limitations on network bandwidth or for security reasons. All of these users will benefit from an alternative text description of an image.

A B 2 1 C

---

5. Merge the codes together (without any spaces), ensuring they are in the correct order.

6. At checkout, click **Have a redemption code?** and enter your unique product string. Click **Apply**, and the price should be free!

Finally, we'd like to thank you for purchasing the print edition of *The HTML and CSS Workshop*! We hope that you finish the course feeling capable of tackling challenges in the real world. Remember that we're here to help if you ever feel like you're not making progress.

If you run into issues during redemption (or have any other feedback) you can reach us at workshops@packt.com.

Table of Contents

Chapter 2: Structure and Layout   61

Chapter 3: Text and Typography   105

Chapter 6: Responsive Web Design and Media Queries 231

Chapter 8: Animations   319

Chapter 9: Accessibility   363

Chapter 12: Web Components

Chapter 13: The Future of HTML and CSS

Preface

About

This section briefly introduces the coverage of this book, the technical skills you'll need to get started, and the software requirements required to complete all of the included activities and exercises.

About the Book

You already know you want to learn HTML and CSS, and a smart way to learn HTML and CSS is to learn by doing. *The HTML and CSS Workshop* focuses on building up your practical skills so that you can build your own static web pages from scratch, or work with existing themes on modern platforms such as WordPress. It's the perfect way to get started with web development. You'll learn from real examples that lead to real results.

Throughout *The HTML and CSS Workshop*, you'll take an engaging step-by-step approach to beginning HTML and CSS development. You won't have to sit through any unnecessary theory. If you're short on time, you can jump into a single exercise each day or spend an entire weekend learning about CSS preprocessors. It's your choice. Learning on your terms, you'll build up and reinforce key skills in a way that feels rewarding.

Every physical copy of *The HTML and CSS Workshop* unlocks access to the interactive edition. With videos detailing all the exercises and activities, you'll always have a guided solution. You can also benchmark yourself against assessments, track your progress, and receive free content updates. It's a premium learning experience that's included with your printed copy. To redeem this, follow the instructions located at the start of your HTML CSS book.

Fast-paced and direct, *The HTML and CSS Workshop* is the ideal companion for an HTML and CSS beginner. You'll build and iterate on your code like a software developer, learning along the way. This process means that you'll find that your new skills stick, embedded as best practice, and this will constitute a solid foundation for the years ahead.

About the Chapters

Chapter 1, Introduction to HTML and CSS, introduces you to the two foundation technologies of the web – HTML and CSS. We'll go over the syntax of both and look at how they are combined to make a web page.

Chapter 2, Structure and Layout, introduces you to the structural elements in HTML, including header, footer, and section tags. You will also learn the three main CSS layout techniques – float, flex, and grid.

Chapter 3, Text and Typography, introduces you to text-based elements, such as paragraphs, headings, and lists. We will go over how to style text-based elements in web pages and let your creativity run wild.

Chapter 4, Forms, introduces you to the creation of web forms, starting with the key HTML elements used in forms and then learning how to style them with CSS. Along the way, you will also learn how to style your forms with validation styling.

Chapter 5, Themes, Colors, and Polish, introduces you to the practical world of CSS. Applying what you've learned so far, you will craft multiple themes to apply to a website in order to change the look and feel.

Chapter 6, Responsive Web Design and Media Queries, introduces you to the world of mobile-first web design, building web pages that can adapt to different screen sizes in order to give the user a better browsing experience, regardless of their browser size. This chapter will teach you how to use modern responsive web design techniques using a CSS feature called Media Queries.

Chapter 7, Media – Audio, Video, and Canvas, introduces you to HTML's audio, video, and canvas elements, all of which can be used to make our web pages a richer, more interactive experience.

Chapter 8, Animations, introduces you to CSS animation, which can be used to animate elements on a page or application to add moving elements that can add richness to your page and tell a complex story.

Chapter 9, Accessibility, introduces you to the important topic of accessibility. The web is meant for everyone, and we can use a variety of simple techniques to make it accessible.

Chapter 10, Preprocessors and Tooling, introduces you to the world of CSS preprocessing and explains how to write a CSS preprocessing language called SCSS (Sassy Cascading Style Sheets). With these new skills, you will be able to compile SCSS into CSS, creating more maintainable code by writing less in order to achieve more.

Chapter 11, Maintainable CSS, introduces you to a variety of different techniques that we can use to create CSS that is easy to maintain, including writing semantic markup using Block Element Modifier (BEM), making reusable components in CSS, grouping CSS rules, and structuring SCSS files correctly.

Chapter 12, Web Components, introduces you to several technologies – custom elements, HTML templates, and the shadow DOM – that can be combined into composable, reusable, and shareable web components.

Chapter 13, The Future of HTML and CSS, introduces you to the cutting edge of web technologies through the CSS Paint API and looks at how we can make use of progressive enhancements to open up future technologies to us now.

Conventions

Code words in text, database table names, folder names, filenames, file extensions, pathnames, dummy URLs, user input, onscreen text, and Twitter handles are shown as follows: "The **head** element is home to most machine-read information in an HTML document."

A block of code is set as follows:

```
<!doctype html>
<html lang="en">
    <head>
        <title>Page Title</title>
    </head>
</html>
```

New terms and important words are shown like this: "We can also represent the HTML document as a **Document Object Model (DOM)**."

Long code snippets are truncated and the corresponding names of the code files on GitHub are placed at the top of the truncated code. The permalinks to the entire code are placed below the code snippet. It should look as follows:

Example 6.01.html

```
13   @media (orientation: landscape) {
14      p.warning {
15         display: block;
16      }
17   }
```

The complete code for this example is available at: https://packt.live/2JXLsyM

Before You Begin

Each great journey begins with a humble step. Our upcoming adventure in the land of HTML and CSS is no exception. Before we can do awesome things with data, we need to be prepared with a productive environment. In this section, we shall see how to do that.

Installing Chrome

In this book, we have used Chrome version 76, which was the latest version at the time of publication. Make sure you have downloaded and installed the latest version of Chrome to ensure an optimal experience.

To install Chrome on Windows, follow the instructions given here: https://www.google.com/chrome/

For other platforms such as macOS and Linux, scroll down to the bottom left-hand corner of this website: https://www.google.com/chrome/ and click on Other Platforms, and choose accordingly.

Installing Visual Studio Code

Here are the steps to install Visual Studio Code (VSCode):

1. Download the latest VSCode from https://packt.live/2BIlniA:

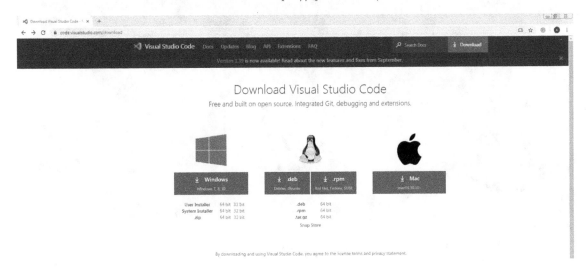

Figure 0.1: Downloading VSCode

2. Open the downloaded file, follow the installation steps, and complete the installation process.

Installing the "Open in Default Browser" Extension

1. Open your VSCode, click on the **Extensions** icon, and type in `Open In Default Browser` in the search bar, as shown in the following screenshot:

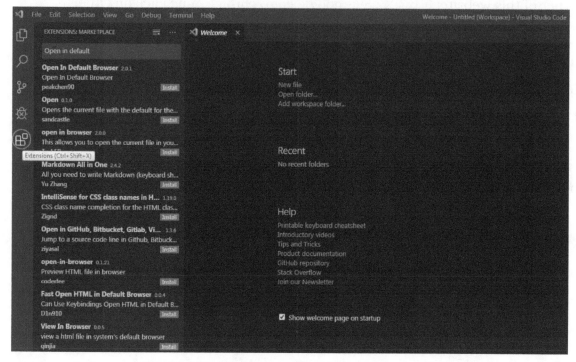

Figure 0.2: Open in Default Browser extension search

2. Click on **Install** to complete the installation process, as shown in the following screenshot:

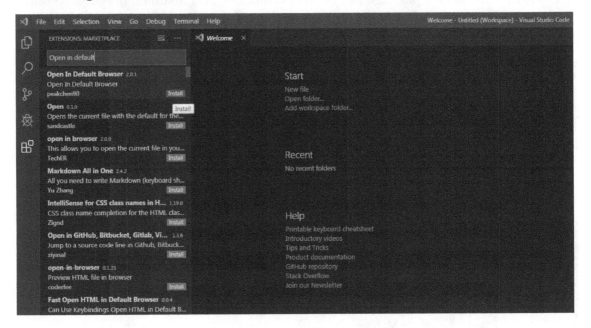

Figure 0.3: Installing the extension

Installing the web sever

Some exercises in the book will require the use of a personal web server to run the scripts and view the pages you'll be creating. If you do not have a personal web server already set up, you can use the following steps to get one installed.

To begin, you will need to go to https://www.apachefriends.org/index.html and download the XAMPP package for your operating system. XAMPP is available for Windows, Mac, and Linux.

Instructions for Windows

1. Once the web server package has been downloaded and installed on your system, open the XAMPP Control Panel and check the status of your web server. If **Apache** is highlighted in green, then the web server is running:

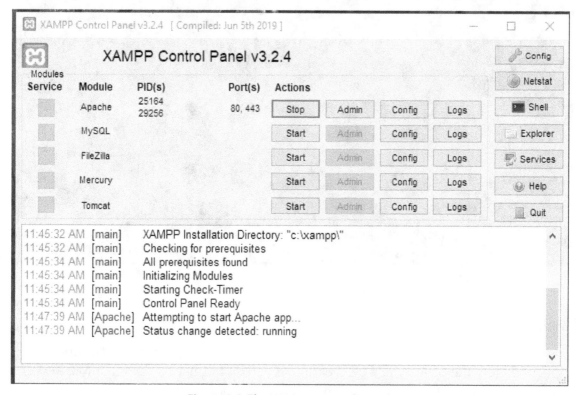

Figure 0.4: The XAMPP Control Panel

If it is not highlighted in green, then you will need to start the Apache web server:

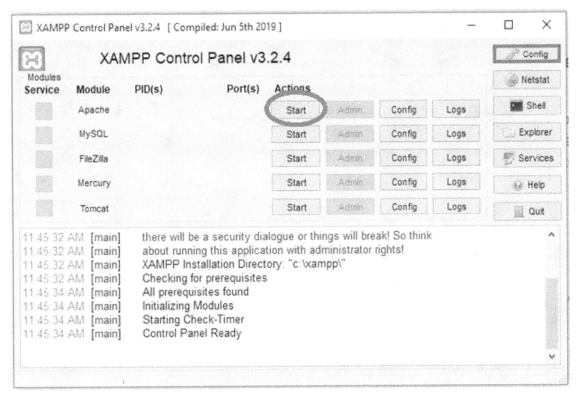

Figure 0.5: Starting the Apache Web Server for Windows

2. Click the first **Start** button in the control panel that corresponds to Apache. The web server will then initialize and start so you can use it.

3. To check if your web server is running correctly, open your browser, type **localhost** into the browser's URL bar and then hit *Enter*:

Figure 0.6: Navigating to localhost in the browser

If your installation was successful and Apache is running, you will get the XAMPP welcome page:

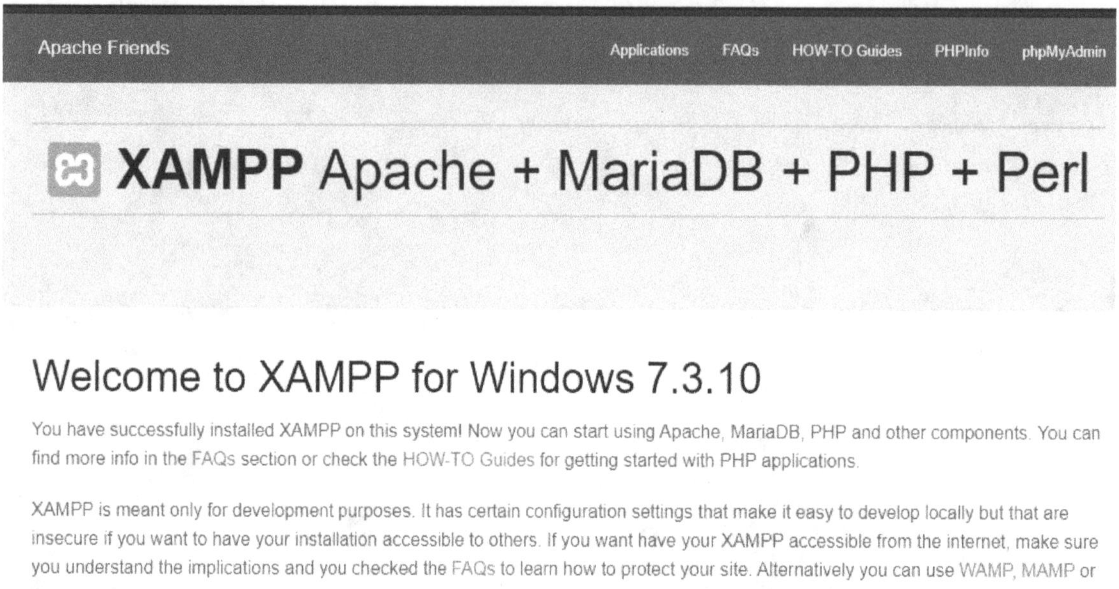

Apache Friends   Applications   FAQs   HOW-TO Guides   PHPInfo   phpMyAdmin

XAMPP Apache + MariaDB + PHP + Perl

Welcome to XAMPP for Windows 7.3.10

You have successfully installed XAMPP on this system! Now you can start using Apache, MariaDB, PHP and other components. You can find more info in the FAQs section or check the HOW-TO Guides for getting started with PHP applications.

XAMPP is meant only for development purposes. It has certain configuration settings that make it easy to develop locally but that are insecure if you want to have your installation accessible to others. If you want have your XAMPP accessible from the internet, make sure you understand the implications and you checked the FAQs to learn how to protect your site. Alternatively you can use WAMP, MAMP or

Figure 0.7: Viewing the XAMPP Welcome Page

4. Next, open your XAMPP directory so you understand where to place your files when you want to use the web server to view your pages. From the XAMPP Control Panel, click the **Explorer** tab:

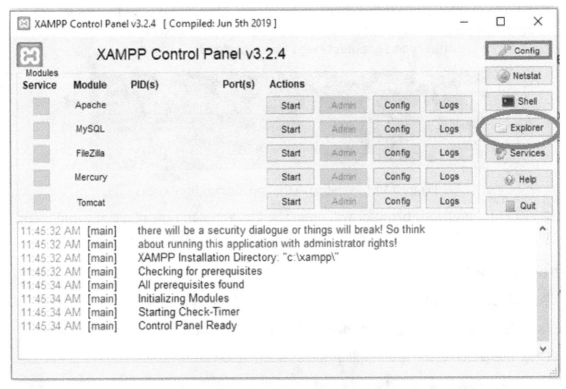

Figure 0.8: Navigating to the File Explorer

5. This will open the Windows Explorer and show the files used by XAMPP for running your web server. Locate the folder named **htdocs**. This is the folder that holds any files that you want to view in the browser:

This PC > Local Disk (C:) > xampp > htdocs

Name	Date
dashboard	10/19/2019 1:42 I
img	10/19/2019 1:42 I
webalizer	10/19/2019 1:42 I
xampp	10/19/2019 1:42 I
favicon.ico	7/16/2015 10:32 ,
index.php	7/16/2015 10:32 ,
applications.html	8/27/2019 9:02 A
bitnami.css	8/27/2019 9:02 A
sample.zip	3/6/2020 1:24 PM

Figure 0.9: Viewing the Web Server Folders and Files

6. Any files you place into this folder can be accessed through your localhost. For example, we see the file **applications.html**. If we want to access this file in the browser, we would type **localhost/applications.html**:

Figure 0.10: Navigating to the applications.html Page

7. Press *Enter* and your browser will open the file. You can now place any folders or files you wish to open using the web server, by placing them into this directory.

Instructions for Mac

1. Download MAC file from this link: https://www.apachefriends.org/index.html.

2. Double click the downloaded .dmg file (for example: **xampp-osx-7.4.5-0-vm.dmg**)

3. Drag the DMG image into application as seen below:

Figure 0.11: XAMPP Installation screen

4. To start XAMPP, go to **Finder** -> **Applications** -> **XAMPP** OR press CMD + *Space* (Spotlight search) and type "**XAMPP**". You will see the following result:

Figure 0.12: MAC Finder screen with Applications folder view

5. Click on the installer. You will get the following dialog box:

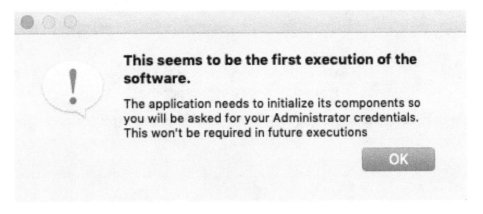

Figure 0.13: First time installation screen

6. Click **OK** on this screen and then it will ask your password in the next screen. Once you enter your password, the next screen will appear which looks like this:

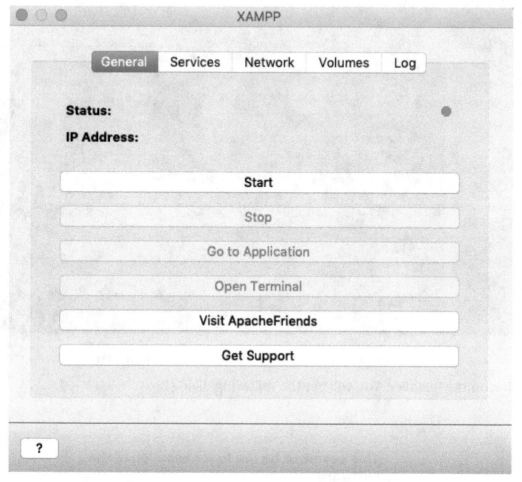

Figure 0.14: XAMPP Control panel

7. Click on the **Start** button to start the service (observe the red dot and IP address is empty)

8. After you click on the **Start** button, you will see green-dot and IP address as shown in the following figure:

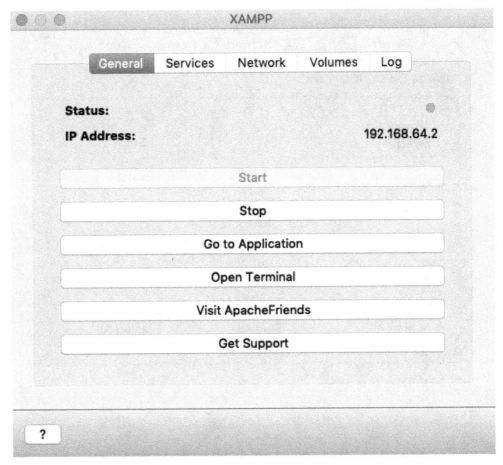

Figure 0.15: XAMPP control panel with active "General" Tab

9. To see status of all the services, click on the **"Services"** tab and check status:

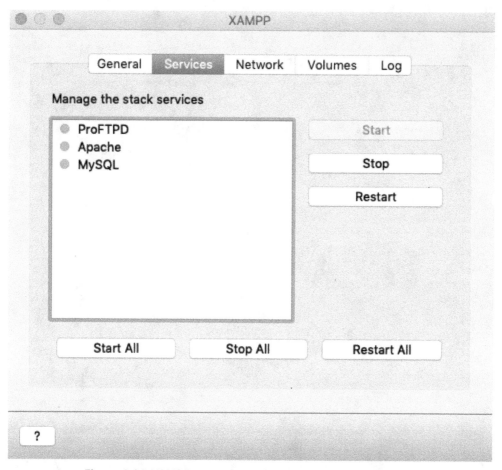

Figure 0.16: XAMPP control panel with active "Services" Tab

10. Now, click back on "**General**" tab and click on "**Go To Application**" button OR copy the IP address and go to your browser and hit enter. It will open default (welcome) page of XAMPP in your default web-browser. Go through this page, it will give you basic information:

Welcome to XAMPP for 7.4.5-0

You have successfully installed XAMPP on this system! Now you can start using Apache, MariaDB, PHP and other components. You can find more info in the FAQs section or check the HOW-TO Guides for getting started with PHP applications.

XAMPP is meant only for development purposes. It has certain configuration settings that make it easy to develop locally but that are insecure if you want to have your installation accessible to others. If you want have your XAMPP accessible from the internet, make sure you understand the implications and you checked the FAQs to learn how to protect your site. Alternatively you can use WAMP, MAMP or LAMP which are similar packages which are more suitable for production.

Start the XAMPP Control Panel to check the server status.

Community

XAMPP has been around for more than 10 years – there is a huge community behind it. You can get involved by joining our Forums, adding yourself to the Mailing List, and liking us on Facebook, following our exploits on Twitter, or adding us to your Google+ circles.

Contribute to XAMPP translation at translate.apachefriends.org.

Can you help translate XAMPP for other community members? We need your help to translate XAMPP into different languages. We have set up a site, translate.apachefriends.org, where users can contribute translations.

Install applications on XAMPP using Bitnami

Apache Friends and Bitnami are cooperating to make dozens of open source applications available on XAMPP, for free. Bitnami-packaged applications include Wordpress, Drupal, Joomla! and dozens of others and can be deployed with one-click installers. Visit the Bitnami XAMPP page for details on the currently available apps.

Figure 0.17: XAMPP Welcome screen

11. If you wish to use localhost instead of IP, then click on "**Network**" tab and enable the below port:

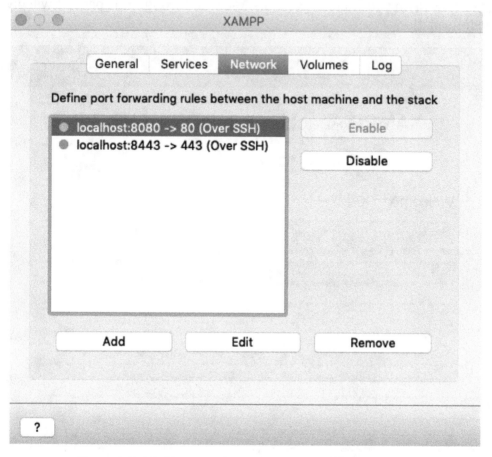

Figure 0.18: XAMPP control panel with active "Network" Tab

12. Now go to your web-browser and type-in **localhost:8080** (as visible in the above figure) and you will see the same "Welcome" screen.

13. To place your files into XAMPP, click on the **"Volumes"** tab and click on the **"Mount"** button. Once clicked, the other two options **"Unmount"** & **"Explore"** will be activated for you. Now click on **"Explore"** button to go to the LAMPP folder:

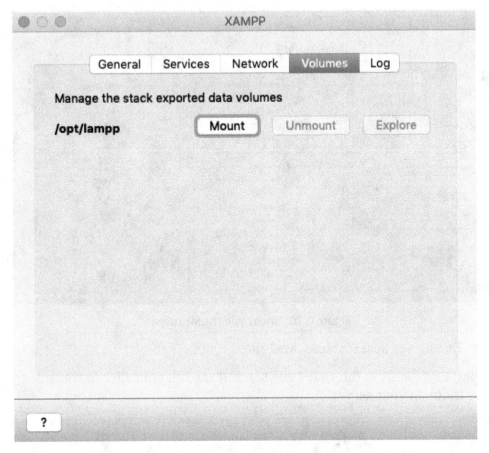

Figure 0.19: XAMPP control panel with active "Volumes" Tab

14. When you click on "**Explore**" button, you see Finder window will open and search for "**htdocs**" folder as shown below:

Figure 0.20: Finder with htdocs view

There you can see **applications.html** file.

15. Navigate that file from the web-browser and save all your files in this folder to access them on your server:

Figure 0.21: XAPP welcome screen

16. To stop the server, go to the XAMPP server and in "**General**" tab, click "**Stop**" button.

Instructions for Linux

1. Download the Linux installer and find that in your **Downloads** folder.

2. Open terminal into **Downloads** and type the following codes:

```
chmod 755 xampp-linux-x64-7.4.5-0-installer.run
sudo ./xampp-linux-x64-7.4.5-0-installer.run
```

3. Enter your password and you will see the following screen:

Figure 0.22: Linux setup (installation) screen

4. Click on the **Next** button and follow other instructions, until you see the installation process starting as shown in the following figure:

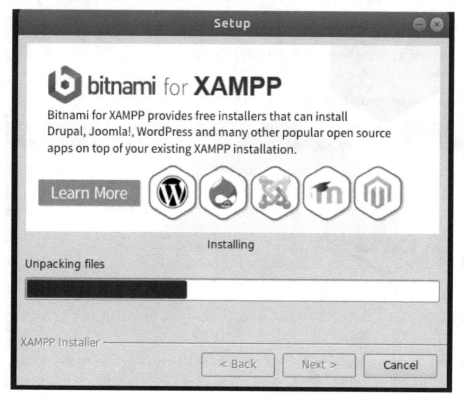

Figure 0.23: Linux setup (installation) screen

5. Wait for this process to complete. Once done, the following screen will appear which states that the setup has finished:

Figure 0.24: Linux setup (installation) screen

6. Click on the **Finish** button to launch the XAMPP. XAMPP Control panel will look like this:

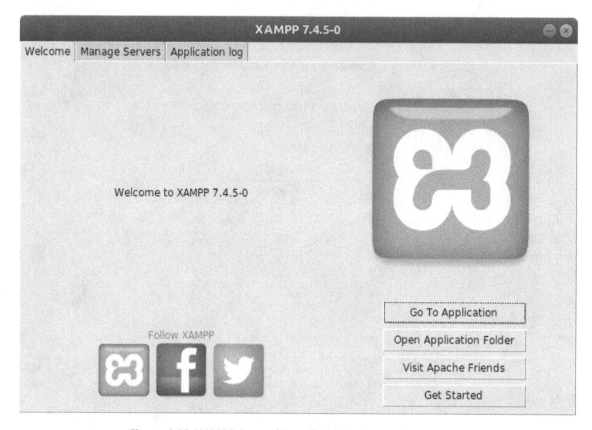

Figure 0.25: XAMPP Control Panel with active "Welcome" tab

7. Start the server first by selecting "**Manage Servers**" tab and clicking on **Start All** button:

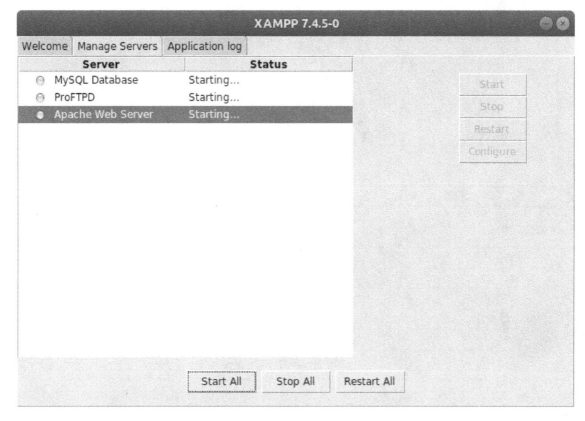

Figure 0.26: XAMPP Control Panel with active "Manage Servers" tab

8. Once they are started, you will see green light in front of each of them.

9. From the **Welcome** tab, click on the "**Open Application Folder**", which will open "**/opt/lampp**" folder. Search for "**htdocs**" folder and click on it. You will see **applications.html** file.

10. Run localhost from your favorite web-browser and type-in "**http://localhost/ applications.html**". You will see Apache Welcome screen. From the control panel you can start/stop any services.

Installing the Code Bundle

Download the code files from GitHub at https://packt.live/2N7M3yC and place them in a new folder called **C:\Code**. Refer to these code files for the complete code bundle.

If you face any trouble with installation or with getting the code up and running, please reach out to the team at workshops@packt.com.

Introduction to HTML and CSS

Overview

By the end of this chapter, you will be able to describe the roles of HTML and CSS in a web page; explain the HTML DOM and CSSOM; explain how a web page renders; create a web page from scratch; and use CSS selectors and calculate CSS specificity. This chapter introduces two core technologies of the web – HTML and CSS. We will look at how they work, how we can write them, and how we can use them to build web pages.

Introduction

Whether you want to create a simple web page to advertise a business, blog about your hobbies and interests, maintain an online community, or even create your own social media network, HTML and CSS are the foundational technologies upon which you can build for the web and are a way for you to get your ideas out there to as wide an audience as possible.

When a web browser navigates to a web page, it will receive and parse an HTML document, which may include text, pictures, links, and other media (for instance, sound and video).

HTML structures this content. It gives it context, describes the content and tells the browser what to do with it. CSS tells the browser how to present the content. A set of styling rules lets the browser know how to render elements within the HTML document. HTML and CSS together give the browser the information it needs to render the web page for the user.

Navigate to a website and what you see is the rendered output of content marked up with HTML and styled with CSS. As a browser user, you have access to the source code of a web page. In Chrome, for example, you can view a page's source code with the keyboard shortcut *Ctrl* + U on a PC or *Cmd* + U on a Mac, or right-click and choose **View Page Source**. Try it yourself. As an example, the following two figures show what the Packt website's Web Development portal looks like when rendered in the browser and as source code respectively.

Ultimately, by learning how to write the HTML and CSS found in that source code, we can create a modern website:

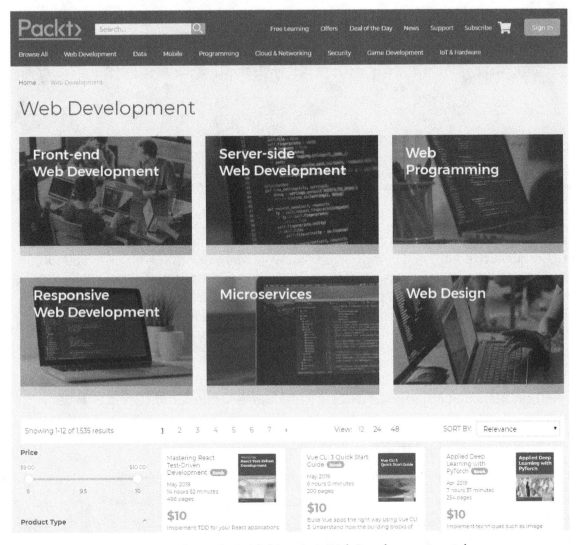

Figure 1.1: The Packt Publishing site's Web Development portal

The following figure shows the source code of the Packt website:

Figure 1.2: The HTML source code of the Packt site

In this chapter, we will look at how a web page renders by following the process from initial request to completed composition. We will create our first web page and look at how a web browser will parse HTML and CSS to render that page. We will look at how browser developer tools, such as those included with the Chrome browser, can help us to identify and edit nodes within an HTML document and the CSS applied to those nodes.

HTML

HyperText Markup Language (**HTML**) is a markup language used to describe the structure of a web page.

Consider a snippet of text with no markup:

```
HTML HyperText Markup Language (HTML) is a markup language used to describe the
structure of a web page. We can use it to differentiate such content as headings
lists links images Want to https://www.packtpub.com/web-development Learn more
about web development.
```

The above snippet of text may make some sense to you, but it may also raise some questions. Why does the snippet begin with the word HTML? Why is there a URL in the middle of a sentence? Is this one paragraph?

Using HTML, we can differentiate several bits of content to give them greater meaning. We could mark the word HTML as a heading, `<h1>HTML</h1>`; we could mark a link to another web page using the URL `<a href="https://www.packtpub.com/web-development">Learn more about web development</a>`.

Throughout this chapter, we will be looking at the HTML5 version of the HTML language. We will look at the syntax of HTML in the next section.

Syntax

The syntax of HTML is made up of **tags** (with angle brackets, <>) and attributes. HTML provides a set of tags that can be used to mark the beginning and end of a bit of content. The opening tag, closing tag, and all content within those bounds represent an HTML element. The following figure shows the HTML element representation without attributes:

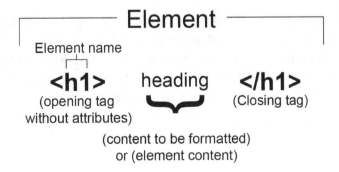

Figure 1.3: HTML element representation without tag attributes

The following figure shows the HTML element representation with tag attributes:

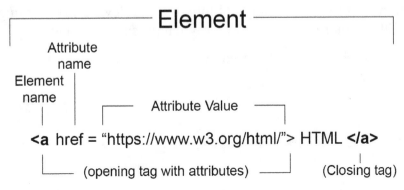

Figure 1.4: HTML element representation with tag attributes

A tag has a name (for instance, **p**, **img**, **h1**, **h2**, **h3**, **br**, or **hr**) and that name combined with attributes will describe how the browser should handle the content. Many tags have a start and an end tag with some content in between, but there are also tags that don't expect any content, and these can be self-closing.

An opening tag can have any number of **attributes** associated with it. These are modifiers of the element. An attribute is a name-value pair. For example, `href="https://www.packtpub.com/web-development"` is an attribute with the name `href` and the value `https://www.packtpub.com/web-development`. An `href` attribute represents a hypertext reference or a URL, and when this attribute is added to an anchor element, `<a>`, it creates a hyperlink that the user can click in the browser to navigate to that URL.

To provide information within an HTML document to be ignored by the parser and not shown to the end user, you can add **comments**. These are useful for notes and documentation to aid anyone who might read or amend the source of the HTML document. A comment begins with `<!--` and ends with `-->`. Comments, in HTML, can be single or multiline. The following are some examples:

```
<!-- Comment on a single line -->


<!--

    This comment is over multiple lines.
    Comments can be used to inform and for detailed documentation.
-->
```

You can use comments to provide helpful hints to other developers working on the web page but they will be ignored by the browser when parsing the page.

Let's see what the previous snippet of text content looks like when it is given some meaning with HTML:

```html
<h1>HTML</h1>
<p>
HyperText Markup Language (HTML) is a markup language used to describe the
structure of a web page.
</p>
<p>
We can use it to differentiate such content as:
</p>
<ul>
    <li>headings</li>
    <li>lists</li>
    <li>links</li>
    <li>images</li>
</ul>
<p>
Want to <a href="https://www.packtpub.com/web-development">learn more about web
development?</a>
</p>
```

If we were to look at this HTML code rendered in a browser, it would look like the following figure:

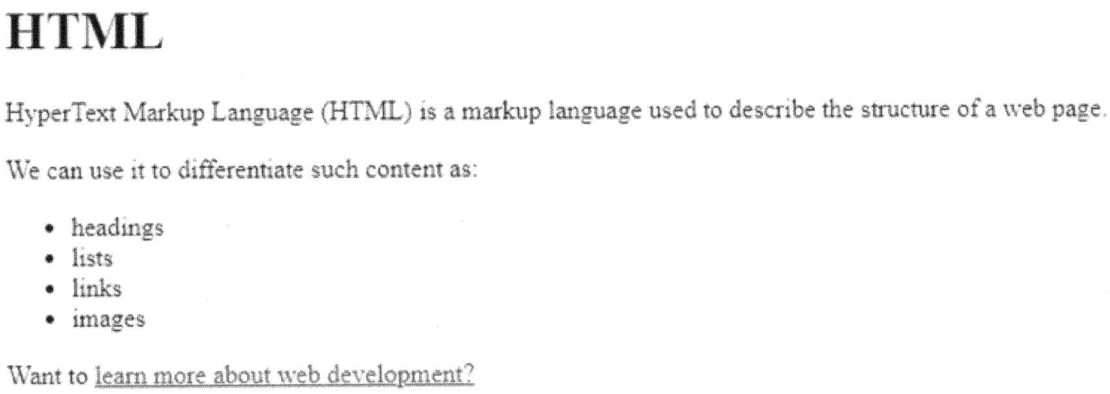

HTML

HyperText Markup Language (HTML) is a markup language used to describe the structure of a web page.

We can use it to differentiate such content as:

- headings
- lists
- links
- images

Want to learn more about web development?

Figure 1.5: HTML rendered in the Google Chrome web browser

The first line shows the text content "HTML" with a start tag, **<h1>**, and an end tag, **</h1>**. This tells the browser to treat the text content as an **h1** heading element.

The next line of our code snippet has a **<p>** start tag, which means the content until the corresponding end tag, **</p>** (on the last line), will be treated as a **paragraph** element. We then have another paragraph and then an unordered list element that starts with the **<ul>** start tag and ends with the **</ul>** end tag. The unordered list has four child elements, which are all list item elements (from the **<li>** start tag to the **</li>** end tag).

The last element in the example is another paragraph element, which combines text content and an anchor element. The anchor element, starting from the **<a>** start tag and ending at the **</a>** end tag, has the text content **learn more about web development?** and an **href** attribute. The **href** attribute turns the anchor element into a hyperlink, which a user can click to navigate to the URL given as the value of the **href** attribute.

As with our example, the contents of a paragraph element might be text but can also be other HTML elements, such as an anchor tag, **<a>**. The relationship between the anchor and paragraph elements is a parent-child relationship.

Content Types

HTML5 provides many elements for describing the metadata, content, and structure of an HTML document, and you will learn more about the meaning of specific elements throughout the following chapters.

When starting out with HTML, it can be easy to find the number and variety of elements overwhelming. That is why it may be helpful to think about HTML in terms of content types. We can categorize HTML elements as one of the following content types:

- Metadata

- Flow

- Sectioning

- Phrasing

- Heading

- Embedded

- Interactive

The following table has the description and example of different content types:

Type	Description	Example
Metadata	Content hosted in the head of an HTML document. Doesn't appear in the web page directly but is used to describe a web page and its relationship to other external resources.	`<meta name="viewport" content="width=device-width,initial-scale=1.0">`

Flow	Text and all elements that can appear as content in the body of a HTML document.	`<body>`
		`  <h1>Heading</h1>`
		`  <p>Some content...</p>`
		`</body>`

Sectioning	Used to structure the content of a web page and to help with layout. Elements in this category are described in Chapter 2, Structure and Layout.	`<aside></aside>`
		`<article class="blog-post">`
		`  <section></section>`
		`</article>`

Phrasing	Elements such as those used for marking up content within a paragraph element. Chapter 3, Text and Typography, will be largely concerned with this content type.	`<p><em>Emphasized text</em> and some normal text.</p>`
Heading	Elements used to define the headings of a section of an HTML document. The elements h1-6 represent headings with h1 having the highest ranking.	`<h1>Main Heading</h1>` `<h2>Subheading</h2>`
Embedded	Embedded content includes media, such as video, audio, and images. We will look at these in Chapter 7, Media – Audio, Video, and Canvas.	`<img src="media/kitten.png" alt="A cute kitten">`
Interactive	Elements that a user can interact with, which includes media elements with controls, form inputs, buttons, and links.	`<input type="password" name="password" required>`

Figure 1.6: Table describing the different content types

To see how we can use these categories, we will introduce an HTML5 element and see how it will fit into these category types. We will look at the **<img>** element.

If we want to embed an image in our web page, the simplest way is to use the **img** element (for more on images, see *Chapter 7, Media, Audio, Video, and Canvas*). If we want to create an **img** element, an example of the code looks like this: **<img src="media/ kitten.png" alt="A cute kitten">**.

We set the **src** attribute on the **img** element to an image URL; this is the source of the image that will be embedded in the web page.

Unless your image has no value other than as a decoration, it is a very good idea to include an **alt** attribute. The **alt** attribute provides an alternative description of the image as text, which can then be used by screen readers if an image does not load, or in a non-graphical browser.

An **img** element is a form of **embedded** content because it embeds an image in an HTML document. It can appear in the body of an HTML document as the child element of the **body** element, so it would be categorized as **flow** content.

An image can be included as content in a paragraph, so it is a type of **phrasing** content. For example, we could have inline images appear in the flow of a paragraph:

```
<p>Kittens are everywhere on the internet. The best thing about kittens is that
they are cute. Look here's a kitten now: <img src="media/kitten.jpg" alt="A cute
kitten">. See, cute isn't it?</p>
```

This code would render the following figure, with the image embedded in the paragraph and the rest of the text flowing around it:

Kittens are everywhere on the internet. The best thing about kittens is that they are

cute. Look here's a kitten now: . See, cute isn't it?

Figure 1.7: Image with text flowing around it

In certain circumstances, an **img** element is a type of **interactive** content. For this to be the case, the image must have a **usemap** attribute. The **usemap** attribute allows you to specify an image map, which defines areas of an image that are treated as hyperlinks. This makes the image interactive.

An **img** element does not act as **metadata** and it does not provide a **sectioning** structure to an HTML document. Nor is it a **heading**.

There are lots more elements in the HTML5 standard. Elements can appear in more than one category and there is some overlap between the relationships of the categories. Some of these elements are very common and are used often, but some of these elements have very specific purposes and you may never come across a use case for them.

The content types can be useful for grouping elements into more manageable chunks, for getting an overview of the choices HTML gives you and the restrictions it puts on the content of an element, and for understanding the content types generally before we drill down into the uses of specific elements in later chapters.

For further reference, we can see where each available element is categorized in the W3C's documentation on HTML5: https://packt.live/2OvPGRi.

The HTML Document

An HTML document represents a hierarchical tree structure, rather like a family tree. Starting from a **root** element, the relationship between an element and its contents can be seen as that of a parent element and a child element. An element that is at the same level of the hierarchy as another element can be considered a sibling to that element, and we can describe elements within a branch of the tree as ancestors and descendants. This structure can be represented as a tree diagram to get a better idea of the relationship between elements.

Take, for example, this simple HTML document:

```html
<html>
  <head>
    <title>HTML Document structure</title>
  </head>
  <body>
    <div>
      <h1>Heading</h1>
      <p>First paragraph of text.</p>
      <p>Second paragraph of text.</p>
    </div>
  </body>
</html>
```

Here, we have an HTML element, the root of the document, which hosts a **head** element containing a **title** element, and a **body** element, containing some content including a **div** element with an **h1** heading element and some paragraph elements. It can be represented as a tree diagram as follows:

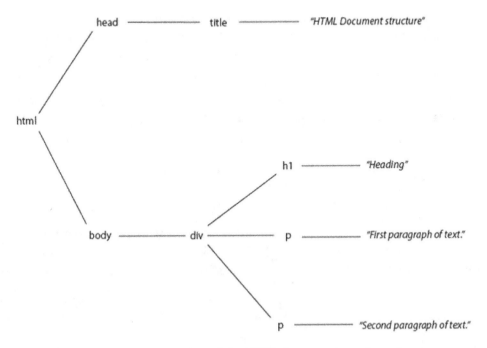

Figure 1.8: A representation of the HTML document as a tree diagram

In the browser, this code would render the following web page:

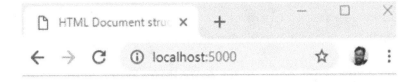

Heading

First paragraph of text.

Second paragraph of text.

Figure 1.9: HTML rendered in the Google Chrome web browser

The **<html>** element is the parent of **<head>** and **<body>**, which (as children of the same parent) are siblings. **<body>** has one child, a **<div>** element, and that has three children: an **<h1>** element and two **<p>** elements. The **<h1>** element is a descendant of the **<body>** element but not of the **<head>** element.

Understanding this structure will become more important when we look at CSS selectors and how we target parts of the HTML document later in this chapter.

The HTML DOM

We can also represent the HTML document as a **Document Object Model** (**DOM**). The DOM is an in-memory representation of our HTML document as a tree of objects. The tree is represented as a set of nodes and their connections to other nodes.

A **node** is associated with an object. The object stores properties, methods, and events associated with the HTML element. The node represents where that element sits in the document tree structure by storing references to its parent and child nodes.

When we want to change the document programmatically, as is often the case when we use JavaScript, we do so through the DOM. We can traverse the document programmatically and make changes to elements by changing properties on the objects.

As an example, we can use JavaScript's DOM API to create an anchor element, change properties on the element, and then add it to a paragraph with the **parent** class attribute:

```
<script>
   const anchorElement = document.createElement('a');
   anchorElement.href = '#';
   anchorElement.textContent = 'Click me!';
   const p = document.querySelector('.parent');
   p.appendChild(anchorElement);
</script>
```

The DOM represents the HTML document in a way that means we can traverse the tree and make changes to it programmatically. This allows modern web pages to be more than static documents.

We will mostly concentrate on the HTML and CSS parts of the web page but will see how we can use JavaScript to enhance and manipulate the HTML document in later chapters.

The Doctype Declaration

To let the browser, know which version of HTML to expect on our web page, we provide a doctype declaration. There are multiple versions of HTML and, at the time of writing, we are on version 5.2. Each version of HTML has a different declaration.

We will be working with HTML5 in this book so the appropriate doctype is as follows:

```
<!DOCTYPE html>
```

> **Note**
>
> The doctype declaration is not case-sensitive, so variations such as <!doctype html> and <!DOCTYPE HTML> are equally valid.

Without a **doctype** declaration, the browser can still try to parse an HTML document but it will do so in something called **quirks** mode. The effects of quirks mode can be different depending on the browser and whether the HTML document complies with the standard or not. This mode is there for backward compatibility and to handle old web pages and may cause your web page to render in unexpected ways. It is recommended to always add a doctype.

One of the nice things about HTML5 is that it really simplifies doctype declaration. Before HTML5, there were two commonly used variations of web markup – HTML4 and XHTML1 – and they both had strict, transitional, and frameset versions of their doctype declarations. For example, the HTML 4 strict declaration looked like this: `<!DOCTYPE HTML PUBLIC "-//W3C//DTD HTML 4.01//EN" "http://www.w3.org/TR/html4/strict.dtd">`.

The HTML5 doctype simplifies things and makes it a lot easier to get started creating a web page.

Structuring an HTML Document

A requirement of HTML5 is that all HTML documents are structured with a root element, the `html` element, and two child elements, the `head` element, and the `body` element.

HTML

The **html** element is the root element of an HTML document. In an HTML document, the only thing that appears outside the **html** element is the doctype declaration, which appears above the **html** element. Before we put any content on our web page, the code looks like this:

```
<!doctype html>
<html lang="en"></html>
```

The **html** element is the top-level element and all other elements must be a descendant of it. The **html** element has two children – one **head** element and a **body** element, which must follow the **head** element.

It is strongly recommended that you add a **lang** attribute to your **html** element to allow the browser, screen readers, and other technologies, such as translation tools, to better understand the text content of your web page.

Head

The **head** element contains metadata. It stores information about the HTML document that is used by machines – browsers, web crawlers, and search engines – to process the document. This section of an HTML document is not rendered for human users to read.

The minimum content the **head** element can have is a **title** element. Within a web page, the code would look like this:

```
<!doctype html>
<html lang="en">
    <head>
        <title>Page Title</title>
    </head>
</html>
```

Body

An HTML element is expected to have one **body** element and it is expected to be the second child of the **html** element following the **head** element.

Our minimal HTML document would therefore have the following code:

```
<!doctype html>
<html lang="en">
    <head>
        <title>Page Title</title>
```

```
    </head>
    <body>
    </body>
</html>
```

The **body** element represents the content of the document. Not everything in the body will necessarily be rendered in the browser, but all human-readable content should be hosted in the **body** element. This includes headers and footers, articles, and navigation.

You will learn more about the elements that can be the content of the **body** element throughout the following chapters.

Our First Web Page

In our first example, we will create a very simple web page. This will help us to understand the structure of an HTML document and where we put different types of content.

Exercise 1.01: Creating a Web Page

In this exercise, we will create our first web page. This is the foundation upon which the future chapters will build.

> **Note**
>
> Before beginning the exercises in the book, please make sure you have followed the instructions given in the *Preface* regarding installing VSCode and the extension **Open In Default Browser**.

The steps are as follows:

1. To start, we want to create a new folder, named **Chapter01**, in a directory of your choice. Then open that folder in Visual Studio Code (**File** > **Open Folder...**).

2. Next, we will create a new plain text file by clicking **File** > **New File**. Then, save it in HTML format, by clicking **File** > **Save As...** and enter the **File name: Exercise 1.01.html**. Finally, click on **Save**.

3. In **Exercise 1.01.html**, we start by writing the doctype declaration for HTML5:

```
<!DOCTYPE html>
```

4. Next, we add an HTML tag (the root element of the HTML document):

```
<html lang="en">
</html>
```

5. In between the opening and closing tags of the **html** element, we add a **head** tag. This is where we can put metadata content. For now, the **head** tag will contain a title:

```
<head>
    <title>HTML and CSS</title>
</head>
```

6. Below the head tag and above the closing **html** tag, we can then add a **body** tag. This is where we will put the majority of our content. For now, we will render a heading and a paragraph:

```
<body>
    <h1>HTML and CSS</h1>
    <p>How to create a modern, responsive website with HTML and
CSS</p>
</body>
```

If you now right-click on the filename in VSCode on the left-hand side of the screen and select **Open In Default Browser**, you will see the following web page in your browser:

HTML and CSS

How to create a modern, responsive website with HTML and CSS

Figure 1.10: The web page as displayed in the Chrome web browser

Metadata

The **head** element is home to most machine-read information in an HTML document. We will look at some commonly used metadata elements and how they enhance a web page and how they can optimize a web page for search engines and modern browsers.

The following elements are considered metadata content: **base**, **link**, **meta**, **noscript**, **script**, **style**, and **title**.

We've already added a **title** element in the previous exercise. This is the name of your web page and it appears in the tab of most modern browsers as well as in a search engine's results as the heading for the web page's listing.

The **link** element lets us determine the relationships between our document and external resources. A common use for this element is to link to an external style sheet. We will look at that use case in the section on CSS later in this chapter. There are several other uses for the **link** element. These include linking to icons and informing the browser to preload assets.

The **base** element lets you set a base URL. This will be used as the base for all relative URLs in the HTML document. For example, we could set the base **href** and then link to a style sheet with a relative URL:

```
<base href="http://www.example.com">
<link rel="stylesheet" href="/style.css">
```

This would result in our browser trying to download a style sheet from **http://www.example.com/style.css**.

The **meta** element acts as a catch-all for other metadata not represented by the other metadata content elements. For example, we can use the **meta** element to provide a description or information about the author of the web page.

Another use for the **meta** element is to provide information about the HTML document, such as the character encoding used. This can be very important as text characters will render differently or not at all if not set correctly. For example, we normally set the character encoding to UTF-8:

```
<meta charset="utf-8">
```

This character encoding declaration tells the browser the character set of the document. UTF-8 is the default and is recommended. This gives information to the browser but does not ensure the document conforms to the character encoding. It is also necessary to save the document with the correct character encoding. Again, UTF-8 is often the default but this varies with different text **editors**.

It is important that the character encoding declaration appears early in the document as most browsers will try to determine the character encoding from the first 1,024 bytes of a file. The **noindex** attribute value is set for the web pages that need not be indexed, whereas the **nofollow** attribute is set for preventing the web crawler from following links.

Another **meta** element that is very useful for working with mobile browsers and different display sizes is the **viewport** element:

```
<meta name="viewport" content="width=device-width, initial-scale=1">
```

The viewport element is not standard but is widely supported by browsers and will help a browser define the size of the web page and the scale to which it is zoomed on smaller display sizes. The units of viewport height and viewport width are **vh** and **vw** respectively; for example, 1vh = 1% of the viewport width. We will dive deeper into the viewport element and other aspects of responsive web development in *Chapter 6, Responsive Web Design and Media Queries*.

The **script** element lets us embed code in our HTML document. Typically, the code is JavaScript code, which will execute when the browser finishes parsing the content of the **script** element.

The **noscript** element allows us to provide a fallback for browsers without scripting capabilities or where those capabilities are switched off by the user.

We will look at the **style** element in more detail when we look at CSS later in this chapter.

These elements won't appear on the web page as content the user sees in the browser. What they do is give web developers a lot of power to tell a browser how to handle the HTML document and how it relates to its environment. The web is a complex environment and we can describe our web page for other interested parties (such as search engines and web crawlers) using metadata.

Exercise 1.02: Adding Metadata

In this exercise, we will add metadata to a web page to make it stand out in search engine results. The page will be a recipe page for a cookery website called Cook School. We want the page's metadata to reflect both the importance of the individual recipe and the larger website so it will appear in relevant searches.

To achieve this, we will add metadata – a title, a description, and some information for search engine robots. On the web, this information could then help users find a blog post online via a search engine.

Here are the steps we will follow:

1. Open the **Chapter01** folder in VSCode (**File** > **Open Folder**...) and create a new plain text file by clicking **File** > **New File**. Then, save it in HTML format by clicking **File** > **Save As**...and enter the **File name: Exercise 1.02.html**. Next, we will start with a basic HTML document:

```
<!DOCTYPE html>
<html lang="en">
    <head>
        <!-- Metadata will go in the head -->
    </head>
    <body>
        <!-- Cupcake recipe will go in the body -->
    </body>
</html>
```

2. Let's add a title for the recipe page that will be relevant to users who have navigated to the page or who are looking at a search engine results page. We will add this to the **head** element:

```
<title>Cook School - Amazing Homemade Cupcakes</title>
```

3. Just after the opening **<head>** element, we will add a metadata element, **<meta>**, to let the browser know which character encoding to use:

```
<meta charset="utf-8">
```

4. Next, we are going to add a description **meta** element below the **title** element:

```
<meta name="description" content="Learn to bake delicious, homemade cupcakes
with this great recipe from Cook School.">
```

5. We will add another **meta** element. This time, it is the robots **meta** element, which is used to make search engine crawling and indexing behave in a certain way. For example, if you didn't want a page to be indexed by a search engine, you could set the value to **noindex**. We will set a value of **nofollow**, which means a web crawler will not follow links from the page:

```
<meta name="robots" content="nofollow">
```

If you don't set this tag, the default value will be **index** and **follow**. This is normally what you want but you might not want a search engine to follow links in comments or index a particular page.

6. The viewport **meta** element, which is very useful for working with mobile browsers and different display sizes, is added just below the **title** element in the **head** element:

```
<meta name="viewport" content="width=device-width, initial-scale=1">
```

7. To finish, let's add some content to the **body** element that correlates with the metadata we've added:

```
<h1>Cook School</h1>
<article>
    <h2>Amazing Homemade Cupcakes</h2>
    <p>Here are the steps to serving up amazing cupcakes:</p>
    <ol>
        <li>Buy cupcakes from a shop</li>
        <li>Remove packaging</li>
        <li>Pretend you baked them</li>
    </ol>
</article>
```

If you now right-click on the filename in VSCode on the left-hand side of the screen and select **Open In Default Browser**, you will see the following web page in your browser:

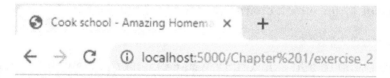

Cook School

Amazing Homemade Cupcakes

Here are the steps to serving up amazing cupcakes:

1. Buy cupcakes from a shop
2. Remove packaging
3. Pretend you baked them

Figure 1.11: The web page as displayed in the Chrome web browser

The important thing in the relationship between content and metadata is that they should make sense together. There is no point in adding keywords or writing a description of cars if the article is about cupcakes. The metadata should describe and relate to your actual content.

There are many search engines out there and they all do things a bit differently, and often with their own bespoke meta elements. If you would like to know more about how Google's search engine understands the meta tags from your web page's source code, some information is available at https://packt.live/35fRZOF.

Mistakes in HTML

Most browsers do their best to render a web page even if the HTML does not comply perfectly with the W3C's HTML5 standard. One area where HTML5 differs from the previous versions, including XHTML, is that the standard gives detailed instructions for browser developers on how to handle mistakes and issues in an HTML5 document. HTML5 tries to standardize how browsers handle problems and there is a lot of flexibility built into the standard (such as optional omitted end tags).

With that said, there are still many ways that an HTML document might be wrong due to typos, omitted tags, or the incorrect use of tags. Let's look at a few potential mistakes.

A common problem is a missing or mistyped closing tag. Here, we can see an example: in the following snippet of HTML, we have a paragraph with two anchor elements, both of which are pointing at different pages of the Packt website:

```
<p>
    Learn about <a href="https://www.packtpub.com/web-development">web
    development</a>. Try out some of the <a
    href="https://www.packtpub.com/free-learning">Free learning on the
    Packt site.
</p>
<p>
    Lorem ipsum...
</p>
```

There is one problem with this code. The second link does not have a closing tag and so the anchor element never closes. This makes the rest of the paragraph and the next paragraph in the document the same anchor tag. Anything beneath the opening anchor tag (**<a>**) until another closing anchor tag (**</a>**) would become an active link due to this mistake.

We can see the result in the following figure where the link text runs on to the second paragraph:

Learn about web development. Try out some of the Free learning on the Packt site.

Lorem ipsum...

Figure 1.12: Missing closing tag on an anchor element

Some issues are not to do with syntax error but are regarding the **semantic** constraints of HTML5. In other words, an element might have a specific role or meaning and having more than one instance might not make sense.

For example, the **main** element describes the main content of an HTML document. There should never be more than one **main** element visible on a web page at any one time.

The following code would not be valid:

```
<body>
    <main id="main1"><!-- main content here … --></main>
    <main id="main2"><!-- more main content here ... --></main>
</body>
```

However, if we were to hide one of the instances of the **main** element and only render that one when we hide the other, we would be using the main element in a way that is acceptable. The browser could still determine what the main content of our web page is. For example, the following code would be valid:

```
<body>
    <main id="main1"><!-- main content here … --></main>
    <main id="main2" hidden><!-- more main content here ... --></main>
</body>
```

Look carefully and you will see that we have added the **hidden** attribute to the second instance of the **main** element. This means there is only one visible **main** element in the web page.

You will learn more about **main** and other structural elements in the next chapter.

Sometimes, mistakes are caused by not knowing the specification. Take, for example, Boolean attributes such as the **disabled** attribute. We can apply this attribute to some interactive elements such as form inputs and buttons.

A button element creates a clickable button UI on a web page. We can use this element to trigger form submissions or to change the web page. We can use the **disabled** attribute with this element to stop it from submitting or taking any action.

If we add the **disabled** attribute like this, **<button disabled="false">Click me!</button>**, we might expect this element to be enabled. We've set the **disabled** attribute to **false**, after all. However, the specification for the **disabled** attribute says that the state of the element is decided by whether the attribute is present or not and the value is not regarded. To enable the element, you must remove the **disabled** attribute.

Because of the ways most modern browsers try to correct problems in HTML5 documents, it might not be immediately obvious what the benefits of making your HTML document valid are. However, it is good to keep in mind that, while developing for the web, you could have an audience on a variety of browsers – not all of them the most modern. Equally, it is still very easy to make mistakes that will cause obvious rendering issues. The best way to solve these is to make sure your document is valid and therefore working as expected.

There are tools available to help you check that your web page is valid. In the forthcoming exercise, we will look at how we can use an online tool to validate a web page.

Validating HTML

Mistakes in HTML5 can cause our web pages to render in ways we are not expecting. They can also cause problems for screen reader technologies that rely on the semantic meaning of HTML elements. Thankfully, there are some really useful tools out there to automate the validation of our HTML document and to keep us from making costly mistakes.

In this section, we will introduce the W3C's Markup Validation Service, an online tool that will validate a web page for us. We will then try out the tool with the help of exercises.

W3C's Markup Validation Service is an online tool that lets us validate a web page. The tool is available at https://packt.live/323qgOI. Navigating to that URL, we will see the tool as in the following figure:

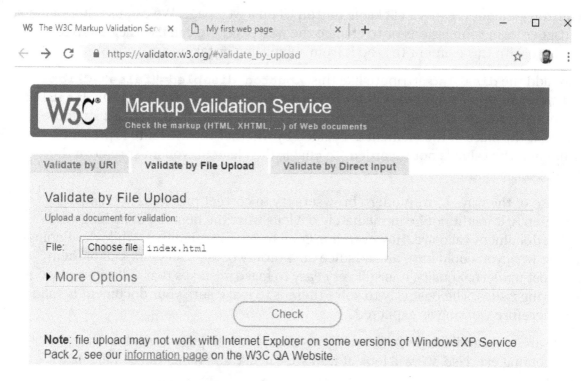

Figure 1.13: The W3C's Markup Validation Service

There are several options available, depending on how you wish to provide the validator a web page to validate. The options are:

- **Validate by URI** – choose a publicly accessible URL you wish to validate

- **Validate by File Upload** – validate a file uploaded from your computer

- **Validate by Direct Input** – copy and paste code to validate

As well as these input options, you have some more advanced options under the **More Options** heading, as seen in the following figure:

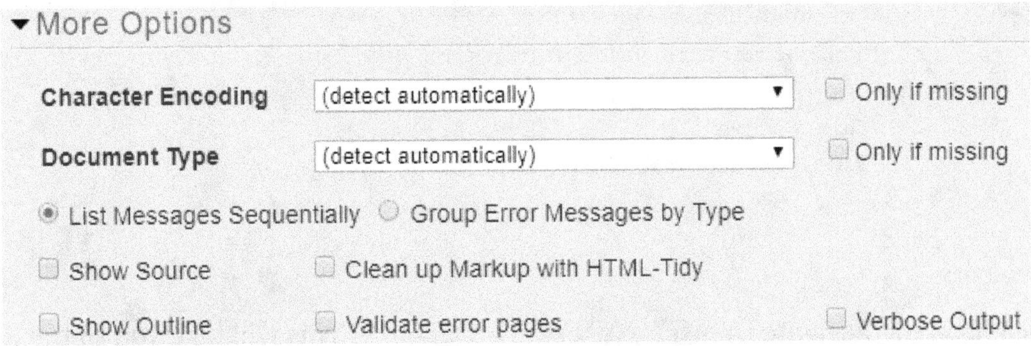

Figure 1.14: The More Options panel of the W3C's Markup Validation Service

With these options, you can:

- Set the character encoding and document type
- Group error messages
- Show source code
- Set the level of output (verbose)

Once you hit the **Check** button on the validator, it will run the tool and your results will appear in the results panel. You will either see a message telling you your document is valid or you will see a list of errors describing where the document is invalid and possible reasons.

We will see examples of both these results in the next two exercises.

Exercise 1.03: Validation

In this exercise, we will validate a web page using the W3C's Markup Validation Service. The steps are as follows:

1. For this exercise, we need a web page to validate. Create a new file within the **Chapter01** folder by clicking **File > New File**. Then, save it in HTML format, by clicking **File > Save As...**and enter the **File name: Exercise 1.03.html**. Copy the following content into the file:

```
<!DOCTYPE html>
<html lang="en">
    <head>
        <meta charset="utf-8">
        <title>Valid HTML document</title>
        <meta name="viewport" content="width=device-width,
          initial-scale=1">
    </head>
    <body>
```

```
<h1>Valid HTML document</h1>
<p>This document is valid HTML5.</p>
<!--
    This document will not throw errors in
    W3C's Markup Validation Service
-->
</body>
</html>
```

2. In a browser, navigate to https://packt.live/323qgOI. This will take you to the W3C's online Markup Validation Service.

3. You can validate web pages by URI, file upload, or copy and paste. We will use the file upload method. Click the **Validate by File Upload** tab.

4. Click the **Choose file** button to select the **Exercise 1.03.html** file.

5. Click the **Check** button.

If all went well, we should see a results page similar to that shown in the following figure, where there is green highlighted text saying **Document checking completed. No errors or warnings to show**. This means we have a valid HTML document:

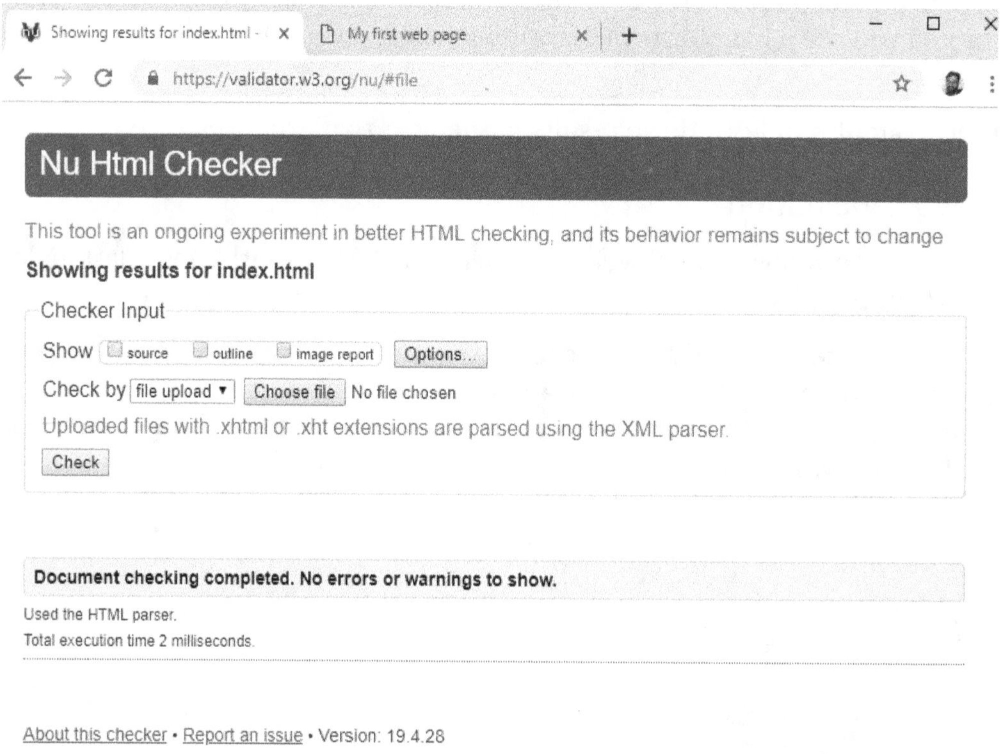

Figure 1.15: W3C's Markup Validation Service results page with no errors

Exercise 1.04: Validation Errors

Now we will see what happens if the HTML document does not validate. Here are the steps:

1. Firstly, we need a web page with some errors in its source code. Create a new file under **Chapter01** folder by clicking **File** > **New File**. Then, save it in HTML format, by clicking **File** > **Save As...** and enter the **File name: Exercise 1.04.html**.

2. In **Exercise 1.04.html**, we will add the following content:

```html
<html lang="en">
    <head>

    </head>
    <body>
        <p>Hello world!</p>
        <title>My first web page</title>
    </body>
</html>
```

This HTML document has some problems that should cause validation errors when this page is run through the W3C's validation service.

3. Repeat the steps from *Exercise 1.03, Validation*, uploading the **Exercise 1.04.html** file in place of the valid file.

The results should look like the following figure. Each error is flagged with a line number. In our case, there should be three errors because the doctype is missing, a **<title>** element is expected in the **<head>** element of the document, and **<title>** is not allowed as a child of the **<body>** element.

The error messages can point out issues with the content model, such as the **<title>** element missing in a **<head>** element, as well as issues where an opening tag does not have a corresponding closing tag:

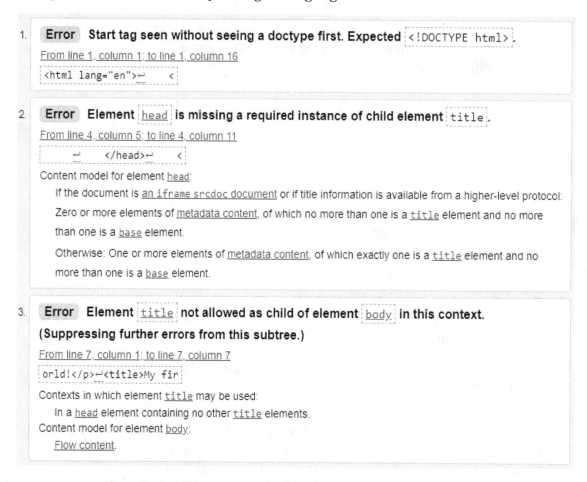

Figure 1.16: Validation errors in the W3C's Markup Validation Service

The HTML5 standard is pretty robust and tries to handle things such as omitted closing tags, but it is still possible to cause errors with typos or missed tags, and so the validator is a very useful tool. A valid HTML document is likely to be better optimized for performance, cause fewer bugs in JavaScript or CSS, and be easier for **web crawlers**, **search engines**, and browsers to understand.

Activity 1.01: Video Store Page Template

We've been tasked with creating a website for an online on-demand film store called *Films On Demand*. We don't have designs yet but want to set up web page boilerplate that we can use for all the pages on the site.

We will use comments as placeholders to know what needs to change for each page that is built on top of the boilerplate template. For visible content in the body element, we will use lorem ipsum to get an idea of how content will flow.

The following figure shows the expected output for this activity:

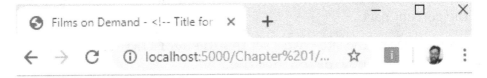

Lorem ipsum

Lorem ipsum dolor sit amet, consectetur adipiscing elit. Nullam quis scelerisque mauris. Curabitur aliquam ligula in erat placerat finibus. Mauris leo neque, malesuada et augue at, consectetur rhoncus libero. Suspendisse vitae dictum dolor. Vestibulum hendrerit iaculis ipsum, ac ornare ligula. Vestibulum efficitur mattis urna vitae ultrices. Nunc condimentum blandit tellus ut mattis. Morbi eget gravida leo. Mauris ornare lorem a mattis ultricies. Nullam convallis tincidunt nunc, eget rhoncus nulla tincidunt sed. Nulla consequat tellus lectus, in porta nulla facilisis eu. Donec bibendum nisi felis, sit amet cursus nisl suscipit ut. Pellentesque bibendum id libero at cursus. Donec ac viverra tellus. Proin sed dolor quis justo convallis auctor sit amet nec orci. Orci varius natoque penatibus et magnis dis parturient montes, nascetur ridiculus mus.

Figure 1.17: Expected output for the video store page template

The steps are as follows:

1. Create a file named `Activity 1.01.html`.

2. We want the page to be a valid HTML5 document. So, we will need to add the following:

   The correct doctype definition.

   Elements to structure the document: the `html` element, the `head` element, and the `body` element.

   A title element that combines the *Films on Demand* brand with some specifics about the current page.

   Metadata to describe the site – we'll set this to `Buy films from our great selection. Watch movies on demand.`

   Metadata for the page character set and a viewport tag to help make the site render better on mobile browsers.

3. We want to add placeholders for a heading (an `h1` element) for the page, which we will populate with lorem ipsum and a paragraph for the content flow, which we will also populate with the following lorem ipsum text:

   "Lorem ipsum dolor sit amet, consectetur adipiscing elit. Nullam quis scelerisque mauris. Curabitur aliquam ligula in erat placerat finibus. Mauris leo neque, malesuada et augue at, consectetur rhoncus libero. Suspendisse vitae dictum dolor. Vestibulum hendrerit iaculis ipsum, ac ornare ligula. Vestibulum efficitur mattis urna vitae ultrices. Nunc condimentum blandit tellus ut mattis. Morbi eget gravida leo. Mauris ornare lorem a mattis ultricies. Nullam convallis tincidunt nunc, eget rhoncus nulla tincidunt sed.

   Nulla consequat tellus lectus, in porta nulla facilisis eu. Donec bibendum nisi felis, sit amet cursus nisl suscipit ut. Pellentesque bibendum id libero at cursus. Donec ac viverra tellus. Proin sed dolor quis justo convallis auctor sit amet nec orci. Orci varius natoque penatibus et magnis dis parturient montes, nascetur ridiculus mus."

   > **Note**
   >
   > The solution to this activity can be found on page 584.

CSS

Cascading Style Sheets (CSS) is a style sheet language used to describe the presentation of a web page. The language is designed to **separate concerns**. It allows the design, layout, and presentation of a web page to be defined separately from content semantics and structure. This separation helps keeps source code readable and it is important because a designer can update styles separately from a developer who is creating the page structure or a web editor who is changing content on a page.

A set of CSS rules in a style sheet determines how an HTML document is displayed to the user. It can determine whether elements in the document are rendered at all, whether they appear in some context but not others, how they are laid out on the web page, whether they are rendered in a different order to the order in which they appear within a document, and their aesthetic appearance.

We will begin by looking at the syntax of CSS.

Syntax

A **CSS declaration** is made of two parts: a property and a value. The property is the name for some aspect of style you want to change; the value is what you want to set it to.

Here is an example of a CSS declaration:

```
color: red;
```

The property is `color` and the value is **red**. In CSS, `color` is the property name for the foreground `color` value of an element. That essentially means the color of the text and any text decoration (such as underline or strikethrough). It also sets a `currentcolor` value.

For this declaration to have any effect on an HTML document, it must be applied to one or more elements in the document. We do this with a selector. For example, you can select all the **<p>** elements in a web page with the **p** selector. So, if you wanted to make the color of all text in all paragraph elements red, you would use the following CSS ruleset:

```
p {
  color: red;
}
```

The result of this CSS ruleset applied to an HTML document can be seen in the following figure:

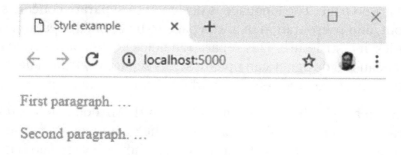

Figure 1.18: Result of a CSS rule applied to <p> elements in HTML

The curly braces represent a declaration block and that means more than one CSS declaration can be added to this block. If you wanted to make the text in all paragraph elements red, bold, and underlined, you could do that with the following ruleset:

```
p {
    color: red;
    font-weight: bold;
    text-decoration: underline;
}
```

The result of this CSS ruleset applied to an HTML document can be seen in the following figure:

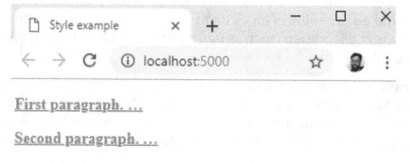

Figure 1.19: Several CSS declarations applied to <p> elements in HTML

Multiple selectors can share a CSS ruleset. We can target these with a comma-separated list. For example, to apply the color red to **p** elements, **h1** elements, and **h2** elements, we could use the following ruleset:

```
p, h1, h2 {
    color: red;
}
```

Multiple CSS rulesets form a style sheet. The order of these CSS rules in a style sheet is very important as this is partly how the cascade or specificity of a rule is determined. A more specific rule will be ranked higher than a less specific rule and a higher-ranked rule will be the style shown to the end user. We will look at cascade and specificity later in this chapter:

CSS Rule Set

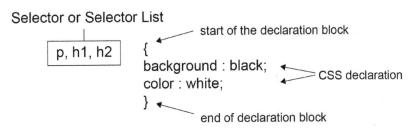

Figure 1.20: A CSS ruleset explained

Adding Styles to a Web Page

There are several ways to add your styles to a web page:

- You can use inline styles, which are applied directly to an element in the HTML document using the **style** attribute. The value of the **style** attribute is a CSS declaration block, meaning you can apply a semicolon-separated list of CSS declarations to the element.

- You can use a **style** element to add style information to an HTML document. The **style** element can be a child of either the **head** element or **body** element of a document. The **head** element tends to be preferable as the styles will be applied to your page more quickly.

- You can provide a style sheet as an external resource using the **link** element. One of the rationalities behind style sheets is the separation of concerns, which is why this approach is often recommended.

We will try out each of these methods in the following exercises.

Exercise 1.05: Adding Styles

In this exercise, we will be styling web pages by adding styles within the HTML document itself.

Here are the steps:

1. Open the **Chapter01** folder in VSCode (**File** > **Open Folder...**) and we will create a new plain text file by clicking **File** > **New File**. Then, save it in HTML format, by clicking **File** > **Save As...** and enter the **File name: Exercise 1.05.html** and start with the following web page:

```
<!DOCTYPE html>
<html lang="en">
    <head>
        <meta charset="utf-8">
        <title>Adding styles</title>
    </head>
    <body>
        <h1>Adding styles</h1>
        <p>First paragraph</p>
        <p>Second paragraph</p>
    </body>
</html>
```

Without any styles, this web page would look like the following:

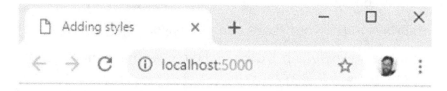

Adding styles

First paragraph

Second paragraph

Figure 1.21: The unstyled web page

2. We've decided to use a nicer font and make some of the text red. To do this, we will add a **style** element to the **head** element. We can add the following code under the **title** element:

```
<style>
  h1 {
      font-family: Arial, Helvetica, sans-serif;
      font-size: 24px;
      margin: 0;
      padding-bottom: 6px;
  }
  p {
      color: red;
  }
</style>
```

The results of this code change are that we should now have styles applied to the **h1** element and to both of the **p** elements, all of which will have red text. The result will look similar to the following figure:

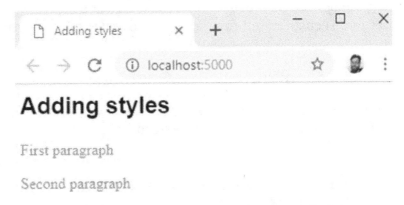

Figure 1.22: The web page with styles applied

3. Finally, we will give the first paragraph element a different style by overriding the style set in the **head** element. Let's add an inline style to the first paragraph, setting the color to blue and adding a **line-through** text decoration as follows:

```
<p style="color: blue; text-decoration: line-
    through">First paragraph</p>
```

If you now right-click on the filename in VSCode on the left-hand side of the screen and select **Open In Default Browser**, you will see the following web page in your browser:

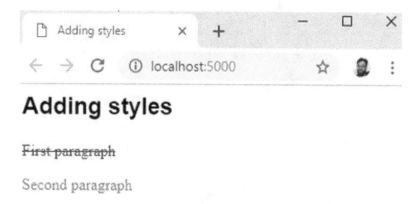

Figure 1.23: The web page with inline style applied

Something to note is that the inline style, applied to the first paragraph, takes precedence over the more general CSS rule applied to all **p** elements in the **style** element in the **head** element of the document. We will look at specificity and the rules of the cascade later in this chapter, but first, let's try moving these style rules into an external file.

Exercise 1.06: Styles in an External File

In this exercise, we will separate the concerns of presentation and structure of *Exercise 1.05, Adding Styles*, by moving all the styles to an external file.

The steps are as follows:

1. Open the **Chapter01** folder in VSCode (**File** > **Open Folder...**) and create a new plain text file by clicking **File** > **New File**. Then, save it in HTML format, by clicking **File** > **Save As...** and enter the **File name: Exercise 1.06.html**.

2. Add the same web page as in *Exercise 1.05, Adding Styles*, to the file:

```
<!DOCTYPE html>
<html lang="en">
    <head>
        <meta charset="utf-8">
        <title>Adding styles</title>
    </head>
    <body>
        <h1>Adding styles</h1>
        <p>First paragraph</p>
```

```
        <p>Second paragraph</p>
    </body>
</html>
```

3. We will add a **link** element to reference a **.css** file below the **title** element:

```
<link href="styles/Exercise_1.06.css" rel="stylesheet">
```

4. Create a **styles** directory in the same directory as **Exercise 1.06.html**. Next, create a file named **Exercise_1.06.css** within the styles directory.

5. Now, add the following styles to **Exercise_1.06.css**:

```
h1 {
    font-family: Arial, Helvetica, sans-serif;
    font-size: 24px;
    margin: 0;
    padding-bottom: 6px;
}


p {
    color: red;
}
```

6. To get the equivalent styles that we had at the end of *Exercise 1.05, Adding Styles*, without using an inline style, we have to have a specific way of targeting the first **p** element. We will use the **:first-of-type** pseudo-class. You will learn more about CSS selectors later in this chapter. For now, add this CSS rule to the bottom of **Exercise_1.06.css**:

```
p:first-of-type {
    color: blue;
    text-decoration: line-through;
}
```

The result will be as seen in *Figure 1.23* – the same result as in *Exercise 1.05, Adding Styles*. The difference is that we have removed all references to styles from the HTML document into their own external resources. We have successfully separated concerns.

Both these methods add styles to the HTML document when it loads. Similar to the HTML DOM, we can manipulate CSS programmatically with JavaScript. This is because the styles are also represented as an object model called the CSSOM.

CSSOM

The CSS Object Model (CSSOM) is similar to the HTML DOM, which we described earlier. The CSSOM is an in-memory representation of the styles in the document as they are computed on elements. It is a tree-structure with nodes that mirror those in the HTML DOM, and the objects associated have a list of style properties where all CSS rules have been applied.

The CSSOM represents all the styles that have been created in the document as objects with properties that we can change with JavaScript. We can access these styles and change the values of style properties.

We mostly access these styles via the **style** property of a DOM element, as here:

```
const boldElement = document.querySelector('.aBoldElement');
boldElement.style.fontWeight = 'bold';
```

In JavaScript, we can also access the CSSOM with the **getComputedStyle** method on the **window** object; for example:

```
const boldElement = document.querySelector('.aBoldElement');
window.getComputedStyle(boldElement);
```

This will return a computed styles object for an element with the **aBoldElement** class attribute. This method returns a read-only styles object with all computed styles for the element.

In the next section, we will look at the different CSS selectors we can use to apply our styles to a web page.

CSS Selectors

To target elements in the HTML document with CSS, we use selectors. There are a lot of options available to help you select a wide range of elements or very specific elements in certain states.

Selectors are a powerful tool and we will look at them in some detail as the different options available can help with both web page performance and making your CSS more maintainable.

For example, you can use a selector to target the first letter of a heading, like you might expect to see in a medieval book:

```
h1::first-letter {
    font-size: 5rem;
}
```

Or you could use a selector to invert the colors of every odd paragraph in an article:

```
p {
   color: white;
   background-color: black;
}
p:nth-of-type(odd) {
   color: black;
   background-color: white;
}
```

We will explore a variety of the options available to us when creating selectors.

Element, ID, and Class

Three commonly used selectors are:

- Element type: For example, to select all **p** elements in an HTML document, we use the **p** selector in a CSS ruleset. Other examples are **h1**, **ul**, and **div**.

- A class attribute: The class selector starts with a dot. For example, given the HTML snippet **<h1 class="heading">Heading</h1>**, you could target that element with the **.heading** selector. Other examples are **.post** and **.sub-heading**.

- An **id** attribute: The **id** selector starts with a hash symbol. For example, given the HTML snippet **<div id="login"> <!-- login content --> </div>**, you could target this element with the **#login** selector. Other examples include **#page-footer** and **#site-logo**.

The Universal Selector (*)

To select all elements throughout an HTML document, you can use the universal selector, which is the asterisk symbol (*). Here is an example snippet of CSS that is often added to web pages; a value is set on the **html** element and then inherited by all descendant elements:

```
html {
   box-sizing: border-box;
}


*, *:before, *:after {
   box-sizing: inherit;
}
```

herit keyword and the universal selector, we can pass a value on to all the of the **html** element. This snippet will universally apply the border-box lements and their pseudo-elements (that's the reason for **:before** and **:after**). You'll learn more about the box model and layout in the next chapter.

Attribute Selectors

Attribute selectors let you select elements based on the presence of an attribute or based on the value of an attribute. The syntax is square brackets, **[]**, with the suitable attribute inside. There are several variations that you can use to make matches:

- **[attribute]** will select all elements with an attribute present; for example, **[href]** will select all elements with an **href** attribute.

- **[attribute=value]** will select all elements with an attribute with an exact value; for example, **[lang="en"]** will select all elements with a **lang** attribute set to **en**.

- **[attribute^=value]** will select all elements with an attribute with a value that begins with the matching value; for example, **[href^="https://"]** will select all elements with an **href** attribute beginning with **https://**, which links to a secure URL.

- **[attribute$=value]** will select elements with an attribute with a value that ends with the matching value; for example, **[href$=".com"]** will select all elements with an **href** attribute that ends with **.com**.

- **[attribute*=value]** will select elements with an attribute with a value that has a match somewhere in the string; for example, **[href*="co.uk"]** will select all elements with an **href** attribute matching **.co.uk**. **http://www.example. co.uk?test=true** would be a match, as would **https://www.example.co.uk**.

Pseudo-classes

To select an element when it is in a particular state, we have several pseudo-classes defined. The syntax of a pseudo-class is a colon, **:**, followed by a keyword.

There are a great number of pseudo-classes, but most developers' first experience of them is when styling links. A link has several states associated with it:

- When an anchor element has an **href** attribute, it will have the **:link** pseudo-class applied to it.

- When a user hovers over the link, the **:hover** pseudo-class is applied to it.

- When the link has been visited, it has the **:visited** pseudo-class applied to it.

- When the link is being clicked, it has the **:active** pseudo-class applied to it.

Here is an example of applying styling to the various pseudo-class states of an anchor element:

```
a:link, a:visited {
    color: deepskyblue;
    text-decoration: none;
}


a:hover, a:active {
    color: hotpink;
    text-decoration: dashed underline;
}
```

In the following figure, we can see the first link with the `:link` or `:visited` styles applied and the second link with the `:hover` or `:active` styles applied:

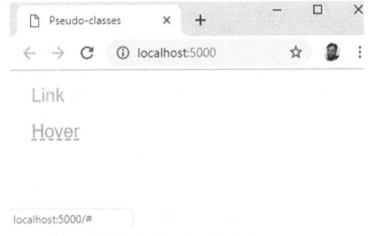

Figure 1.24: Link with and without the hover state

The cascade can cause some issues with styling links. The order in which you specify your CSS rules for each state of the link is important. If, for example, we applied the **a:hover** rule before the **a:link** rule in the previous example, we would not see the hover effect. A mnemonic exists for remembering the order: **love-hate**. The **l** is for `:link`, the **v** is for `:visited`, the **h** is for `:hover`, and the **a** is for `:active`.

Some other useful pseudo-classes for selecting elements in a particular interactive state include `:checked`, `:disabled`, and `:focus`.

There are several pseudo-classes that help us select a pattern of children nested under an element. These include `:first-child`, `:last-child`, `:nth-child`, `:nth-last-child`, `:first-of-type`, `:last-of-type`, `:nth-of-type`, and `:nth-last-of-type`.

For example, we can use `:nth-child` with an unordered list to give a different style to list items based on their position in the list:

```
<style>
  ul {
    font-family: Arial, Helvetica, sans-serif;
    margin: 0;
    padding: 0;
  }

  li {
    display: block;
    padding: 16px;
  }

  li:nth-child(3n-1) {
    background: skyblue;
    color: white;
    font-weight: bold;
  }

  li:nth-child(3n) {
    background: deepskyblue;
    color: white;
    font-weight: bolder;
  }
</style>

<!-- unordered list in HTML document -->
<ul>
    <li>Item 1</li>
    <li>Item 2</li>
    <li>Item 3</li>
    <li>Item 4</li>
    <li>Item 5</li>
    <li>Item 6</li>
    <li>Item 7</li>
</ul>
```

The following figure shows the result. The `:nth-child` pseudo-class gives you a lot of flexibility because you can use keywords such as **odd** and **even** or functional notation such as **3n - 1**:

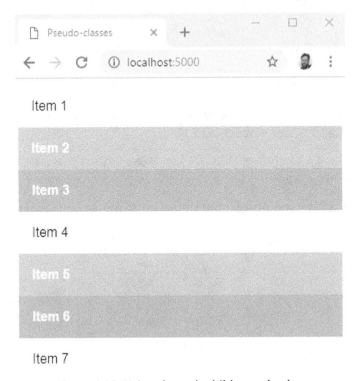

Figure 1.25: Using the :nth-child pseudo-class

Pseudo-elements

Pseudo-element selectors are preceded by two colons (`::`) and they are used to select part of an element. The available pseudo-elements include `::after`, `::before`, `::first-letter`, `::first-line`, `::selection`, and `::backdrop`.

These pseudo-elements give us a handle we can use to add stylistic elements without adding to the HTML document. This can be a good thing if the pseudo-element has no semantic value and is purely presentational, but it should be used with care.

Combining Selectors

What makes CSS selectors particularly powerful is that we can combine them in several ways to refine our selections. For example, we can select a subset of `li` elements in an HTML document that also has a `.primary` class selector with `li.primary`.

We also have several options, sometimes called combinators, for making selections based on the relationships of elements:

- To select all the `li` elements that are **descendants** of a `ul` element, we could use `ul li`.

- To select all the `li` elements that are **direct children** of a `ul` element with the `primary` class, we might use `ul.primary > li`. This would select only the direct children of `ul.primary` and not any `li` elements that are nested.

- To select an `li` element that is the **next sibling** of `li` elements with the `selected` class, we could use `li.selected + li`.

- To select all of the `li` elements that are the **next siblings** of `li` elements with the `selected` class, we could use `li.selected ~ li`.

The following figure shows the difference between using `li.selected + li` and `li.selected ~ li`:

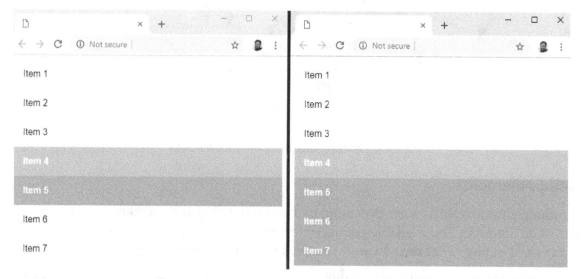

Figure 1.26: Selecting the next adjacent sibling compared to selecting all of the next siblings

Let's try out some of the selectors we've learned about in an exercise.

Exercise 1.07: Selecting Elements

In this exercise, we will differentiate list items by styling the odd items. We will use a class selector to style a selected item and a next-siblings combinator to style the elements after the selected item.

The steps are as follows:

1. Open the **Chapter01** folder in VSCode (**File** > **Open Folder**...) and we will create a new plain text file by clicking **File** > **New File**. Then, save it in HTML format, by clicking **File** > **Save As...** and enter the **File name: Exercise 1.07.html**. Now, copy the following simple web page with a **ul** list element and nine list items:

```
<!DOCTYPE html>
<html lang="en">
    <head>
        <meta charset="utf-8">
        <title>Selectors</title>
    </head>
    <body>
        <ul>
            <li>Item 1</li>
            <li>Item 2</li>
            <li>Item 3</li>
            <li>Item 4</li>
            <li>Item 5</li>
            <li>Item 6</li>
            <li>Item 7</li>
            <li>Item 8</li>
            <li>Item 9</li>
        </ul>
    </body>
</html>
```

2. So that we can style a selected item differently, we will add a **selected** class to the fifth list item:

```
<li class="selected">Item 5</li>
```

3. Next, we will add a **style** element to the **head** element with the following CSS:

```
<head>
    <meta charset="utf-8">
    <title>Selectors</title>
    <style>
      ul {
          font-family: Arial, Helvetica, sans-serif;
          margin: 0;
          padding: 0;
      }

      li {
          display: block;
          padding: 16px;
      }
    </style>
</head>
```

This will remove some of the default styling of the unordered list in the browser. It will remove margins and padding on the list and set the font style to Arial (with Helvetica and sans-serif as a fallback).

4. Next, we will style the odd list items with the **:nth-child** pseudo-class. We can use the **odd** keyword for this. With this style, any **odd** list item will have a blue background and white text:

```
li:nth-child(odd) {
    background-color: deepskyblue;
    color: white;
    font-weight: bold;
}
```

This gives us the stripy effect that we can see in the following figure:

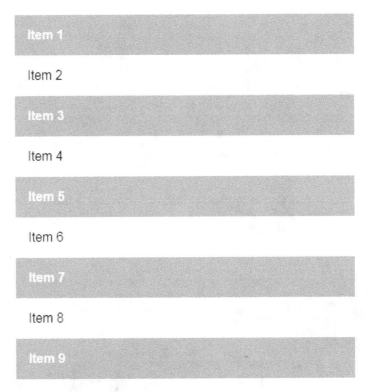

Figure 1.27: Stripy list using :nth-child(odd)

5. We can style the selected class selector:

```
li.selected {
    background-color: hotpink;
}
```

This overrides the striped effect for the selected item as seen in the following figure:

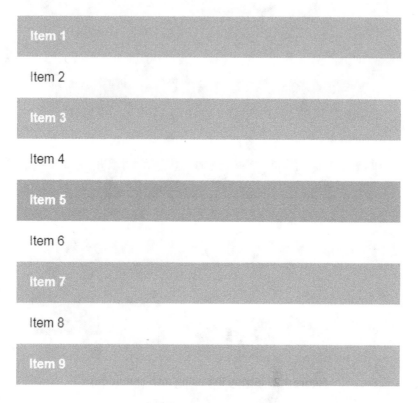

Figure 1.28: Stripy list with a selected item

6. Finally, we will style the **odd** list items after the selected item using the all-next-siblings combinator:

```
li.selected ~ li:nth-child(odd) {
    background-color: orange;
}
```

If you now right-click on the filename in VSCode on the left-hand side of the screen and select **Open In Default Browser**, you will see the following web page in your browser:

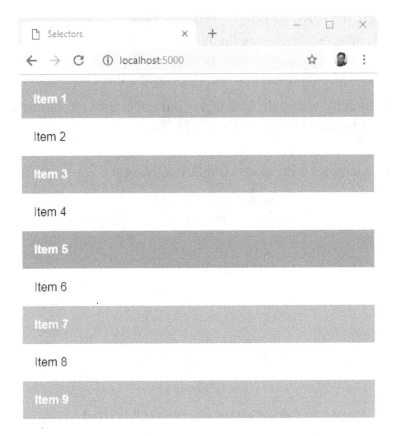

Figure 1.29: Combining selectors to style a list

Style sheets can have a large number of style rules and combinations of selectors. It is good to understand why one rule takes precedence over another one. This is where CSS specificity comes in.

CSS Specificity

If we have two CSS declarations that have an effect on the same style property of an element, how do we know which of those declarations will take precedent?

There are several factors that decide the ranking of a CSS declaration and whether it is the style the browser will apply. The term for these factors is **specificity**.

A style attribute that adds inline styles to an element has the highest specificity value. An ID selector has a greater specificity value than a class selector and a class selector or attribute selector has a greater specificity value than an element type. We can calculate the specificity value by giving points to each of these specificity values.

The most common way of representing this is as a comma-separated list of integers, where the leftmost integer represents the highest specificity. In other words, the leftmost value is the inline style attribute; next is an ID selector; next is a class selector, pseudo-class, or attribute selector; and the rightmost value is an element.

An inline style would have the value 1, 0, 0, 0. An ID selector would have the value 0, 1, 0, 0. A class selector would have the value 0, 0, 1, 0, and an **h1** element selector would have the value 0, 0, 0, 1.

Let's look at a few examples with more complex selectors:

- **li.selected a[href]** has two element selectors (**li** and **a**), a class selector (**.selected**) and an attribute selector (**[href]**), so its specificity value would be 0, 0, 2, 2:

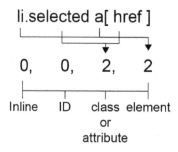

Figure 1.30: Calculating the specificity of li.selected a[href]

- **#newItem #mainHeading span.smallPrint** has two ID selectors, a class selector (**.smallPrint**), and a span element, so its specificity value would be 0, 2, 1, 1:

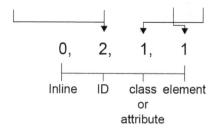

Figure 1.31: Calculating the specificity of #newItem #mainHeading span.smallPrint

Comparing the two selectors, we can see that the selector in the second example is more specific than the selector in the first example.

The Special Case of !important

The **!important** value can be appended to the value of any CSS declaration. It is a special keyword that can be applied to selectors to override the specificity value. It trumps any other specificity value. In terms of specificity value, it would add another column to become the value **1, 0, 0, 0, 0**. It would even take precedence over an inline style.

For example, we might want to create a style rule that is reusable and lets us hide content on a web page. Something like this:

```
.hide {
   display: none;
}
```

If we apply this class to an element, we want that element to be hidden and to not be rendered on the web page. We might use this to later reveal the element using JavaScript. However, consider the following example:

```
<style>
 div.media {
   display: block;
   width: 100%;
   float: left;
 }
 .hide {
   display: none;
 }
</style>
<div class="media hide">
  ...Some content
</div>
```

We might expect our **div** element to be hidden because the **.hide** class appears second in the style sheet. However, if we apply the specificity calculations we've learned about, we can see that **div.media** scores **0, 0, 1, 1** and **.hide** only scores **0, 0, 1, 0**. The **div.media** rule for the **display** property with **block** value will override the **none** value of the **.hide** class. We can't really use this instance of the **.hide** class as we don't know whether it will have any effect.

Now consider the same `.hide` class but using the `!important` keyword:

```
.hide {
  display: none !important;
}
```

Adding the `!important` keyword will make this `.hide` class much more reusable and useful as we can pretty much guarantee that it will hide content as we desire.

We've learned a lot about CSS in this part of the chapter. Let's apply some of this knowledge to an activity.

Activity 1.02: Styling the Video Store Template Page

In the previous activity, we were tasked with creating boilerplate HTML for a web page for the *Films On Demand* website. In this activity, we are going to add some style to that template page.

The following figure shows the expected output for this activity:

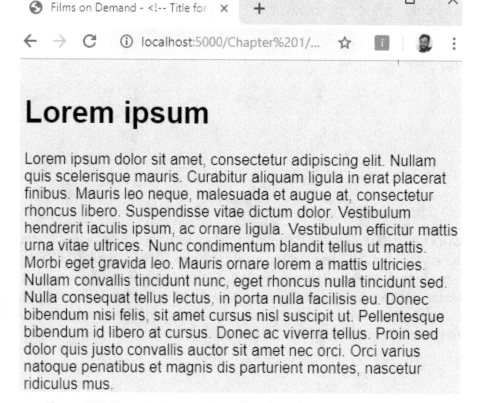

Figure 1.32: Expected output for styling the video store template page

The steps are as follows:

1. We will start with the template from *Activity 1.01, Video Store Page Template*, which we will save as **Activity 1.02.html**:

```
<!DOCTYPE html>
<html lang="en">
    <head>
        <meta charset="utf-8">
        <title>Films on Demand - <!-- Title for page goes here --
        ></title>
        <meta name="description" content="Buy films from our great
          selection. Watch movies on demand.">
        <meta name="viewport" content="width=device-width,
          initial-scale=1">
    </head>
    <body>
        <h1>Lorem ipsum</h1>

        <p>
        Lorem ipsum dolor sit amet, consectetur adipiscing elit.
        Nullam quis scelerisque mauris. Curabitur aliquam ligula
        in erat placerat finibus. Mauris leo neque, malesuada et
        augue at, consectetur rhoncus libero. Suspendisse vitae
        dictum dolor. Vestibulum hendrerit iaculis ipsum, ac
        ornare ligula. Vestibulum efficitur mattis urna vitae
        ultrices. Nunc condimentum blandit tellus ut mattis.
        Morbi eget gravida leo. Mauris ornare lorem a mattis
        ultricies. Nullam convallis tincidunt nunc, eget
        rhoncus nulla tincidunt sed. Nulla consequat tellus
        lectus, in porta nulla facilisis eu. Donec bibendum
        nisi felis, sit amet cursus nisl suscipit ut.
        Pellentesque bibendum id libero at cursus. Donec ac
        viverra tellus. Proin sed dolor quis justo convallis
        auctor sit amet nec orci. Orci varius natoque
        penatibus et magnis dis parturient montes, nascetur
        ridiculus mus.

        </p>
    </body>
</html>
```

2. We are going to link to an external CSS file. One of the difficulties with styling web pages is handling differences between browsers. We are going to do this by adding a file to normalize our default styles. We will use the open source **normalize.css** for this. Download the file from https://packt.live/3fijzzn. Add the file to a **styles** folder and link to it from the **Activity 1.02.html** web page.

3. We are going to add a **style** element to the **head** element of **Activity 1.02. html**. In the **style** element, we want to set some styles used across all pages. We want to do the following:

Set **box-sizing** to **border-box** for all elements using the universal selector (*****).

Add a font family with the **Arial, Helvetica, sans-serif** values and a font size of **16px** to the whole page.

Add the background color **#eeeae4** for the whole page. To do this, we will add a **div** element wrapper with the **pageWrapper** ID, where we will set the background color and padding of **16px**, and a **full-page** class, where we will set the minimum height to **100vh** (**100%** of the viewport height).

Add an **h1** element selector that sets the **margin** to **0** and adds **padding** of **16px** to the bottom of the **h1** element.

> **Note**
>
> The solution to this activity can be found on page 587.

We've learned a lot about HTML and CSS in this chapter. In the next section, we will learn a little bit about the Chrome developer tools and how we can use them to better understand our web page.

Dev Tools

Most browsers come with some tools to help web developers create and change web pages. One of the most useful of these tools is the built-in developer tools that come with the Chrome browser.

You can access the developer tools on any web page with the *Command + Option + I* keyboard shortcut (on Mac) and *F12* or *Control + Shift + I* (on Linux and Windows).

On opening the developer tools, you should see something similar to the following figure:

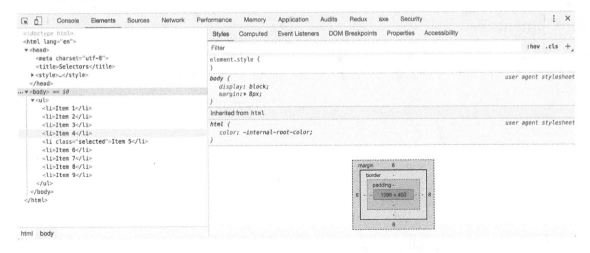

Figure 1.33: Chrome developer tools (Elements tab)

There are a lot of options available in the developer tools. We will not cover all of them here but will focus on the top bar and the **Elements** tab in this chapter.

The Top Bar

The top bar gives you access to several options and tools, including access to all the other sections of the developer tools via the tabs:

Figure 1.34: Chrome DevTools top bar

The top bar has the following tools and options:

- Select tool – You can use this tool to select an element from within the web page.

- 🗗 Devices toolbar – Changes the view so you can select the view size of various devices.

- Tabs – We can access various tools from the top bar menu such as **Console**, **Elements**, **Sources**, **Network**, and **Performance**. We will focus on the **Elements** tab.

- ⋮ Configuration – Gives you access to various settings for the developer tools.

- ✕ Close – Closes the developer tools.

While developing the HTML and CSS for a web page, one of the tabs we use the most is the Elements tab.

The Elements Tab

The Elements tab has two areas – the left-hand panel shows the current HTML DOM and the right-hand panel shows details about the selected element.

You can use the left-hand panel to select an element from the HTML DOM and you can also edit an element to change its attributes:

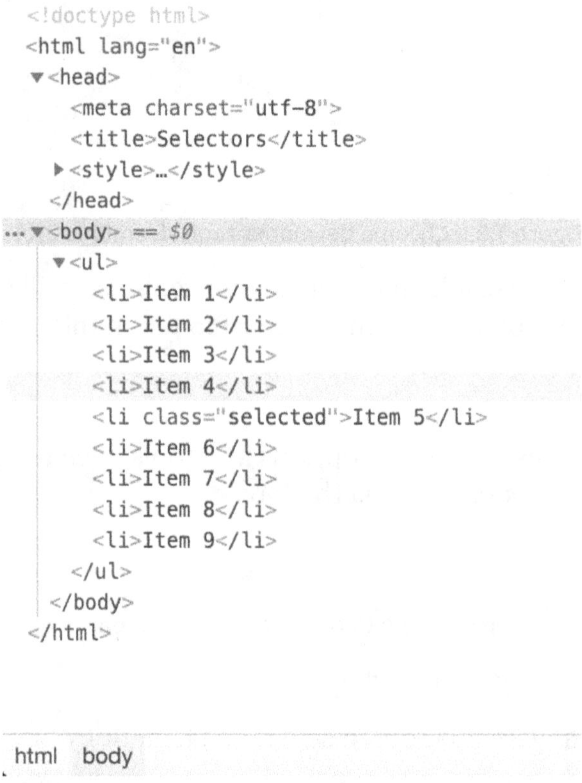

```
<!doctype html>
<html lang="en">
▼<head>
    <meta charset="utf-8">
    <title>Selectors</title>
   ▶<style>…</style>
   </head>
···▼<body> == $0
   ▼<ul>
       <li>Item 1</li>
       <li>Item 2</li>
       <li>Item 3</li>
       <li>Item 4</li>
       <li class="selected">Item 5</li>
       <li>Item 6</li>
       <li>Item 7</li>
       <li>Item 8</li>
       <li>Item 9</li>
    </ul>
   </body>
 </html>
```

html   body

Figure 1.35: Selectable HTML DOM in the left-hand panel of the Elements tab

Once you have an element selected, you can see a lot of information about the element in the right-hand panel. This information is divided into tabs and the first tab shows the styles:

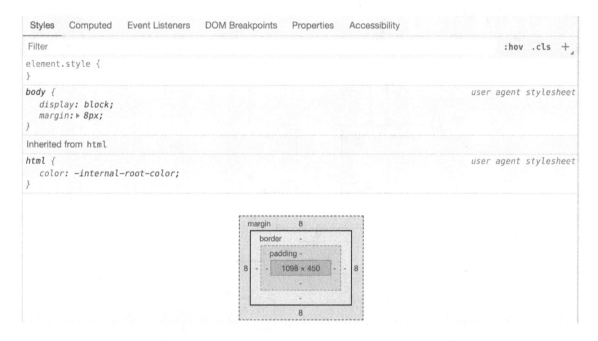

Figure 1.36: The Styles tab of the right-hand panel of the Elements tab

Essentially, the **Styles** tab shows a representation of the CSSOM. You can look at the styles associated with the element, including the cascade of styles, showing you where a style is inherited from and where it has been set but overridden. The topmost element is the most specific. You can also use this tab to edit the styles of an element and to check different states, such as the hover state.

The **Computed** tab shows styles as they have been computed for the element.

There are also tabs to show you the event listeners and properties associated with an element.

Having had a glimpse at the power of Chrome's developer tools, let's finally consider how the browser works with HTML and CSS technologies to render a web page.

How a Web Page Renders

How does a web page render in the browser? We've learned a lot about HTML and CSS in this chapter, but let's see how these technologies are put together by the browser to render our web page.

The following figure shows a flowchart of the process, which is further explained below:

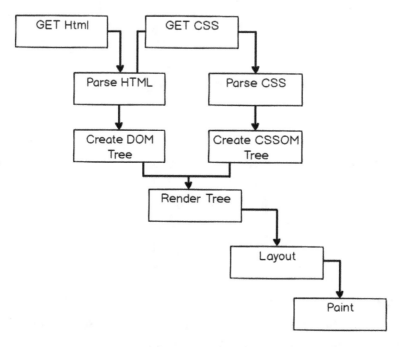

Figure 1.37: Flow chart of the web page render process

To summarize the process:

1. The user starts by navigating to a URL, possibly via a hyperlink or by typing the URL into the address bar of their browser.

2. The browser will make a **GET** request to the disk or a network. It will read the raw bytes from that location and convert them to characters (based on character encoding such as UTF-8).

3. The browser then parses these characters according to the HTML standard to find the tokens that are familiar as HTML elements, such as <html> and <body>.

4. Another **parse** is then made to take these tokens and construct objects with their properties and based on the rules appropriate to that token. At this point, the objects are defined.

5. Finally, the browser can define the relationships between these objects and construct the HTML DOM for the web page.

6. At this point, we have a DOM but not a rendered web page. The next task is to construct the CSSOM. Again, the browser will load any style sheet resources it needs to, which were found while parsing the document. It will then construct the styles associated with nodes in the tree structure, which gives us the CSSOM.

7. With the information gathered in the DOM and the CSSOM, the browser can create a render tree. The render tree is constructed by combining information from the CSSOM and the HTML DOM. Nodes in the HTML DOM that will not be rendered (for instance, those with the `display: none;` style) are excluded from the render tree. Those that are rendered are enriched with their computed style rules.

8. Now the browser has all the information it needs, it can begin to calculate the positions of elements in the rendered viewport. This is called the layout stage. The browser lays elements out based on their size and position within the browser viewport. This stage is often also called reflow. It means the browser must recalculate the positions of elements in the viewport when elements are added to or removed from the page or when the viewport size is changed.

9. Finally, the browser will rasterize or paint each element on the page, depending on their styles, shadows, and filters to render the page the user will see.

That is a brief and simplified summary of the rendering of a web page. Think about how many resources might be loaded on a relatively complicated website and with JavaScript running events and we can see that much of this process happens frequently and not in such a linear manner. We can start to see the complexities of what a browser is doing when it renders your web page.

Summary

In this chapter, we have looked at how we write HTML and CSS and their roles in structuring and presenting a web page. We've looked at how to make sure our HTML document is working as we expect and is understandable to a web browser, and we've looked at how to target elements in the HTML document so we can style them.

To demonstrate our understanding, we have created a web page that we will use as a template for the pages of a website and we have added CSS to normalize the style and add some initial styles for the site.

In the next chapter, we will look at the options for structuring and laying out a web page and elements within it and how we can style these for different web page layouts.

2

Structure and Layout

Overview

By the end of this chapter, you will be able to use the correct HTML5 elements to markup a web page; style a web page using float, flex, and grid layouts; describe how the box model works; and build a home page and a product page layout. This chapter introduces the essential HTML elements that are required in order to build a web page. Finally, this knowledge will be utilized by carrying out a number of exercises to create a few well-structured web pages.

Introduction

In the previous chapter, we learned about the basics of HTML and CSS. In this chapter, we will consolidate this basic understanding and look at how web pages are structured with HTML and CSS. When creating web pages using HTML, it is imperative that you use the correct elements. This is because HTML is read by both humans and machines, and so the content of a web page should be associated with the most appropriate element. Additionally, any error in the code might be difficult to track if the code base is too large.

The HTML language offers a vast array of different tags that we can place at our disposal. In this chapter, we will focus on the structural elements that are used to divide the web page up into its key parts. You may be familiar with the concept of a page header or footer, and these would be examples of structural elements. We will be looking at these amongst many other HTML structural elements.

In this chapter, we will focus our attention on the **HTML5** version of the language, which is the most current version of HTML. HTML5 offers us additional tags that enable us to make our markup more meaningful. The developer experience is more enjoyable compared to writing **XHTML** as the HTML5 language is less strict with regard to syntax.

> **Note**
>
> In this chapter, we will use the terms "tags" and "elements" synonymously.

Web pages are typically styled using CSS. Once we have our web pages marked up correctly, we need to know how to style these into a range of layouts. CSS offers us a range of options for laying out our pages, but the three most common methods are **float**, **flex**, and **grid**-based. In this chapter, we will explore each of these techniques in turn.

Just knowing the various layout methods is not enough to style web pages. We will investigate the box model, which is foundational to understanding how HTML elements are styled. We will break this down into the individual layers – **content box**, **padding**, **border**, and **margin**. With this knowledge in hand, you will be free to develop a host of different web page layouts.

We will now take a look at the structural elements provided by HTML and examine what the key elements are one by one.

Structural Elements

HTML5 provides us with a variety of tags that we can use when dividing our page into different parts. When browsing the web, you would have noticed that web pages typically have a few common things to them. For example, a web page will typically have a logo and page navigation area at the top of the page. We would call this area of the page the **header**. You may also have noticed that the bottom of the page may include a list of links and copyright information. We would call this area the **footer**. The following diagram shows the representation of a few of the main elements of a web page:

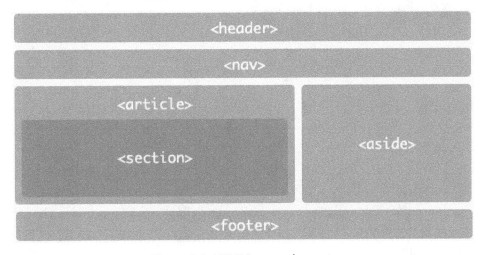

Figure 2.1: HTML5 page elements

In this topic, we will be looking at the following HTML5 page elements:

- **header**
- **footer**
- **section**
- **article**
- **nav**
- **aside**
- **div**

The header Tag

The **header** tag is used to describe the header or top area of a web page. Typically, inside this tag, you would have the page heading, a logo, and, possibly, the page navigation. Prior to HTML5, you would use a **div** tag with a class name so that the header could be styled, and its intention was clear to developers. HTML5 improves on this by giving us a tag specifically for this very task. You will learn more about this improvement under the section, *Semantic Markup* in *Chapter 3*, *Text and Typography*. Now, examine the following codes that show the differences between the old and new way of writing the markup for the **header** area:

```
<!-- old way -->
<div class="header">
   <!-- heading, logo, nav goes here -->
</div>
<!-- new way -->
<header>
   <!-- heading, logo, nav goes here -->
<header>
```

Now, let's open the Packt website at https://www.packtpub.com/ to see how a header is represented in an actual website. In the following diagram, you can see that the header element is highlighted via a box, illustrating where a header element is typically placed on a web page:

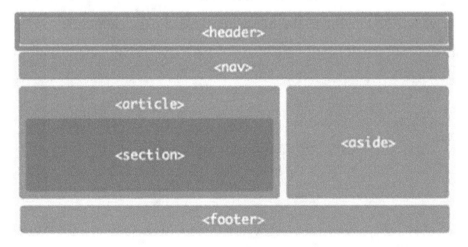

Figure 2.2: The header element

In the following figure, you can see that the header element is highlighted via a box. As this is an example taken from the Packt website, you will notice that it contains items such as the company logo, search bar, and the **Sign In** button:

Figure 2.3: The header element on the Packt site

The footer Tag

The **footer** tag is very similar to the **header** tag but is used at the bottom of a web page. You would typically have the copyright information and website links inside the footer. Similarly, with the header tag in the previous version of HTML, you would use a **div** tag with a class name. Since the use of footers on web pages is so common, HTML5 provides a new tag solely for this purpose. The following codes show the differences between the old and new way of writing the markup for the **footer** area:

```
<!-- old way -->
<div class="footer">
  <!-- copyright, list of links go here -->
</div>
<!-- new way -->
<footer>
  <!-- copyright, list of links go here -->
<footer>
```

In the following figure, you can see that the **footer** element is highlighted via a box, illustrating where a **footer** element is typically placed on a web page:

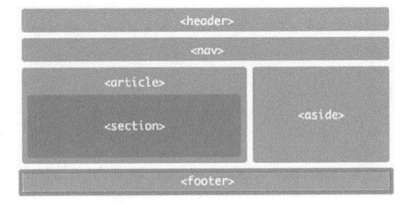

Figure 2.4: The footer element

In the following figure, you can see that the **footer** element is highlighted via a box. As this is an example taken from the Packt website, you will notice it contains items such as useful links and social media icons:

Figure 2.5: The footer element on Packt site

The section Tag

The **section** tag is different from the header and footer tags as it can be used in many different places on a web page. Some examples of when you would use a **section** tag could be for the main content area of a page or to group a list of related images together. You use this tag anytime you want to divide some of the markup into a logical section of the page. Again, prior to HTML5, you would most likely use a **div** tag with a class name to divide a section of the page. The following codes show the differences between the old and new way of writing the markup for the **section** area:

```
<!-- old way -->
<div class="main-content-section">
  <!-- main content -->
</div>

<!-- new way -->
<section>
  <!-- main content -->
</section>
```

In the following figure, you can see that the section element is highlighted via a box, illustrating where a section element is typically placed on a web page:

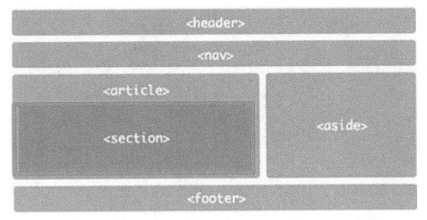

Figure 2.6: The section element

The article Tag

The **article** tag is used for the self-contained part of a web page. Some examples of an article could be an individual news article or blog post. You can have multiple articles on a page, but each must be self-contained and not dependent on any other context within the page. It is common to see the **article** tag used in conjunction with **section** tags to divide up an article into discrete sections. The following code shows this:

```
<article>
    <section>
      <!-- primary blog content -->
    </section>
    <section>
      <!-- secondary blog content -->
    </section>
</article>
```

In the following figure, you can see that the article element is highlighted via a box, illustrating where an article element is typically placed on a web page:

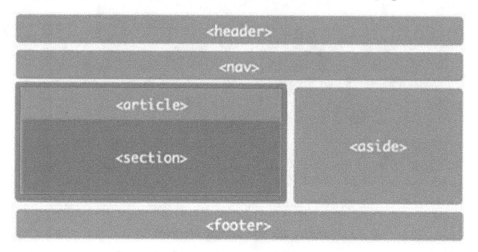

Figure 2.7: The article element

The nav Tag

Inside the navigation area, you will have a list of page links for the different pages of the website. Prior to HTML5, you would again use a **div** tag with a class name. The following codes show the differences between the old and new way of writing the markup for the navigation area:

```
<!-- old way -->
<div class="navigation">
  <!-- list of links go here -->
</div>
<!-- new way -->
<nav>
  <!-- list of links go here -->
</nav>
```

In the following figure, you can see that the **nav** element is highlighted via a box, illustrating where a **nav** element is typically placed on a web page:

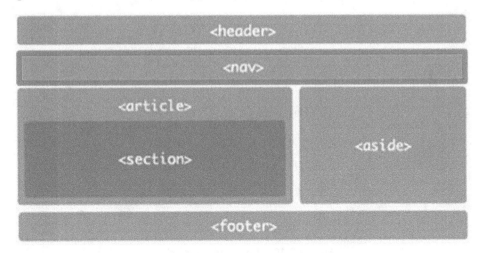

Figure 2.8: The nav element

In the following figure, you can see that the **nav** element is highlighted via a box. As this is an example taken from the Packt website, you will notice it contains a list of page links:

Figure 2.9: The nav element on the Packt site

The aside Tag

The **aside** tag is used to show content that is indirectly related to the main content of a document. You will typically see this tag used for sidebars or for showing notes relating to some content. Again, before the advent of HTML5, developers would use a **div** tag with a class name for this type of content. The following codes show the differences between the old and new way of writing the markup for the **aside** element:

```
<!-- old way -->
<div class="sidebar">
  <!-- indirectly related content goes here -->
</div>
<!--new way -->
<aside>
  <!-- indirectly related content goes here -->
</aside>
```

In the following figure, you can see that the **aside** element is highlighted via a box, illustrating where an aside element is typically placed on a web page:

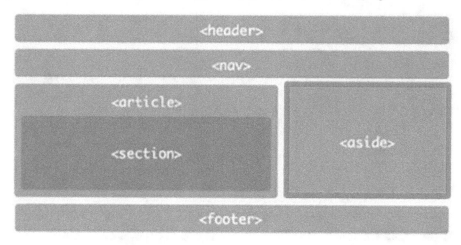

Figure 2.10: The aside element

The div Tag

The **div** tag is probably the most widely used tag on the World Wide Web. In fact, if you view the source code of your favorite website, most of the HTML elements you see will be **div** elements. This tag actually stands for division and is used to divide or group content together. Although HTML5 provides specialist elements for the most common types of page groups, you will still find many uses for using **div** tags. It might help to think of this element as a generic way to group the markup into logical parts. The following are a few example codes of how a **div** tag may be used:

```
<div class="sidebar">
  <!-- indirectly related content goes here -->
</div>
<div class="navigation">
  <div class="navigation-inner"><!-- navigation links go here --></div>
</div>
```

That concludes our tour of the structural HTML elements that are important to us. We will now apply some of this theory with the help of an exercise.

A News Article Web Page

Now that we have an understanding of the structural elements provided by HTML5, let's put our newly acquired knowledge into practice by writing the structural HTML for a news article page. You can get a sense of what this type of page will look like by visiting a popular online news website such as https://packt.live/2nx1HLh or https://packt.live/2AYuA6e and clicking on an article.

Exercise 2.01: Marking up the Page

In this exercise, we will create the markup for our HTML5 page. Our aim will be to produce a page with output, similar to that of *Figure 2.10* without the `<section>` element in it.

Let's complete the exercise with the following steps:

1. Create a folder named **Chapter02** and within this folder create a file named **Exercise 2.01.html** in VSCode.

2. We will use the following starter HTML document, which contains some basic styling for our structural elements. Don't worry if you don't understand the CSS just yet; you will by the end of this book:

```
<!DOCTYPE html>
<html lang = "en">
    <head>
        <title>News article page</title>
        <style>
          header,
          nav,
          article,
          aside,
          footer {
            background: #659494;
            border-radius: 5px;
            color: white;
            font-family: arial, san-serif;
            font-size: 30px;
```

```
          text-align: center;
          padding: 30px;
          margin-bottom: 20px;
        }
      header:before,
      nav:before,
      article:before,
      aside:before,
      footer:before {
        content: '<';
      }
      header:after,
      nav:after,
      article:after,
      aside:after,
      footer:after {
        content: '>';
      }
      article {
        float: left;
        margin-right: 20px;
        width: 60%;
      }
      aside {
        float: left;
        width: calc(40% - 140px);
      }
      footer {
        clear: both;
      }
    </style>
  </head>
  <body>
    <!-- your code will go here -->
  </body>
</html>
```

3. First, let's add our first structural element, which is the **header** tag. We will place it in between the opening and closing body tags. In this example, we will just add some text as content but, when building a real web page, you would include things such as logos, search bars, and links:

```
<body>
    <header>header</header>
</body>
```

4. After our **header** tag comes the navigation area, which is used for including links to different pages of the website. Once again, we will just add some text for the content but, when building a real web page, you would include a list of links:

```
<body>
    <header>header</header>
    <nav>nav</nav>
</body>
```

5. For the main news article content, we will use an **article** tag. Once again, we will just add some text for the content but, when building a real web page, you would include the content of the articles:

```
<body>
    <header>header</header>
    <nav>nav</nav>
    <article>article</article>
</body>
```

6. To the right of the **article** tag, we have an **aside** tag, which will typically contain content such as advertising images or related content links:

```
<body>
    <header>header</header>
    <nav>nav</nav>
    <article>article</article>
    <aside>aside</aside>
</body>
```

7. Finally, we can finish off the markup for our web page by adding the **footer** tag at the bottom of the page. For now, we will just add some text as content but, in real life, you would include elements such as copyright information, and links to other pages:

```
<body>
    <header>header</header>
    <nav>nav</nav>
    <article>article</article>
    <aside>aside</aside>
    <footer>footer</footer>
</body>
```

If you now right-click on the filename in VSCode on the left-hand side of the screen and select **Open In Default Browser**, you will see the following web page in your browser:

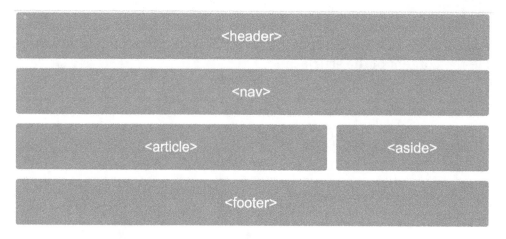

Figure 2.11: Output for the product page

If you look at this page in your browser, you may not be impressed with what you see, but you actually have the foundations in place for a web page.

Wireframes

When working on commercial projects, it is common for web page designs to be provided to web developers in the form of a **wireframe**. A wireframe is a low-fidelity design that provides enough information about a page for the developer to start coding. Usually, they will not include much visual design information and are focused on the main structure of a page. The following figure is an example of a wireframe for a new home page:

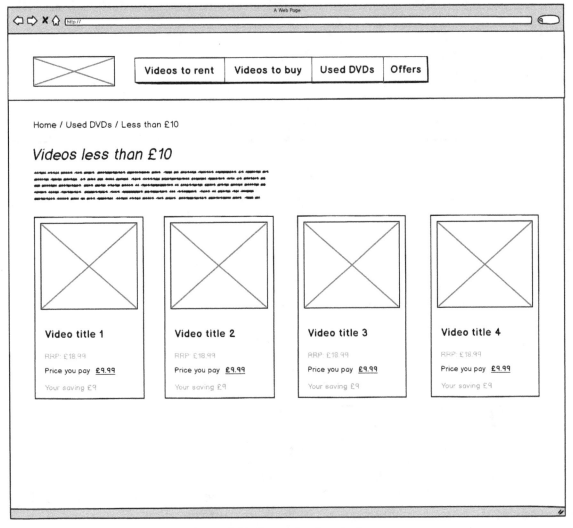

Figure 2.12: Example of a wireframe

Activity 2.01: Video Store Home Page

Suppose you are a frontend developer working for a tech start-up. You have been asked to build a home page for the online video store. You have been given the following wireframe from the UX designer:

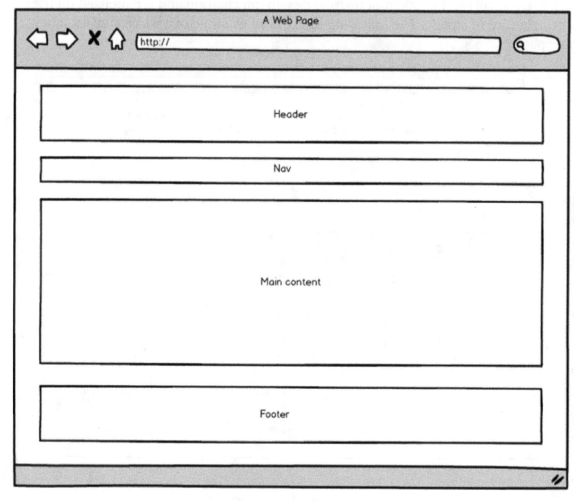

Figure 2.13: Wireframe as per the UX designer's expectation

Using your newly acquired HTML5 knowledge, you can start to convert the wireframe into working HTML code. At this stage, you should just be concerned with writing the structural HTML tags and shouldn't worry about content right now.

The aim will be to achieve a web page similar the following output screenshot:

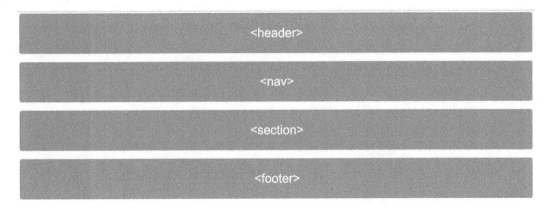

Figure 2.14: Expected output of video store home page

The steps are as follows:

1. Create a file named **Activity 2.01.html** within the **Chapter02** folder in VSCode.

2. Use the following code as a page skeleton. Again, do not worry about not understanding the styling part of the code:

```
<!DOCTYPE html>
<html lang = "en">
    <head>
        <title>Video store home page</title>
        <style>
          header,
          nav,
          section,
          footer {
            background: #659494;
            border-radius: 5px;
            color: white;
            font-family: arial, san-serif;
            font-size: 30px;
            text-align: center;
            padding: 30px;
            margin-bottom: 20px;
          }

          header:before,
          nav:before,
          section:before,
          footer:before {
```

```
        content: '<';
      }

    header:after,
    nav:after,
    section:after,
    footer:after {
      content: '>';
    }
  </style>
</head>
<body>
  <!-- your code will go here -->
</body>
</html>
```

3. Start adding the HTML5 structural elements inside the **body** tag one by one, the same as we did in *Exercise 2.01, Marking up the Page*.

4. As with *Exercise 2.01, Marking up the Page*, we will just add the tag name for content such as **header** and **footer**.

If you now right-click on the filename in VSCode on the left-hand side of the screen and select **Open In Default Browser**, you will see the web page in your browser.

Hopefully, you are now getting a feel for the process of putting basic web pages together. We will build on this knowledge in the coming exercises.

> **Note**
>
> The solution to this activity can be found on page 592.

We are now ready to start making our web pages more realistic by learning some CSS page layout techniques.

CSS Page Layouts

CSS provides us with a range of possibilities for laying out web pages. We will be looking into the three most common techniques for laying out web pages. These are as follows:

- **float**

- **flex**

- **grid**

Armed with this knowledge, combined with your knowledge of HTML structural tags, you will be able to code a range of web page layouts. The concepts learned in this part of the chapter will form the core of your frontend development skillset and you will use these techniques over and over throughout your career.

Video Store Product Page

In order to gain a solid understanding of how these three different approaches to layout work, we shall use a video store product listing page as a concrete example. We will work through solutions to the following design using the three most common layout techniques, one by one. For the examples that follow, we will only be concerned with the product section of the page:

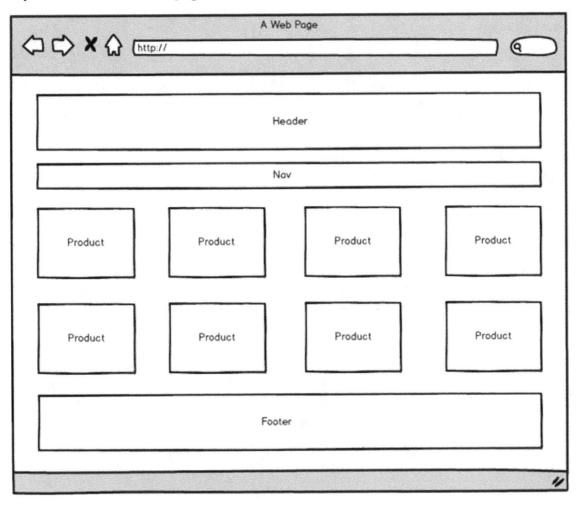

Figure 2.15: Product page wireframe

Float-Based Layouts

The **float**-based CSS layout technique is the oldest of the three. Whilst CSS provides us with improved techniques for layout, the **float**-based layout is still used today. Having a firm grasp of how **float**-based layouts work in practice will set you up for more advanced styling segments in this book.

The float Property

The CSS **float** property, when applied to an element, will place the element to either the left or right of its containing element. Let's examine a few examples of the most common values for this property.

To **float** elements to the right, you would use the **right** value, as shown in the following code:

```
float: right;
```

Whereas, to float elements to the left, you would use the **left** value, as shown in the following code:

```
float: left;
```

The **none** value isn't used as frequently but, with the following code, it can be handy if you wish to override either the left or right values:

```
float: none;
```

The width Property

When we apply the **float** property to elements, we typically will also want to give the element an explicit **width** value as well. We can either give a value in pixels or percentages. The following code shows the input for **width** in pixels, that is, by writing **px** after the value:

```
width: 100px;
```

Whereas the following code shows the input for **width** as a percentage, that is, by entering the **%** symbol after the value:

```
width: 25%;
```

Clearing Floated Elements

As the name suggests, floated elements do, in fact, appear to float in relation to the other non-floated elements on the page. A common issue with floated elements inside a container is illustrated in the following figure:

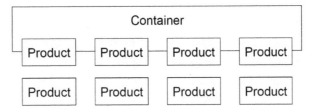

Figure 2.16: Floating elements' illustration

> **Note**
>
> This solution of clearing floated elements has been used for simplicity.

There are many solutions to this issue, but by far the easiest solution is to apply the following CSS to the containing element:

```
section {
    overflow: hidden;
}
```

With the preceding code added to the container, we will now have floated elements contained inside the wrapping element, as illustrated in the following figure:

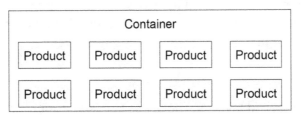

Figure 2.17: Cleared floats' illustration

The following example code shows how you could achieve the preceding layout using **float**:

```
<!-- HTML -->
<section>
  <div>product 1</div>
  <div>product 2</div>
  <div>product 3</div>
  <div>product 4</div>
  <div>product 5</div>
  <div>product 6</div>
  <div>product 7</div>
  <div>product 8</div>
</section>
/* CSS */
section {
  overflow: hidden;
}
div {
  float: left;
  width: 25%;
}
```

Flex-Based Layouts

The **flex**-based CSS layout technique is a new and improved alternative to the **float**-based approach. With **flex**, we have much more flexibility and can easily achieve complex layouts with very little code. With **flex**, we no longer have to worry about clearing floating elements. We will now look into some of the key properties and values in order to let us build the product page layout using **flex**.

The flex Container

When developing **flex**-based layouts, there are two key concepts you must first understand. The first is the **flex** container, which is the element that contains the child elements. To activate a **flex** layout, we must first apply the following code to the container or parent element that holds the individual items:

```
display: flex;
```

We also have to choose how we want the container to handle the layout of the child elements. By default, all child elements will fit into one row. If we want the child elements to show on multiple rows, then we need to add the following code:

```
flex-wrap: wrap;
```

The flex Items

Now that we know how to set the **flex** container up, we can turn to the child elements. The main issue of concern here is the need to specify the width of the child elements. To specify this, we need to add the following code:

```
flex-basis: 25%;
```

You can think of this as being equivalent to the width in our **float**-based example.

The following example code shows how you could achieve the product layout, as shown in *Figure 2.15*, using **flex**:

```
<!-- HTML -->
<section>
   <div>product 1</div>
   <div>product 2</div>
   <div>product 3</div>
   <div>product 4</div>
   <div>product 5</div>
   <div>product 6</div>
   <div>product 7</div>
   <div>product 8</div>
</section>

/* CSS */
section {
   display: flex;
   flex-wrap: wrap;
}
div {
   flex-basis: 25%;
}
```

Grid-Based Layouts

The **grid**-based CSS layout technique is the newest of the three different approaches we will be exploring. This new approach was introduced in order to simplify the page layout and offer developers even more flexibility vis-à-vis the previous two techniques. We will now look into some of the key properties and values to enable us to build the product page layout using a grid-based approach.

The grid Container

When developing **grid**-based layouts, there are two key concepts you must first understand. The first is the grid container, which is the element that contains the child elements. To activate a grid layout, we must first apply the following code to the parent element:

```
display: grid;
```

Now that we have activated the container to use the **grid**-based layout, we need to specify the number, and sizes, of our columns in the grid. The following code would be used to have four equally spaced columns:

```
grid-template-columns: auto auto auto auto;
```

The grid Items

When we used **float** and **flex** layouts, we had to explicitly set the width of the child elements. With **grid**-based layouts, we no longer need to do this, at least for simple layouts.

We will now put our new-found knowledge into practice and build the product cards shown in *Figure 2.15*. We will use the grid layout technique since the product cards are actually within a **grid** layout, comprising four equally spaced columns.

Exercise 2.02: A grid-Based Layout

In this exercise, we will create our CSS page layout with the aim of producing a web page where six products are displayed as shown in the wireframe *Figure* 2.15. The expected output is as follows:

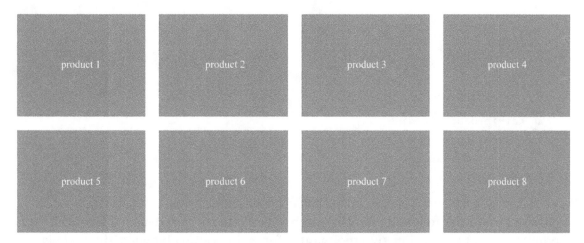

Figure 2.18: Expected output for the grid based layout

Following are the steps to complete this exercise:

1. Let's begin with the following HTML skeleton and create a file called **Exercise 2.02.html** within the **Chapter02** folder in VSCode. Don't worry if you do not understand the CSS used here; you will soon enough:

```html
<!DOCTYPE html>
<html lang = "en">
<head>
  <title>Grid based layout</title>
  <style type="text/css">
    div {
      background: #659494;
      color: white;
      text-align: center;
      margin: 15px;
      padding: 100px;
    }
  </style>
</head>
<body>
</body>
</html>
```

2. Next, we will add the product items using **div** tags, which are placed inside a **section** tag. We will just add a product with a number inside each item, so we know what product each represents:

```
<body>
  <section>
    <div>product 1</div>
    <div>product 2</div>
    <div>product 3</div>
    <div>product 4</div>
    <div>product 5</div>
    <div>product 6</div>
    <div>product 7</div>
    <div>product 8</div>
  </section>
</body>
```

3. Now, let's add the following CSS in order to activate the **grid**-based layout. If you compare this to the other two techniques for laying out web pages, the code is very minimal:

```
section {
  display: grid;
  grid-template-columns: auto auto auto auto;
}
```

If you now right-click on the filename in VSCode on the left-hand side of the screen and select **Open In Default Browser** you will see the output as shown in *Figure 2.18*.

We will now take a detour and look into some fundamental concepts of how CSS styles HTML elements.

The Box Model

So far, all the elements on our pages look almost identical because we have not learned how to adjust the size of each element. We are now ready to progress to more realistic page designs by introducing a foundational layout concept called the **box model**.

Try to picture each HTML element as a box made up of different layers. The different layers are the element's content box, **padding**, border, and margin. We will explore each of these layers one by one. The following figure illustrates how all aspects of the box model relate to one another. You can see that the margin is the outermost part, followed by the element's border and padding between the border and content area:

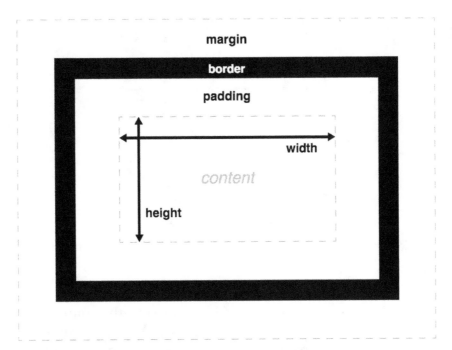

Figure 2.19: The box model

We will now look at each of the box model elements, in turn, starting with the innermost content box.

Content Box

The content box is the part of the element where the actual content lives. This is typically text but could contain other child elements or media elements such as images. The most important CSS properties for this layer are **width** and **height**. Absolute length units such as pixel are not recommended for use on screen, because screen sizes vary so much. Developers should use relative length units to specify a length relative to another length property. Relative length units scale better between different rendering mediums. The following code shows some example values, followed by the corresponding output figure for these properties:

```
width: 200px;
height: 100px;
```

In the following figure, we will see what the content area looks like after CSS is applied to the preceding code:

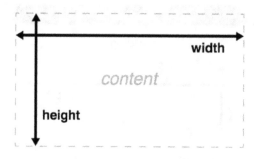

Figure 2.20: The content box

Next, we will work our way out to the next layer of the box model – padding.

The padding Property

The padding area is the layer that provides spacing between the content box and the border. The amount of spacing in this layer can be specified in all directions – top, right, bottom, and left. CSS provides a padding property where you can specify values for the amount of spacing in all directions. If you want to apply the same amount of **padding** in all directions, you can just give a single value. If you want to apply the same values for vertical and horizontal directions, you can specify two values. It also provides direction-specific properties – **padding-top**, **padding-right**, **padding-bottom**, and **padding-left**. The following code shows a number of example values for these properties:

```
/* 50px of padding applied in all directions */
padding: 50px;


/* 50px of padding applied vertically and 0px applied horizontally */
padding: 50px 0;


/* 10px of padding applied to the top */
padding-top: 10px;


/* 10px of padding applied to the right */
padding-right: 10px;


/* 10px of padding applied to the bottom */
padding-bottom: 10px;


/* 10px of padding applied to the left */
padding-left: 10px;
```

The following figure illustrates what the **content** and **padding** areas would look like after CSS is applied to the following code:

```
width: 200px;
height: 100px;
padding: 25px;
```

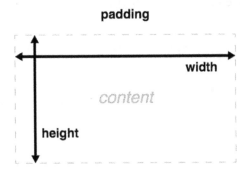

Figure 2.21: Padding

Now that we understand how the content and padding layers relate to one another, we will work our way out to the next layer of the box model – the border.

The border Property

The border area is the layer that sits between the end of the padding area and the beginning of the margin. By default, the border isn't visible; it can only be seen when you explicitly set a value that will allow you to see the border. Similar to the padding property, CSS provides a shorthand property called border, and also the direction-specific properties – **border-top**, **border-right**, **border-bottom**, and **border-left**. All of these properties require three values to be provided; the **width** of the border, the **border** style, and finally, the color of the border. The following code shows some example values for these properties:

```
/* border styles applied in all directions */
border: 5px solid red;

/* border styles applied to the top */
border-top: 5px solid red;

/* border styles applied to the right */
border-right: 15px dotted green;
```

```
/* border styles applied to the bottom */
border-bottom: 10px dashed blue;

/* border styles applied to the left */
border-left: 10px double pink;
```

The following figure illustrates how the four different border styles would appear if applied to an element:

Figure 2.22: Border styles

The **content**, **padding**, and **border** layers is obtained with the following code:

```
width: 200px;
height: 100px;
padding: 25px;
border: 10px solid black;
```

The following figure is the output for the preceding code:

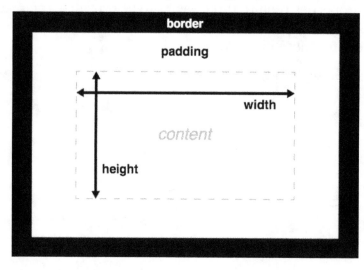

Figure 2.23: Border

Now that we understand how the content, **padding** and **margin** layers relate to one another, we will work our way out to the final layer of the box model – the **margin**.

The margin Property

The margin area is the layer that provides spacing between the edge of the border and out toward other elements on the page. The amount of spacing in this layer can be specified in all directions – top, right, bottom, and left. The CSS provides a margin property where you can specify values for the amount of spacing in all directions. It also provides direction-specific properties – **margin-top**, **margin-right**, **margin-bottom**, and **margin-left**. The following code shows a number of example values for these properties:

```
margin: 50px;
margin: 50px 0;
margin-top: 10px;
margin-right: 10px;
margin-bottom: 10px;
margin-left: 10px;
```

The **content**, **padding**, **border**, and **margin** layers is obtained with the following code:

```
width: 200px;
height: 100px;
padding: 25px;
border: 10px solid black;
margin: 25px;
```

The following figure is the output for the preceding code:

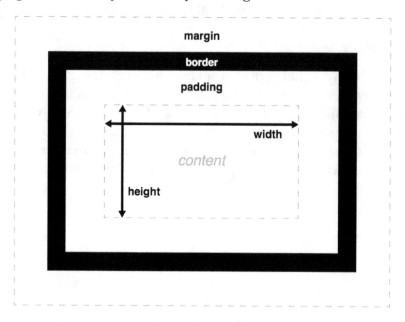

Figure 2.24: Margin

To get some practice looking at how different HTML elements make use of the box model, you can use the webtools inspector in your favorite browser. In Chrome, you can inspect an element and investigate how the box model is used for each element. If you inspect an element and then click the **Computed** tab on the right-hand side, you will see a detailed view. The following figure shows an example of an element from the Packt website revealing the values for properties from the box model:

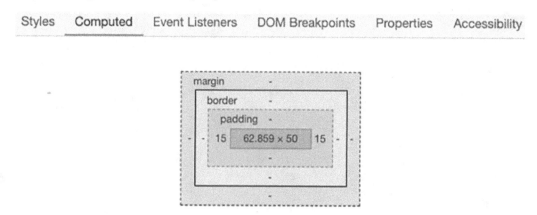

Figure 2.25: Chrome web tools box model inspection view

In the following exercise, we will play around with the different box model properties to get some practice with box model-related CSS properties.

Exercise 2.03: Experimenting with the Box Model

The aim of this exercise will be to create the three boxes as shown in the following output screenshot:

Figure 2.26: Expected boxes

The steps to complete the exercise are as follows:

1. First, let's add the following HTML skeleton to a file called **Exercise 2.03.html** within the **Chapter02** folder in VSCode:

```
<!DOCTYPE html>
<html lang = "en">
<head>
  <title>Experimenting with the box model</title>
  <style type="text/css">
  </style>
</head>
<body>
  <div class="box-1">Box 1</div>
  <div class="box-2">Box 2</div>
  <div class="box-3">Box 3</div>
</body>
</html>
```

2. Now, let's add some CSS to the first box, observing the **width**, **height**, **padding**, and **border** properties we are adding. We will add the CSS in between the opening and closing style tags, as shown in the following code, to render the following figure:

```
<style type="text/css">
   .box-1 {
      float: left;
      width: 200px;
      height: 200px;
      padding: 50px;
      border: 1px solid red;
   }
</style>
```

The following figure shows the output of the preceding code:

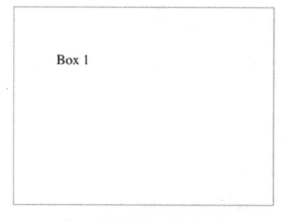

Box 1

Figure 2.27: Output for box 1

3. Now, let's add the CSS to the second box in *Figure 2.25*, observing how the **width**, **height**, **padding**, and **border** properties differ from the first box. We are using percentage-based measurements for the width and height properties, as shown in the following code:

```
.box-2 {
   float: left;
   width: 20%;
   height: 20%;
   padding-top: 50px;
   margin-left: 10px;
   border: 5px solid green;
}
```

The following figure shows the output of the preceding code:

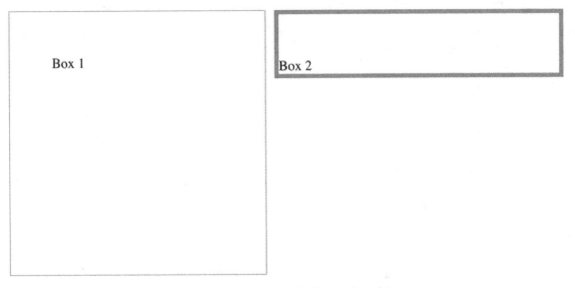

Figure 2.28: Output for boxes 1 and 2

4. Finally, let's add the CSS to the third box in *Figure* 2.25, observing how the **width**, **height**, **padding**, and **border** properties differ from the first and second boxes, as shown in the following code, to render the following figure:

```
.box-3 {
    float: left;
    width: 300px;
    padding: 30px;
    margin: 50px;
    border-top: 50px solid blue;
}
```

The following figure shows the output of the preceding code:

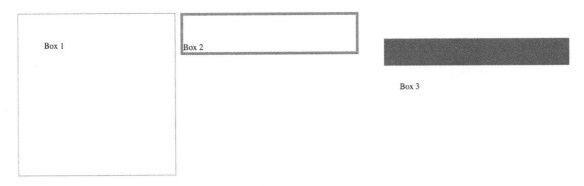

Figure 2.29: Output for boxes 1, 2, and 3

If you now right-click on the filename in VSCode on the left-hand side of the screen and select **Open In Default Browser**, you will see the web page in your browser.

This should give you a sense of what's possible with the box model. Feel free to change the various different properties and experiment with different combinations.

Putting It All Together

We now know how to correctly markup a web page with the correct HTML5 structural tags. We also know how to use the three most popular CSS layout techniques. Finally, we have an understanding of how the box model works. We will now build the two complete web pages, combining all of the things we have learned so far in this chapter.

Exercise 2.04: Home Page Revisited

In this exercise, we will be using the wireframe in *Figure* 2.13 for a home page design used in *Activity 2.01, Video Store Home Page*. We will build a version of this page, incorporating the concepts from the box model topic. Our aim will be to build a page as shown in the wireframe *Figure* 2.15:

The steps to complete this exercise are as follows:

1. Create a new file called **Exercise 2.04.html** within the **Chapter02** folder in VSCode.

2. Use the following HTML code as a start file. Again, don't worry if some of the CSS doesn't make sense to you. We will look into this part of the styling in more detail in *Chapter 3, Text and Typography*:

```html
<!DOCTYPE html>
<html lang = "en">
    <head>
        <title>Video store home page</title>
        <style>
          header,
          nav,
          section,
          footer {
             background: #659494;
             border-radius: 5px;
             color: white;
             font-family: arial, san-serif;
             font-size: 30px;
             text-align: center;
          }
```

```
        header:before,
        nav:before,
        section:before,
        footer:before {
          content: '<';
        }

        header:after,
        nav:after,
        section:after,
        footer:after {
          content: '>';
        }
      </style>
    </head>
    <body>
      <header>header</header>
      <nav>nav</nav>
      <section>main content</section>
      <footer>footer</footer>
    </body>
  </html>
```

3. Now, let's add some styling for the structural elements. Notice how we have used what we have learned from *The Box Model* topic to include **border**, **padding**, and **margin** with our structural elements. We will use a border to visually define the outer edge of the element, along with some padding to add spacing between the text and the outer edge of the element and a bottom **margin** to provide vertical spacing between the elements. We will add this just before the closing style tag:

```
/* CSS code above */
header,
nav,
section,
footer {
  border: 1px solid gray;
  padding: 50px;
  margin-bottom: 25px;
}

</style>
```

If you now right-click on the filename in VSCode on the left-hand side of the screen and select **Open In Default Browser**, you will see the web page in your browser as shown in the following figure:

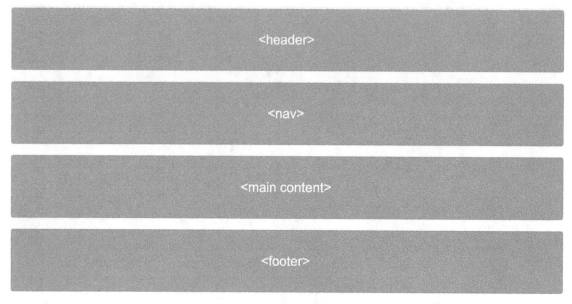

Figure 2.30: Output of home page

You should see a web page resembling the one shown in the home page wireframe.

Exercise 2.05: Video Store Product Page Revisited

In this exercise, we will be using the wireframe for a product page design as in *Figure 2.15*. We will build a more realistic version incorporating the box model. Our aim will be to build a page as shown in the wireframe *Figure 2.15*.

The steps to complete the exercise are as follows:

1. Create a new file called **Exercise 2.05.html** within the **Chapter02** folder in VSCode with the following code:

```
<!DOCTYPE html>
<html lang = "en">
<head>
  <title>Video store product page</title>
  <style>
  </style>
</head>
<body>
</body>
</html>
```

2. In order to add styling, add the following code in between the **style** tags:

```
header, nav, section, footer {
  background: #659494;
  border-radius: 5px;
  color: white;
  font-family: arial, san-serif;
  font-size: 30px;
  text-align: center;
}
header:before, nav:before, footer:before {
  content: '<';
}
header:after, nav:after, footer:after {
  content: '>';
}
```

3. We will now add the HTML for the page elements, which are **header**, **nav**, **section**, and **footer**. The product items will be **div** elements inside the **section** element, as shown in the following code:

```
<body>
  <header>header</header>
  <nav>nav</nav>
  <section>
    <div>product 1</div>
    <div>product 2</div>
    <div>product 3</div>
    <div>product 4</div>
    <div>product 5</div>
    <div>product 6</div>
    <div>product 7</div>
    <div>product 8</div>
  </section>
  <footer>footer</footer>
</body>
```

4. Now, let's add some styling for the structural elements. This is the same code as in the previous exercise. We will use a border to visually define the outer edge of the element, along with some padding to add spacing between the text and the outer edge of the element and a bottom margin to provide vertical spacing between elements. Again, we will add the CSS just before the closing **style** tag:

```css
/* CSS code above */
header,
nav,
section,
footer {
   border: 1px solid gray;
   padding: 20px;
   margin-bottom: 25px;
}
</style>
```

5. We will now need to add some styling for the product cards. We will use the **grid** layout technique, as this will allow our code to be as concise as possible:

```css
/* CSS code above */
section {
   display: grid;
   grid-template-columns: auto auto auto auto;
}
section div {
   border: 2px solid white;
   padding: 30px;
   margin: 10px;
}
</style>
```

If you now right-click on the filename in VSCode on the left-hand side of the screen and select **Open In Default Browser**, you will see the web page in your browser as shown in the following figure:

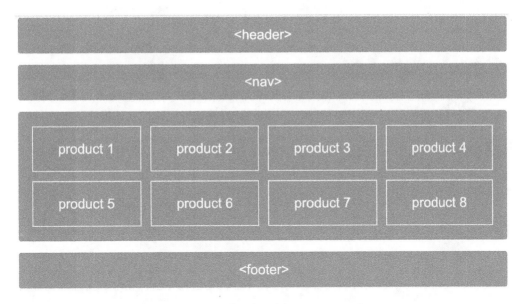

Figure 2.31: Output for the video store product page

You should now see a web page resembling the one shown in the product page wireframe.

Activity 2.02: Online Clothes Store Home Page

Suppose you are a freelance web designer/developer and have just landed a new client. For your first project, the client wants a web home page developed for their online clothes store.

Using the skills learned in this chapter, design and develop the home page layout for the new online store.

The steps are as follows:

1. Produce a wireframe, either by hand or by using a graphics tool, for the new home page layout.

2. Create a file named **Activity 2.02.html** within the **Chapter02** folder in VSCode.

3. Start writing the markup for the page.

4. Now, style the layout with CSS.

The following figure shows the expected output for this activity:

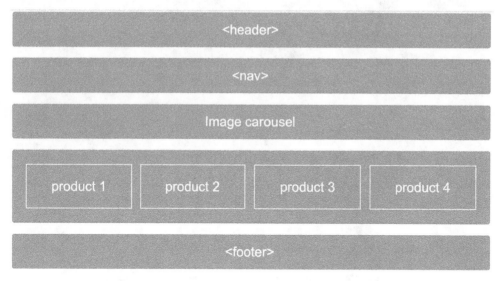

Figure 2.32: Expected output for the online clothes store home page

> **Note**
>
> The solution to this activity can be found on page 594.

Summary

In this chapter, we have begun our journey into building web pages. Knowing the range of HTML tags available to you is crucial in writing well-formed HTML documents. These include header, footer, and section tags.

You should now feel comfortable taking a visual design or wireframe and converting this into the skeleton of an HTML document. We also looked at three common ways of styling a page layout with CSS. These involved the use of **float**-, **flex**-, and **grid**-based layout techniques. We then looked into what makes up the box model and used this knowledge to build the home and product pages of the video store.

In the next chapter, we will learn about the non-structural HTML elements used for content on a web page. We will then look into a number of common styling approaches to these elements using CSS.

3

Text and Typography

Overview

By the end of this chapter, you will be able to identify the most suitable heading element for a web page; use the most common text-based HTML elements; develop common web page features such as navigation and breadcrumbs; explain the importance of semantic markup; and convert a design into semantic markup. This chapter introduces the text-based elements used for web page content. We will first take a tour through the text-based HTML elements. We will then look into the styling of these elements and finally put this into practice with some exercises to do with building common web page components.

Introduction

In the previous chapter, you may have been frustrated that our web pages only dealt with the page structure and didn't contain any actual content. In this chapter, we will look at the HTML elements, that are used for text-based page content. You are probably familiar with most of these elements from using word processing applications when writing documents. These HTML elements include elements such as headings, paragraphs, and lists.

As you are aware, when building web pages, the markup is only one aspect. We also need to style elements of the page. We will take a look at issues concerning the styling of these text-based elements. We will look into concerns such as cross-browser consistency. Finally, we will walk through some exercises on developing common web page components such as navigation bars and breadcrumbs.

Text and typography are very important as a visually appealing web page will ensure that a web user spends more time on it (as this may be good for business conversion).

A highly functional, fast, and efficient website may not attract users if the text/typography is of poor quality. Thus, these are crucial elements while designing a web page, and this chapter equips you with the necessary tool in HTML/CSS to create visually appealing and aesthetic web pages.

Text-Based Elements

HTML provides us with a variety of elements that are used for text-based content. While browsing the web, you might notice that web pages typically have similar text-based content. Most web pages will contain a page heading. The content will typically comprise headings, paragraphs, and lists. HTML equips you with tools to format such elements within a web page.

In this section, we will be looking at the following HTML text-based elements:

- Headings
- Paragraphs
- Inline text elements
- Lists

Headings

Heading elements in HTML offer six levels of hierarchy, ranging from **h1** to **h6**. Now, **h1** is typically only used once on a page as it is the topmost heading for the document as a whole. The following code snippet shows how all of these headings are used and what they look like in the browser by default:

```
<h1>Heading level 1</h1>
<h2>Heading level 2</h2>
<h3>Heading level 3</h3>
<h4>Heading level 4</h4>
<h5>Heading level 5</h5>
<h6>Heading level 6</h6>
```

A web page from the preceding code would appear as follows:

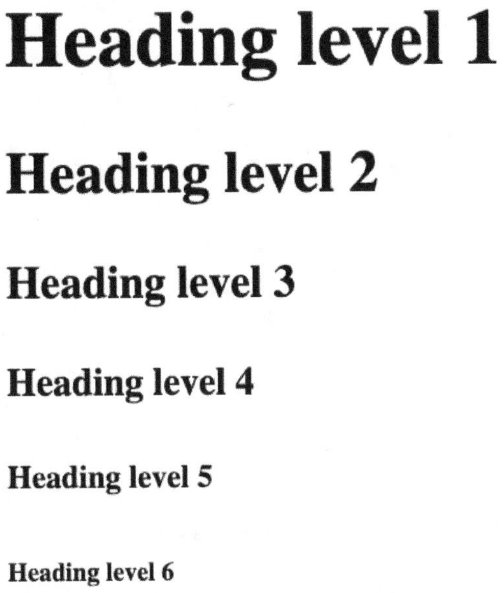

Heading level 1

Heading level 2

Heading level 3

Heading level 4

Heading level 5

Heading level 6

Figure 3.1: Headings shown in a browser

Paragraphs

Paragraphs in HTML can be represented using the **p** tag. On a web page, you might have chunks of the core content of a topic presented to the reader. Such content is included under a **p** tag in HTML. The following code snippet shows how you would include paragraphs in a document and what they look like by default in the browser:

```
<p>
Horatio says 'tis but our fantasy, And will not let belief take hold of him
Touching this dreaded sight, twice seen of us. Therefore I have entreated him
along,With us to watch the minutes of this night, That, if again this apparition.
come, He may approve our eyes and speak to it.
</p>

<p>Tush, tush, 'twill not appear.</p>

<p>
Sit down awhile, And let us once again assail your ears,
That are so fortified against our story,
What we two nights have seen.
</p>
```

> **Note**
>
> The text used for the preceding code is a work of Shakespeare sourced from the following: https://packt.live/2Cmvni4.

A web page for the preceding code will appear as follows:

Horatio says 'tis but our fantasy, And will not let belief take hold of him Touching this dreaded sight, twice seen of us. Therefore I have entreated him along, 35 With us to watch the minutes of this night, That, if again this apparition come, He may approve our eyes and speak to it.

Tush, tush, 'twill not appear.

Sit down awhile, And let us once again assail your ears, That are so fortified against our story, What we two nights have seen.

Figure 3.2: Paragraphs shown in a browser

Inline Text Elements

As a designer of web pages, you may often find yourself in a situation where you need to highlight special terms in a paragraph. Fortunately, HTML provides a solution to this.

It is possible to add what are called inline elements around text contained within paragraphs. Think of using a word processor: you are able to make words bold, underlined, italicized, and so on. HTML provides developers with this ability, and we will now look at some of the most common examples.

If you want to emphasize some text, you can use the **em** tag. An example of how you would use this and what it would look like in a browser is shown here:

```
<p>I need to wake up <em>now</em>!</p>
```

A web page for the preceding code would appear as follows:

I need to wake up *now*!

Figure 3.3: The em tag as it appears on a web page

When you want to show some text as having serious importance, you can use the **strong** tag. An example of how you would use this and what it would look like in a browser is shown here:

```
<p>
Before leaving the house <strong>remember to lock the front
door</strong>!
</p>
```

A web page for the preceding code would appear as follows:

Before leaving the house **remember to lock the front door**!

Figure 3.4: The strong tag as it appears on a web page

Perhaps the most important of all the inline text-based elements is the anchor element, which allows you to add hyperlinks. To inline a link, you use an **a** tag wrapped around some text. An example of how you would use this and what it would look like in a browser is shown here:

```
<p>
Please click <a href="http://www.google.com">here</a> to go to google
</p>
```

A web page for the preceding code would appear as follows:

Please click here to go to google

Figure 3.5: Anchor as it appears on a web page

Another important inline element for you to learn to use is the **span** tag. This is similar to the **div** tag but is used for inline elements. The **span** tag is used as a generic way to divide up content and has no inherent meaning, unlike the other inline tags mentioned previously in this chapter. A common use case is when styling a part of an element's content differently to the rest of the content. The following code shows an example of this:

```
/* styles */
.red {
    color: red;
}
.green {
    color: green;
}
.blue {
    color: blue;
}
<!-- markup -->
<p>
   My favorite colors are <span class="red">red</span>,
   <span class="green">green</span> and <span class="blue">blue</span>.
</p>
```

A web page for the preceding code would appear as follows:

My favorite colors are red, green and blue.

Figure 3.6: Paragraph with highlighted words as it appears on a web page

Lists

Another common type of text-based element that you will be very familiar with is the list. In HTML, these come in three different types: an unordered list, an ordered list, and a definition list. We will take a look at the differences between these types of lists and when you should use them.

Let's begin by taking a look at by far the most common type of list, the unordered list, which is expressed in HTML as **ul**, with **li** being used for the list items. You will most likely be very familiar with this type of list in your everyday life. A common example of this type of list could be a shopping list or a list of things you need to pack before going on holiday. What makes this type of list unordered is the fact that the order of the items in the list isn't important. The following code shows an example of this type of list as you would use it in HTML:

```
<!-- Shopping list -->
<ul>
    <li>Ice Cream</li>
    <li>Cookies</li>
    <li>Salad</li>
    <li>Soap</li>
</ul>
```

A web page for the preceding code would appear as follows:

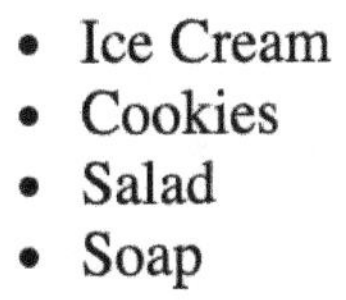

- Ice Cream
- Cookies
- Salad
- Soap

Figure 3.7: Unordered list as it appears on a web page

Following the unordered list, we have the ordered list, which is expressed in HTML as **ol**, with **li** being used for the list items. You are probably also quite familiar with this type of list in your everyday life. A common use case for the ordered list could be a recipe shown in a list of sequential steps. With this type of list, the ordering is important, unlike in the unordered list we just looked at. The following code shows an example of this type of list as you would use it in HTML:

```
<!-- Cheese on toast recipe -->
<ol>
    <li>Place bread under grill until golden brown</li>
    <li>Flip the bread and place cheese slices</li>
    <li>Cook until cheese is golden brown</li>
    <li>Serve immediately</li>
</ol>
```

A web page for the preceding code would appear as follows:

1. Place bread under grill until golden brown
2. Flip the bread and place cheese slices
3. Cook until cheese is golden brown
4. Serve immediately

Figure 3.8: Ordered list as it appears on a web page

If you want to use an ordered or unordered list but don't want to show bullet points or numbers, you have a range of options. Using CSS, you can customize the style of list using the **list-style** property:

```css
/* Alternative styles for unordered lists */
.square {
  list-style: square;
}
.circle {
  list-style: circle;
}
/* Alternative styles for ordered lists */
.upper-alpha {
  list-style: upper-alpha;
}
.upper-roman {
  list-style: upper-roman;
}
```

The following figure shows the output of the preceding code:

- Square list style
- Circle list style
- E. Upper alpha list style
- VII. Upper roman list style

Figure 3.9: Unordered list shown with the different list styles

It is also possible to nest lists and use different list styles for each list. In the following HTML code, you will see that we have different lists being nested:

```html
<ol>
  <li>Numbered</li>
  <ol class="alphabetic">
    <li>Alphabetic</li>
    <ol class="roman">
      <li>Roman</li>
```

```
      <li>Roman</li>
    </ol>
    <li>Alphabetic</li>
  </ol>
  <li>Numbered</li>
</ol>
```

We just need to add two class names to style the two nested lists with alphabetic and Roman list styles as shown here:

```
.alphabetic {
  list-style: upper-alpha;
}
.roman {
  list-style: upper-roman;
}
```

The following figure shows the output of the preceding code:

1. Numbered
    A. Alphabetic
       I. Roman
      II. Roman
    B. Alphabetic
2. Numbered

Figure 3.10: Nested lists shown with different list styles

The third type of list is the definition list, which is expressed in HTML as **dl**. Although this type of list is used less frequently than the two other types of lists, you will probably still be familiar with it. The definition list is used when you want to list out pairs of terms and descriptions. The most common use of this type of list is probably dictionary entries. You have the word you are interested in, which is the term, **dt**, followed by the definition, which is the description, **dd**. The following is an example of this type of list as you would use it in HTML:

```
<!-- Dictionary -->
<dl>
    <dt>HTML</dt>
    <dd>Hypertext markup language</dd>
    <dt>CSS</dt>
    <dd>Cascading style sheets</dd>
</dl>
```

A web page for the preceding code would appear as follows:

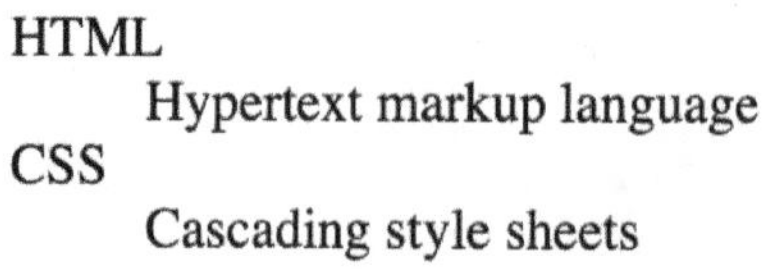

HTML
    Hypertext markup language
CSS
    Cascading style sheets

Figure 3.11: Definition list as it appears on a web page

Exercise 3.01: Combining Text-Based Elements

In this exercise, we will use the following screenshot from the Packt website and write the HTML to match it as closely as possible. This will give us some practice of creating text-based content and recreating the correct HTML for it.

The following figure shows the sample piece of content that we will recreate from the Packt website:

eBook Support

- If you experience a problem with using or installing Adobe Reader, the contact Adobe directly at www.adobe.com/support

- To view the errata for the book, see www.packtpub.com/support and view the pages for the title you have.

- To view your account details or to download a new copy of the book go to www.packtpub.com/account

- To contact us directly if a problem is unresolved use www.packtpub.com/contact

Figure 3.12: Screenshot from the Packt website

Let's complete the exercise with the following steps:

1. First, start by creating a new file in VSCode called **Exercise 3.01.html** within the **Chapter03** folder and use the following code as a starting point:

```
<!DOCTYPE html>
<html lang = "en">
    <head>
        <title>Combining text based elements</title>
    </head>
    <body>
        <!-- your code will go here -->
    </body>
</html>
```

2. Looking at the preceding screenshot, we can see that we will need a heading. Since the heading is not the top-level heading for the page, we will use **h2** in this instance. We will wrap the tag around the text in between the opening and closing **body** tags as follows:

```
<body>
  <h2>eBook Support</h2>
</body>
```

3. Below the heading, we have a list of bullet points. We can assume that these are unordered and, hence, use a **ul** tag for them. Notice that each list item contains a link as well, so we need to include an anchor tag for each. We will place our code just below the **h2** heading as follows:

```
<body>
  <h2>eBook Support</h2>
  <ul>
    <li>If you experience a problem with using or installing Adobe
      Reader, the contact Adobe directly at
      <a href="http://www.adobe.com/support">
        www.adobe.com/support</a>
    </li>
    <li>To view the errata for the book, see
      <a href="http://www.packtpub.com/sup
        port">www.packtpub.com/support</a> and view the pages for
      the title you have.
    </li>
    <li>To view your account details or to download a new copy
      of the book go to
      <a href="http://www.packtpub.com/account">
        www.packtpub.com/account</a>
    </li>
  </ul>
</body>
```

If you now right-click on the filename in VSCode on the left-hand side of the screen and select **Open In Default Browser**, you will see the web page in your browser:

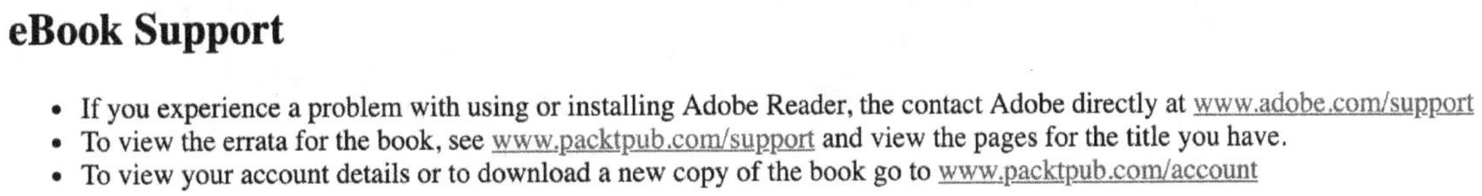

eBook Support

- If you experience a problem with using or installing Adobe Reader, the contact Adobe directly at www.adobe.com/support
- To view the errata for the book, see www.packtpub.com/support and view the pages for the title you have.
- To view your account details or to download a new copy of the book go to www.packtpub.com/account

Figure 3.13: Output of combining text-based elements

You are now getting a feel for the various different text-based HTML elements and when you should use them. Before we start looking into how we will go about styling the HTML elements we have just learned, we will take a look at semantic markup.

Semantic Markup

You will hear the word "semantic" used often when you read or hear about HTML. The core concept behind semantic markup is to ensure that you use the most meaningful HTML element available to describe the content you are marking up. For example, it would be possible for you to wrap the top-level page heading in a **div** tag, however, the **h1** tag conveys the meaning that the content represents, that is, heading level 1. The HTML you write needs to be understandable to both humans and machines, and by using the most meaningful element for each piece of content, you improve the meaning of both.

Ensuring that the HTML you write is as semantic as possible also has additional important benefits. The first being that it will make your web pages more easily searchable by search engines. You will also be helping out users who view your websites using a screen reader.

The following code shows some examples of semantic and non-semantic markup:

```
<!-- Semantic markup -->
<h1>I am a top level page heading</h1>
<p>
  This is a paragraph which contains a word with <strong>strong</strong>
  significance
</p>
<!-- Non semantic markup -->
<div>I am a top level page heading</div>
<div>This is a paragraph which contains a word with <span>strong</span>
  significance</div>
```

Hopefully, you should now understand the differences between semantic and non-semantic markup. Now that we have some knowledge of the most commonly used HTML elements used for content, we can turn to the more fun part of styling.

Styling Text-Based Elements

Until now, we have seen some of the basic text formatting that HTML allows you to implement on a web page. However, depending on the function and the purpose a web page serves, we might need some styling applied to the text-based elements. Here, we will introduce the common issues surrounding the styling of web page content. We will introduce the different units of measurement, including pixel and relative units. We will then walk through some examples of how to style common web components such as breadcrumbs and navigation bars.

CSS Resets

As you begin styling web pages, you will soon realize that different browsers render your pages slightly differently from each other. This can be very frustrating and makes the task of developing websites that look the same across different browsers a nightmare.

Luckily, there is a well-known solution to alleviate at least some of this frustration. A CSS reset is a style sheet whose sole purpose is to level the playing field across browsers. This file will be loaded before any of your page-specific styles are added. The following is an example of the most basic form of CSS reset code:

```
* {
  margin: 0;
  padding: 0;
}
```

What this will ensure is that all HTML elements will have zero margin and padding before you apply your custom styles to your page. This gets around the issue of different browsers by default adding varying amounts of padding and margin to certain elements.

Although using this reset would be better than having no reset at all, there are more sophisticated CSS resets available. The following shows a popular CSS reset developed by Eric Meyer:

```
/* http://meyerweb.com/eric/tools/css/reset/
   v2.0 | 20110126
   License: none (public domain)
*/
html, body, div, span, applet, object, iframe, h1, h2, h3, h4, h5, h6,
p, blockquote, pre, a, abbr, acronym, address, big, cite, code, del,
dfn, em, img, ins, kbd, q, s, samp, small, strike, strong, sub, sup,
tt, var, b, u, i, center,dl, dt, dd, ol, ul, li, fieldset, form,
label, legend, table, caption, tbody, tfoot, thead, tr, th, td,
article, aside, canvas, details, embed, figure, figcaption, footer,
header, hgroup, menu, nav, output, ruby, section, summary,time, mark,
```

```css
audio, video {
    margin: 0;
    padding: 0;
    border: 0;
    font-size: 100%;
    font: inherit;
    vertical-align: baseline;
}
/* HTML5 display-role reset for older browsers */
article, aside, details, figcaption, figure, footer, header, hgroup,
menu, nav, section {
    display: block;
}
body {
    line-height: 1;
}
ol, ul {
    list-style: none;
}
blockquote, q {
    quotes: none;
}
blockquote:before, blockquote:after, q:before, q:after {
    content: '';
    content: none;
}
table {
    border-collapse: collapse;
    border-spacing: 0;
}
```

As you can see, this is more detailed than our first example and would give us a better chance of reducing cross-browser inconsistencies in our web pages.

CSS Text Properties

When styling text-based elements with CSS, there are two main groupings of properties you will see used over and over. These are groups of properties that are based around text and font. We will begin by looking at the most commonly used text-based CSS properties that you will need to become familiar with.

The first property we will look at is **color**, which, as the name suggests, is used to set the text color. You will typically set the color value using either hexadecimal, RGB, or a name.

For hexadecimal values, you specify two hexadecimal integers for the colors red, green, and blue. The values you provide range from **00** to **FF**, with **FF** being the most intense version of the specific color and **00** being the least intense. For example, **#FF0000** will be the highest intensity of the color red.

RGB-based colors are different to hexadecimal values because you provide a value between **0** and **255** for each color, with **0** being the lowest intensity and **255** being the maximum intensity.

The following shows some examples of what this would look like in code form:

```
h1 {
   color: green;
}
p {
   color: #00ff00;
}
span {
   color: rgb(0, 255, 0);
}
```

We will learn more about text colors and background colors in *Chapter 5, Themes, Colors, and Polish*.

You may have noticed that all the text we have seen so far aligns to the left by default. With CSS, we have the power to change this using the `text-align` property. You can style your text to be left-aligned, centered, or right-aligned. The following shows some examples of what this would look like in code form:

```
p {
   text-align: center;
}
```

A web page for the preceding code would appear as follows:

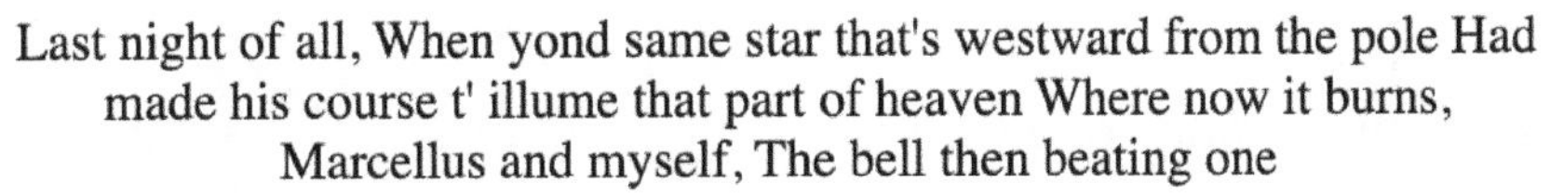

Last night of all, When yond same star that's westward from the pole Had made his course t' illume that part of heaven Where now it burns, Marcellus and myself, The bell then beating one

Figure 3.14: Centrally aligned text as it appears on a web page

Note

The text used for explaining the CSS text properties is a work of Shakespeare sourced from the following: https://packt.live/2Cmvni4.

If you want to underline some text, you can use the **text-decoration** property, which gives you the ability to strikethrough text as well:

```
.underline {
  text-decoration: underline;
}
```

A web page for the preceding code would appear as follows:

Last night of all, When yond same star that's westward from the pole Had made his course t' illume that part of heaven Where now it burns, Marcellus and myself, The bell then beating one

Figure 3.15: Underlined text as it appears on a web page

The following code shows how you would use the **line-through** property:

```
.line-through {
  text-decoration: line-through;
}
```

A web page for the preceding code would appear as follows:

~~Last night of all, When yond same star that's westward from the pole Had made his course t' illume that part of heaven Where now it burns, Marcellus and myself, The bell then beating one~~

Figure 3.16: Strikethrough text as it appears on a web page

Another common styling requirement for text is the ability to control how it is capitalized. For this, we have the **text-transform** property, which gives you the ability to transform the text. By default, the text is set to lowercase, but with this property, you can set the text to all caps or title case. The following code shows how you would use this property:

```
.uppercase {
  text-transform: uppercase;
}
```

A web page for the preceding code would appear as follows:

LAST NIGHT OF ALL, WHEN YOND SAME STAR THAT'S WESTWARD FROM THE POLE HAD MADE HIS COURSE T' ILLUME THAT PART OF HEAVEN WHERE NOW IT BURNS, MARCELLUS AND MYSELF, THE BELL THEN BEATING ONE

Figure 3.17: All-caps text as it appears on a web page

The following code shows how you would use the **lowercase** property:

```
.lowercase {
  text-transform: lowercase;
}
```

A web page for the preceding code would appear as follows:

last night of all, when yond same star that's westward from the pole had made his course t' illume that part of heaven where now it burns, marcellus and myself, the bell then beating one

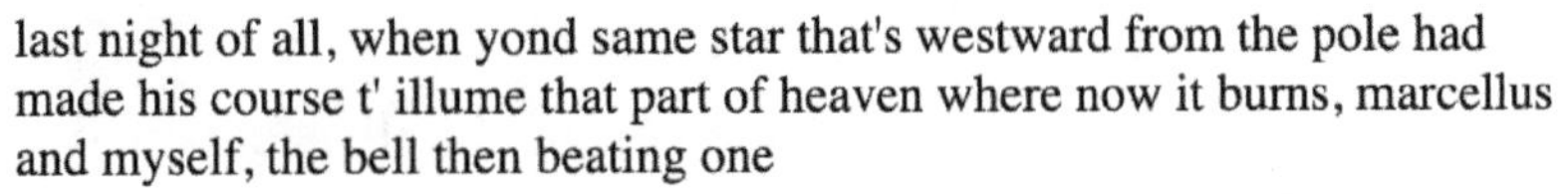

Figure 3.18: Lowercase text as it appears on a web page

The following code shows how you would use the **capitalize** property:

```
.capitalize {
  text-transform: capitalize;
}
```

A web page for the preceding code would appear as follows:

Last Night Of All, When Yond Same Star That's Westward From The Pole Had Made His Course T' Illume That Part Of Heaven Where Now It Burns, Marcellus And Myself, The Bell Then Beating One

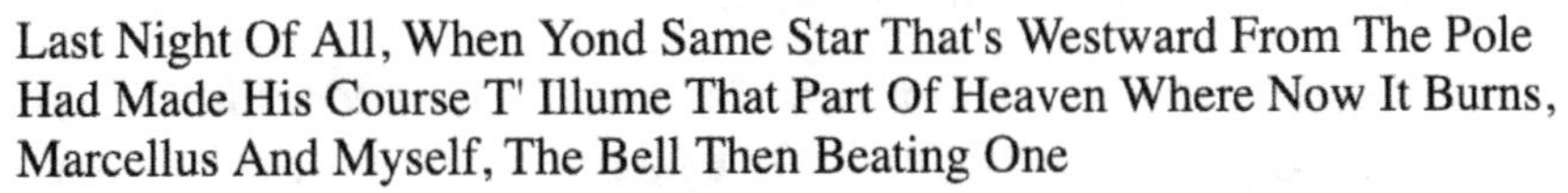

Figure 3.19: Title case text as it appears on a web page

Finally, on our tour of the most used text-based CSS properties, we have the **line-height** property, which is used to control the amount of vertical spacing between lines of text. This is a property you will see used over and over as different types of copy will require different line heights. We will see two extreme examples of how a small and large line-height value affects the readability of the text:

```
.small-line-height {
  line-height: .5;
}
```

A web page for the preceding code would appear as follows:

Lorem ipsum dolor sit amet, consectetur adipiscing elit. Suspendisse odio felis, euismod eu est ac, facilisis euismod odio. Mauris sed vestibulum neque.

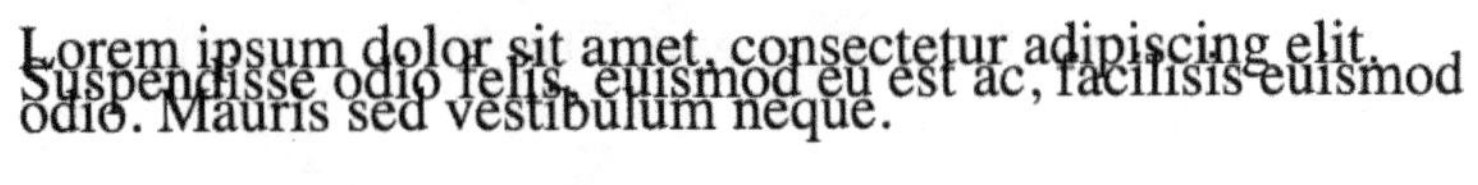

Figure 3.20: Small line-height as it appears on a web page

The following code shows how you would use the **line-height** property:

```
.large-line-height {
   line-height: 1.5;
}
```

A web page for the preceding code would appear as follows:

Lorem ipsum dolor sit amet, consectetur adipiscing elit.

Suspendisse odio felis, euismod eu est ac, facilisis euismod

odio. Mauris sed vestibulum neque.

Figure 3.21: Large line-height as it appears on a web page

CSS Font Properties

The second group of CSS properties we are concerned with is the font-based properties. These are responsible for defining the font family, the size of the font, and the weight. We will now take a quick tour of the most commonly used CSS font properties that you should become familiar with.

The first property we will look at is the **font-family** property, which, as you might have guessed, sets the font family. When using this property, you will usually provide a list of different font families in order of priority. If the web browser doesn't support your first choice of font, it will default to the second or until it finds a font family it can load. The following code shows how this property can be used:

```
body {
   font-family: "Times New Roman", Times, serif;
}
```

The browser will attempt to load the "**Times New Roman**" font from the user's computer first. If it cannot load the font, then it will try "**Times**," and, if this fails, it will load the generic "**serif**" font family.

Now that we know how to set the correct font family for our text, we will need to control the size. For this, we can use the **font-size** property. You can set the value of the size using pixels or relative units such as **ems**. The following shows examples of this property's use:

```
/* pixels */
h1 {
   font-size: 50px;
}
p {
   font-size: 16px;
```

```
}
/* ems */
h1 {
  font-size: 3.125em;
}
p {
  font-size: 1em;
}
```

The benefit of using ems for the unit of measurement is that it allows the user to control the font size. By default, a browser's font size is set to **16px**, but if a user wants to increase their default font size, they can. With **em** units, the font sizes will scale according to the base font size. With pixels used as units, you give the user less flexibility in controlling font sizes.

The last font-based CSS property we will look at is the **font-weight** property. This is used to control the weight of a font, typically to make a font bold as, by default, the weight is set to normal. The following shows you how this would look in code:

```
span {
  font-weight: bold;
}
```

We now have all the knowledge we require to build a realistic-looking web page. We will put this theory into practice by building components for web pages one by one.

The display Property

Before we move onto the next exercise, we need to look into a new CSS property called **display**. By default, elements are either set to **block** or **inline**. Now, **block** elements will take up all horizontal space, while **inline** elements only take up as much horizontal space as their content. An example of a block-level element is a **div** tag and an example of an **inline** element is **span**. Sometimes, you need to style a **block** element as an **inline** element and vice versa. You will do this by using the following CSS:

```
div {
  display: inline;
}
span {
  display: block;
}
```

Video Store Product Page (Revisited)

Remember the video store product page examples from *Chapter 2, Structure and Layout*, where we stepped through several CSS layout techniques? We are going to be using a more detailed version of this page to work through some exercises demonstrating how to code some of the key components on the page. By doing this, we should be able to put most of the theory from this chapter into practice. The following figure is the revised wireframe for the page:

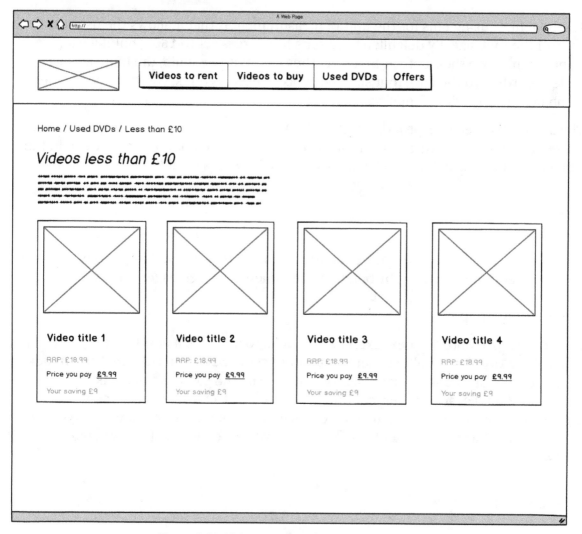

Figure 3.22: Video store product page revisited

Exercise 3.02: Navigation

In this exercise, we will step through the process of writing the HTML and CSS for the navigation component shown in the preceding wireframe. The following figure shows the navigation component in more detail:

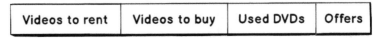

Figure 3.23: Navigation

The following are the steps to complete this exercise:

1. First, let's create a file called **Exercise 3.02.html** within the **Chapter03** folder in VSCode using the following HTML code as a starter file:

```
<!DOCTYPE html>
<html lang = "en">
    <head>
        <title>Exercise 3.02</title>
        <style>
            /* your CSS will go here */
        </style>
    </head>
    <body>
        <!-- your HTML will go here -->
    </body>
</html>
```

2. First, we need to decide on how we should semantically mark up this component. We will start with the **nav** tag, which we learned about in the previous chapter. We will place this in between the opening and closing **body** tags as follows:

```
<body>
    <nav></nav>
</body>
```

3. For the actual list of navigation links, using **ul** would be most appropriate as the order isn't of any significance. We will place the code for **ul** in between our **nav** tags as follows:

```
<body>
    <nav>
        <ul>
            <li>Videos to rent</li>
            <li>Videos to buy</li>
            <li>Used DVDs</li>
```

```
      <li>Offers</li>
    </ul>
  </nav>
</body>
```

4. Since users of the web page will want to be able to click on each of the navigation items, we must add anchors for each item. We will place the anchors around the text contained in our list items as follows:

```
<body>
  <nav>
    <ul>
      <li><a href="">Videos to rent</a></li>
      <li><a href="">Videos to buy</a></li>
      <li><a href="">Used DVDs</a></li>
      <li><a href="">Offers</a></li>
    </ul>
  </nav>
</body>
```

5. We will now have a look in the browser at what this will look like by default without any custom styling. To do this, right-click on the filename in VSCode on the left-hand side of the screen and select **Open In Default Browser**. You should see something that resembles the following screenshot:

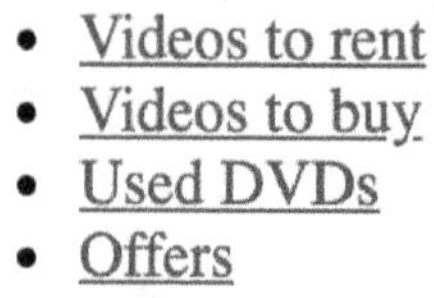

- Videos to rent
- Videos to buy
- Used DVDs
- Offers

Figure 3.24: Default navigation

6. This is very different from what we want the navigation to look like. To correct this, we will now add some CSS to more closely match the navigation component shown in the wireframe. We will start by adding a basic CSS reset, which will be placed in between the opening and closing **style** tags as follows:

```
<style>
  * {
    margin: 0;
    padding: 0;
  }
</style>
```

7. We will use the **flex** layout technique we learned about in *Chapter 2, Structure and Layout*. We will also remove the default bullets shown to the left of the list items:

```
<style>
  *  {
    margin: 0;
    padding: 0;
  }
  nav ul {
    display: flex;
    list-style: none;
  }
</style>
```

8. We will now add some styling to the links so that they are no longer blue and more closely resemble the links shown in the wireframe. We will make the text bold, remove the default underline that gets applied to the links, and give the link some **padding**:

```
nav a {
  color: black;
  font-weight: bold;
  display: block;
  padding: 15px;
  text-decoration: none;
}
```

The following figure shows the output of the web page so far:

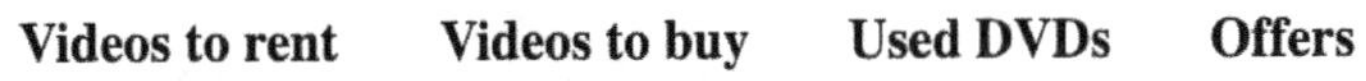

Videos to rent   Videos to buy   Used DVDs   Offers

Figure 3.25: Output without hover

9. Finally, to apply the hover selector on the anchor elements and to set the link text to underline when the user hovers over the text, we will add the following code:

```
nav a:hover {
  text-decoration: underline;
}
```

If you now right-click on the filename in VSCode on the left-hand side of the screen and select **Open In Default Browser**, you will see the navigation component in your browser.

You should now see something that looks like the following screenshot:

Videos to rent   Videos to buy   <u>Used DVDs</u>   Offers

Figure 3.26: Styled navigation

Breadcrumbs

On websites that have lots of pages, it is common for pages to contain what is called **breadcrumbs**. This is a list of links that easily allow the user to see the context of the current page within the website structure and to easily navigate to the parent pages. The following screenshot shows a breadcrumb taken from the Packt website, https://packt.live/3aZBMyo:

Home > All Products > Default Category > All Products > All Books > Default Category > Programming > Default Category > Programming > Application Development > Salesforce Platform Developer I Certification Guide

Figure 3.27: Breadcrumb from the Packt web page

As you can see in the preceding figure, web pages can contain many parent pages or categories. The breadcrumbs of the current page are for a book in the Packt online store. Moving from right to left, the breadcrumb links get more specific, moving from the home page to the closest parent page. The user is free to click on any link, which will take them back to a page related to the current page.

Exercise 3.03: Breadcrumb

In this exercise, we will step through the process of writing the HTML and CSS for the breadcrumb component shown in *Figure 3.22*. The following figure shows it in more detail:

Home / Used DVDs / Less than £10

Figure 3.28: Breadcrumb

The following are the steps to complete this exercise:

1. First, let's create a file called **Exercise 3.03.html** within the **Chapter03** folder in VSCode using the following HTML code as a starter file:

```
<!DOCTYPE html>
<html lang = "en">
    <head>
        <title>Exercise 3.03</title>
        <style>
          /* your CSS will go here */
        </style>
```

```
    </head>
    <body>
        <!-- your HTML will go here -->
    </body>
</html>
```

2. First, we need to decide what the best HTML tag will be for the breadcrumb. Since this is a list of links and the ordering is important, we will use **ol**. We will place this in between the **body** tags as follows:

```
<body>
  <ol class="breadcrumb">
    <li>Home</li>
    <li>Used DVDs</li>
    <li>Less than £10</li>
  </ol>
</body>
```

3. We will then add anchors to all but the last item because the last item represents the current page the user is viewing so doesn't need to be clickable:

```
<body>
  <ol class="breadcrumb">
    <li><a href="">Home</a></li>
    <li><a href="">Used DVDs</a></li>
    <li>Less than £10</li>
  </ol>
</body>
```

4. Let's now take a look at what this will look like in the browser:

1. Home
2. Used DVDs
3. Less than £10

Figure 3.29: Default breadcrumb

5. We will now start adding our styling. Again, we will start with a basic CSS reset and we will add the styles in between the opening and closing **style** tags as follows:

```
<style>
  *  {
    margin: 0;
    padding: 0;
  }
</style>
```

6. We will use the **flex** layout technique again, which we learned about in the previous chapter, and remove the default numbers from the ordered list:

```
<style>
  *   {
      margin: 0;
      padding: 0;
  }
  .breadcrumb {
      display: flex;
      list-style: none;
  }
</style>
```

7. Let's style the list items in the ordered list. We will first add some padding to the list items. Then, we will add a forward slash at the end of all the list items except the last one. We will add some margin to the left to ensure our list items are nicely separated:

```
.breadcrumb li {
    padding: 10px;
}
.breadcrumb li:after {
    content: '/';
    margin-left: 20px;
}
.breadcrumb li:last-child:after {
    content: '';
}
```

8. Finally, we will style the anchors, making sure the color of the text is **black** and only showing an **underline** when the user hovers over the link:

```
.breadcrumb a {
    color: black;
    text-decoration: none;
}
.breadcrumb a:hover {
    text-decoration: underline;
}
```

If you now right-click on the filename in VSCode on the left-hand side of the screen and select **Open In Default Browser**, you will see the breadcrumb component in your browser.

You should now see something similar to the following figure in your browser:

Home   /   Used DVDs   /   Less than £10

Figure 3.30: Styled breadcrumb

Exercise 3.04: Page Heading and Introduction

We will now write the HTML and CSS for the heading and introduction section of the wireframe. The following figure shows it in more detail:

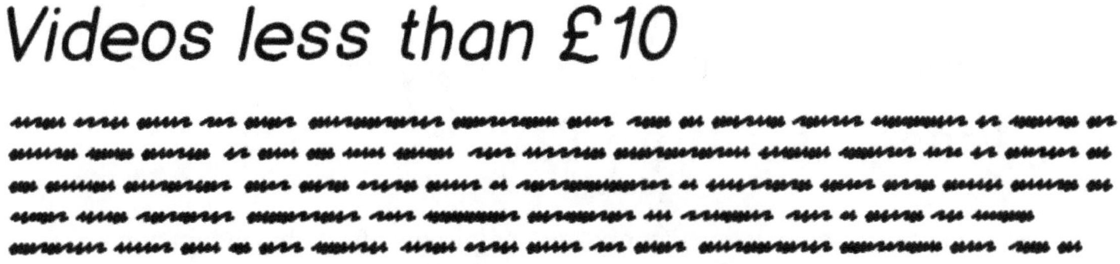

Videos less than £10

Figure 3.31: Introduction section

The steps to complete the exercise are as follows:

1. First, let's create a file called **Exercise 3.04.html** within the **Chapter03** folder in VSCode using the following HTML code as a starter file:

```
<!DOCTYPE html>
<html lang = "en">
    <head>
        <title>Exercise 3.04</title>
        <style>
          /* your CSS will go here */
        </style>
    </head>
    <body>
        <!-- your HTML will go here -->
    </body>
</html>
```

2. First, let's decide on the correct markup to use. Since the heading represents the top-level heading for the page, we will use **h1** and we will use a plain **p** for the introduction. We will place the HTML in between the opening and closing **body** tags:

```html
<body>
  <section class="intro">
  <h1>Videos less than £10</h1>
  <p>
  Lorem ipsum dolor sit amet, consectetur adipiscing elit. In
  bibendum non purus quis vestibulum. Pellentesque ultricies quam
  lacus, ut tristique sapien tristique et.
  </p>
  </section>
</body>
```

3. Let's now take a look at what this looks like in the browser:

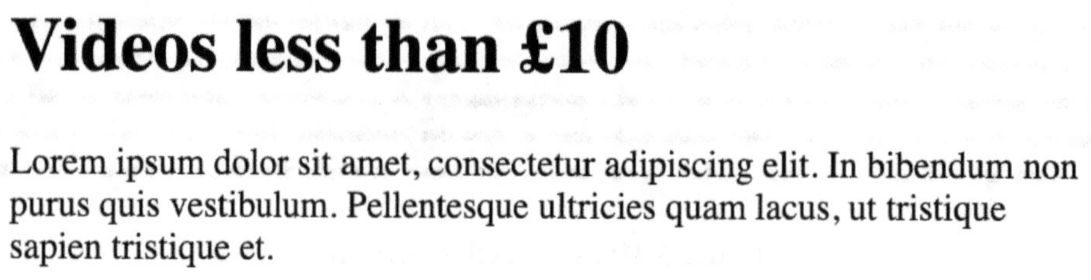

Videos less than £10

Lorem ipsum dolor sit amet, consectetur adipiscing elit. In bibendum non purus quis vestibulum. Pellentesque ultricies quam lacus, ut tristique sapien tristique et.

Figure 3.32: Default introduction section

4. This is actually pretty close to how we want these elements to look. We will just adjust the heading margin and the line height for the paragraph to make the text more readable. We will place the CSS in between the opening and closing **style** tags as follows:

```css
<style>
  * {
    margin: 0;
    padding: 0;
  }
  .intro {
    margin: 30px 0;
    padding-left: 10px;
    width: 50%;
  }
  .intro h1 {
    margin-bottom: 15px;
  }
```

```
    .intro p {
      line-height: 1.5;
    }
  </style>
```

If you now right-click on the filename in VSCode on the left-hand side of the screen and select **Open In Default Browser**, you will see the heading and paragraph.

You should now see something similar to the following figure in your browser:

Videos less than £10

Lorem ipsum dolor sit amet, consectetur adipiscing elit. In bibendum non purus quis vestibulum. Pellentesque ultricies quam lacus, ut tristique sapien tristique et.

Figure 3.33: Styled introduction section

Exercise 3.05: Product Cards

In this exercise, we will step through the process of writing the HTML and CSS for the product card component shown in the wireframe. The following figure shows the product card in more detail:

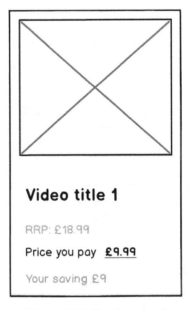

Figure 3.34: Product card

The steps to complete the exercise are as follows:

1. First, let's create a file called **Exercise 3.05.html** within the **Chapter03** folder in VSCode using the following starter file:

```html
<!DOCTYPE html>
<html lang = "en">
    <head>
        <title>Exercise 3.05</title>
        <style>
          /* your CSS will go here */
        </style>
    </head>
    <body>
        <!-- your HTML will go here -->
    </body>
</html>
```

2. Next, let's use a **div** tag with a class name for the outer wrapper of the component. We will place this in between the opening and closing **body** tags as follows:

```html
<body>
  <div class="product-card">
  </div>
</body>
```

3. Now, let's add an image tag; we will add the image URL and some **alt** text:

```html
<body>
  <div class="product-card">
    <img src="https://dummyimage.com/300x300/7EC0EE/000&text=
      Product+Image+1" alt="Product image 1" />
  </div>
</body>
```

4. We will then use **h2** for the heading. Notice how we wrap the text in an anchor so that the user will be able to click on the card:

```html
<body>
  <div class="product-card">
    <img src="https://dummyimage.com/300x300/7EC0EE/000&text=
      Product+Image+1" alt="Product image 1" />
    <h2><a href="">Video title 1</a></h2>
  </div>
</body>
```

5. Now we will add the markup for the pricing information. Note that we will add a class name to allow us to style the information individually:

```
<body>
   <div class="product-card">
    <img src="https://dummyimage.com/300x300/7EC0EE/000&text=
      Product+Image+1" alt="Product image 1" />
    <h2><a href="">Video title 1</a></h2>
    <p class="original-price">RRP: £18.99</p>
    <p class="current-price">Price you pay <span>£9.99</span></p>
    <p class="saving">Your saving £9</p>
   </div>
</body>
```

6. Let's now take a look at what this looks like in the browser without any styling added:

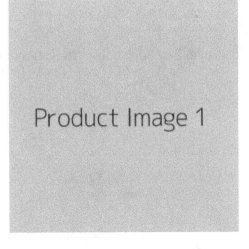

Video title 1

RRP: £18.99

Price you pay £9.99

Your saving £9

Figure 3.35: Default product card

7. Now, let's add some styling for the product card container in between the **style** tags as follows. Notice how we give the product card a black **border** and some **padding**:

```
<style>
  .product-card {
    display: inline-block;
    border: 1px solid black;
    padding: 15px;
  }
</style>
```

8. Then, we will add some styling for the individual elements of the product card. Starting with the image element, we will ensure the width of the image stretches to **100%** by adding the following code:

```
.product-card img {
  width: 100%;
}
```

We will add margin to the level 2 header using the following code:

```
.product-card h2 {
  margin: 30px 0 15px;
}
```

The following code styles the links:

```
.product-card a {
  color: black;
  text-decoration: none;
}
```

With the help of the following code we will style the paragraph element:

```
.product-card p {
  line-height: 1.5;
}
```

We will add the following code to style the original price, current price, and your savings as per our expected wireframe shown in *Figure 3.34*:

```
.original-price {
  color: gray;
  text-transform: uppercase;
}
.current-price span {
```

```
    font-weight: bold;
    text-decoration: underline;
  }
.saving {
  color: green;
  }
```

If you now right-click on the filename in VSCode on the left-hand side of the screen and select **Open In Default Browser**, you will see the product component in your browser.

You should now see something similar to the following figure in your browser:

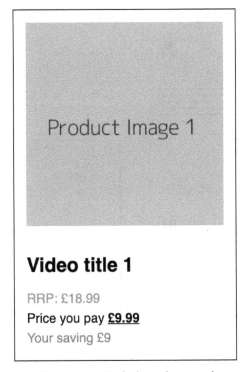

Figure 3.36: Styled product card

Exercise 3.06: Putting It All Together

Now that we have built the individual parts of the product page, we have the fun task of putting them all together to assemble a web page. We will be able to reuse the code we have written already and will just need to make minor tweaks to the CSS to get the page looking good. Our aim will be to produce a web page that resembles the wireframe as shown in the following figure:

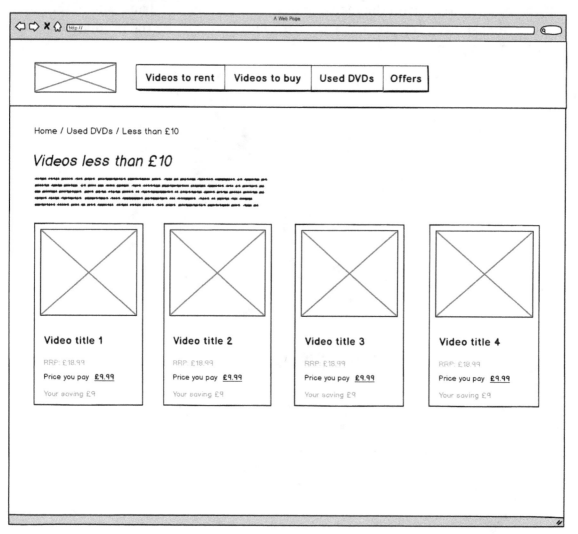

Figure 3.37: Wireframe of the expected output

The steps to complete the exercise are as follows:

1. First, let's create a file called **Exercise 3.06.html** within the **Chapter03** folder in VSCode. We will use the following HTML page template, noting we will use an inline style sheet to make things easier:

```html
<!DOCTYPE html>
<html lang = "en">
<head>
   <title>Video store home page</title>
   <style type="text/css">
     * {
         margin: 0;
         padding: 0;
     }
     body {
         font-family: sans-serif;
         margin: 0 auto;
         width: 1200px;
     }
     header {
         align-items: center;
         display: flex;
         margin-bottom: 25px;
     }
     nav {
         margin-left: 30px;
     }
     .product-cards {
       display: grid;
       grid-template-columns: auto auto auto;
       margin-bottom: 30px;
     }
     /* your styles will go here */
   </style>
</head>
<body>
   <header>
     <img src="https://dummyimage.com/200x100/000/fff&text=Logo"
       alt="" />
     <!-- navigation will go here -->
   </header>
   <section>
```

```
    <!-- breadcrumb will go here -->
  </section>
    <!-- introduction section will go here -->
  <section class="product-cards">
    <!-- product cards will go here -->
  </section>
</body>
</html>
```

2. Now, let's add the navigation CSS, same as the code used in *Exercise 3.02, Navigation*, for the following HTML from the same exercise:

```
<nav>
  <ul>
    <li><a href="">Videos to rent</a></li>
    <li><a href="">Videos to buy</a></li>
    <li><a href="">Used DVDs</a></li>
    <li><a href="">Offers</a></li>
  </ul>
</nav>
```

3. Then, we will add the styles for the breadcrumb component same as that of *Exercise 3.03, Breadcrumb*, for the following HTML from the same exercise:

```
<section>
  <ol class="breadcrumb">
    <li><a href="">Home</a></li>
    <li><a href="">Used DVDs</a></li>
    <li>Less than £10</li>
  </ol>
</section>
```

4. We will then add the introduction section's CSS same as that of *Exercise 3.04, Page Heading and Introduction*, for the following HTML from the same exercise:

```
<section class="intro">
  <h1>Videos less than £10</h1>
    <p>
    Lorem ipsum dolor sit amet, consectetur adipiscing elit.
    In bibendum non purus quis vestibulum. Pellentesque
    ultricies quam lacus, ut tristique sapien
    tristique et.
    </p>
</section>
```

5. Finally, we will add the product card's CSS same as that of *Exercise 3.05*, *Product Cards*, for the following HTML from the same exercise:

```html
<section class="product-cards">
    <div class="product-card">
        <img src="https://dummyimage.com/300x300/000/
          fff&text=Product+Image+1" alt="" />
        <h2><a href="">Video title 1</a></h2>
        <p class="original-price">RRP: £18.99</p>
        <p class="current-price">Price you pay <span>£9.99</span></p>
        <p class="saving">Your saving £9</p>
    </div>
        <!-- Similar to the above product card 1 that is video
        title 1, the product cards for video titles 2, 3, and 4
        can be copied from Exercise 3.05, Product Cards -->
</section>
```

If you now right-click on the filename in VSCode on the left-hand side of the screen and select **Open In Default Browser**, you will see the web page.

You should now see something similar to the following figure in your browser:

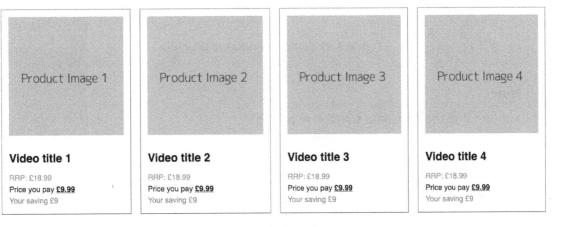

Figure 3.38: Styled product page

Activity 3.01: Converting a Newspaper Article to a Web Page

Using a copy of a recent newspaper, choose a particular article and note down what HTML elements would be used if the paper were a web page. Now create a web page version of the newspaper article using semantic markup and CSS to recreate the layout as closely as possible:

1. Get a copy of a newspaper article and annotate it with a pen to label the individual HTML elements.

2. Create a file named **Activity 3.01.html** within the **Chapter03** folder in VSCode. You can use the starter HTML from a previous exercise as a starting point.

3. Start writing out the HTML for the news article.

4. Now style the text and layout using CSS.

An example of how you could annotate a newspaper article to distinguish the different page elements can be seen in the following figure:

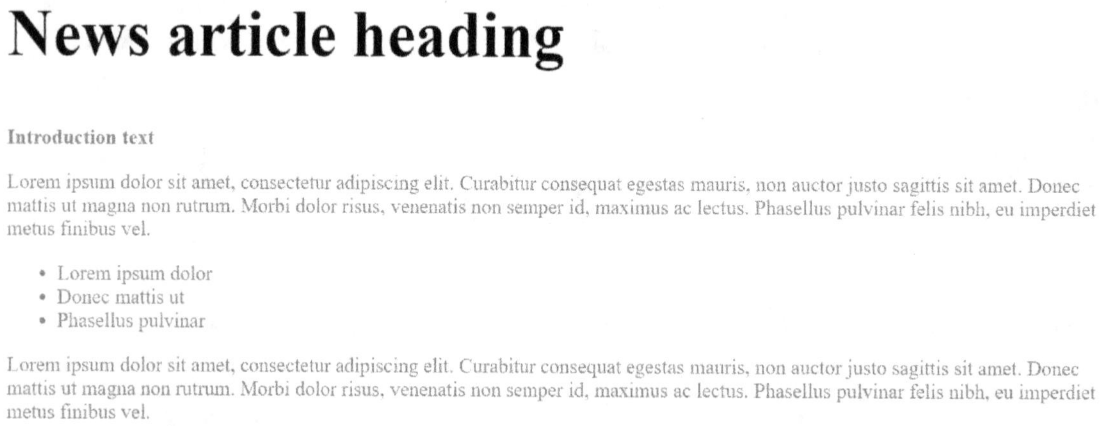

News article heading

Introduction text

Lorem ipsum dolor sit amet, consectetur adipiscing elit. Curabitur consequat egestas mauris, non auctor justo sagittis sit amet. Donec mattis ut magna non rutrum. Morbi dolor risus, venenatis non semper id, maximus ac lectus. Phasellus pulvinar felis nibh, eu imperdiet metus finibus vel.

- Lorem ipsum dolor
- Donec mattis ut
- Phasellus pulvinar

Lorem ipsum dolor sit amet, consectetur adipiscing elit. Curabitur consequat egestas mauris, non auctor justo sagittis sit amet. Donec mattis ut magna non rutrum. Morbi dolor risus, venenatis non semper id, maximus ac lectus. Phasellus pulvinar felis nibh, eu imperdiet metus finibus vel.

Figure 3.39: Sample annotated article

Note

You can find the complete solution in page 597.

Summary

In this chapter, we have continued our journey into building web pages. We first looked at the most common text-based HTML elements such as headings, paragraphs, and lists. We then looked into the most common styling methods available for text-based content. To put this new knowledge into practice, we then walked through building a complete web page.

We took some time to understand the concept and importance of writing semantic HTML. We were also introduced to some common web page components such as navigation and breadcrumbs.

In the next chapters, we will learn how to take our web pages to the next level. We will learn how to make our web pages far more interesting by adding forms, animation, and videos.

4

Forms

Overview

By the end of this chapter, you will be able to use the correct HTML form elements to build an online form; customize form elements to improve the look and feel of your web forms; build online forms; apply form validation styles; and identify when to use checkboxes over radio buttons. This chapter introduces HTML forms and associated elements used within forms. We will first look at the most common HTML form elements used when building forms. We will then take a look at some common techniques for styling forms. We will then put all of this into practice by building signup and checkout forms for a video store.

Introduction

In the previous chapters, we studied how to build web pages that contain static text-based content. From this chapter onward, we will learn how to make web pages much more interesting, starting with forms.

Forms allow users to actually interact with a website. They enable users to sign up for services, order products online, and so on. Forms are arguably one of the most crucial aspects of business websites, as without forms no transactions can take place online. Businesses require online forms to capture user details when creating new user accounts, for instance, to allow users to select flight details when booking a holiday online. Without forms, many online businesses would not be able to function. With this in mind, developing complex forms is an essential skill to add to your toolbelt as a web developer.

In this chapter, we will take a look at the most common elements that are used to build forms with HTML. These HTML elements include text inputs, radio buttons, checkboxes, text areas, submit buttons, and so on. Once we've gained an understanding of the most commonly used form elements, we will look at styling concerns. This will include techniques to make our form elements look visually appealing to a wide range of users. We will put all of this into practice by building different online forms.

Form Elements

HTML provides us with a variety of elements that are used for building forms. While browsing the web, you must have noticed that online forms typically have similar elements. Most forms will contain input fields such as text inputs, checkboxes, and select boxes.

In this section, we will look at the following HTML form elements:

- `form`
- `input`
- `label`
- `textarea`
- `fieldset`
- `select`
- `button`

The form Element

The first element we need to know about when creating forms is the **form** element. This is the outermost element, which contains all other form elements, such as inputs and buttons. The **form** element requires you to pass two attributes, which are the **action** and **method** attributes. The **action** attribute allows the developer to specify the URL where the form data will go to after it has been submitted. The **method** attribute allows the developer to specify whether the form data should be sent via **get** or **post**. You will typically use the **get** method when you are dealing with unsecured data since the data will be present in a query string. On the other hand, the **post** method is typically used when dealing with secure data, or when dealing with a large amount of soft data. The following code snippet shows an example of what an empty form would look like in HTML:

```
<form action="url_to_send_form_data" method="post">
  <!-- form elements go here -->
</form>
```

The input Element

Now that we have our **form** element, we can start building the form using different elements. The first one we will look at is the **input** element. This is the element you would use when creating text input fields, radio buttons, and checkboxes. The **input** element is the most important of all the form elements we will look at and you will find yourself using it over and over again. The **input** element requires two attributes, which are **type** and **name**. The **type** attribute is used to specify what type of input you want, such as radio buttons, checkboxes, or text. The **name** attribute gives the element a unique name that is required when submitting the form. This is so that the form's data can be organized into **key-value** pairs with a unique **name** corresponding with a **value**. It should be noted that the order that you add the attributes in has no significance. The following code snippet shows how to create a text field using the **input** element:

```
<!-- text input -->
<form action="url_to_send_form_data" method="post">
  <div>
    First name: <br />
    <input type="text" name="firstname" />
  </div>
  <div>
    Last name: <br />
    <input type="text" name="lastname" />
  </div>
</form>
```

The following figure shows the output for the preceding code:

First name:

Last name:

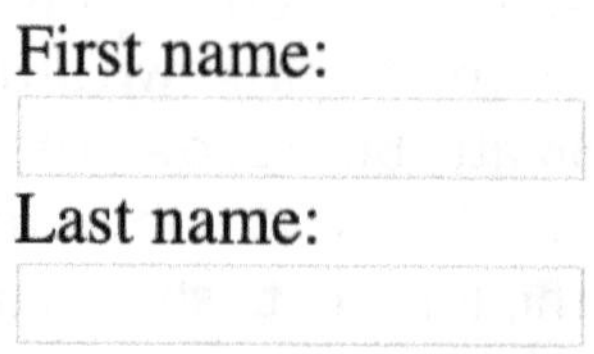

Figure 4.1: Text inputs shown in the browser

Sometimes, when creating text inputs, you will want to limit the number of characters a user can add. A common example of this is when you want to restrict the number of characters for a new username in account signup forms. You can use the **maxlength** attribute and set the maximum number of characters allowed for the **input** field. The following code snippet shows how you would use this attribute:

```
<input type="text" name="username" maxlength="20" />
```

There is also a specialist type of text input that is solely for email addresses. To create an email input, you simply set the **type** to **"email"**. This input type has built-in validation that checks whether the input text is a valid email address. The following code snippet shows how to create an email input:

```
<!-- email input -->
<form action="url_to_send_form_data" method="post">
  <div>
    Email: <br />
    <input type="email" name="email"/>
  </div>
</form>
```

The following figure shows the output for the preceding code:

Email:

john@smith.com

Figure 4.2: Email input shown in the browser

There is a type of text input that is used solely for passwords. To create a password input, you simply set the **type** to **"password"**. This input type will mask the text entered by the user to hide the **password** text. The following code snippet shows how to create a password input:

```
<!-- password input -->
<form action="url_to_send_form_data" method="post">
  <div>
```

```
   Password: <br />
   <input type="password" name="password"/>
 </div>
</form>
```

The following figure shows the output for the preceding code:

Password:

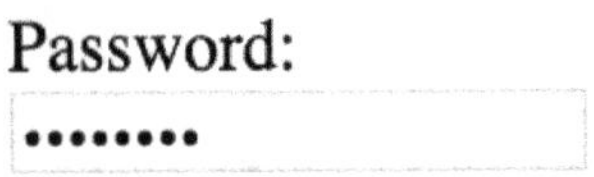

Figure 4.3: Password input shown in the browser

When using checkboxes, you will give all of them a unique value for the **name** attribute and you will need to give each checkbox a unique **value** attribute, as shown in the following code:

```
<!-- checkboxes -->
<form action="url_to_send_form_data" method="post">
  <div>
    <input type="checkbox" name="color1" value="red" /> Red
  </div>
  <div>
    <input type="checkbox" name="color2" value="green" /> Green
  </div>
  <div>
    <input type="checkbox" name="color3" value="blue" /> Blue
  </div>..
</form>
```

The following figure shows the output for the preceding code:

☐ Red
☑ Green
☐ Blue

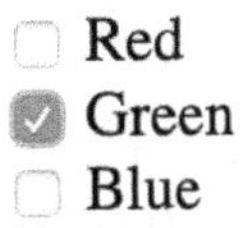

Figure 4.4: Checkboxes shown in the browser

With checkboxes, you can select multiple values at a time. A common use case for checkboxes is when selecting multiple filters to search results.

When using **radio** buttons, you will give all of them the same value for the **name** attribute since there can only be one value selected. You will, however, need to give each radio button a unique **value** attribute, as shown in the following code snippet:

```
<!-- radio buttons -->
<form action="url_to_send_form_data" method="post">
```

```
<div>
  <input type="radio" name="color" value="red" /> Red
</div>
<div>
  <input type="radio" name="color" value="green" /> Green
</div>
<div>
  <input type="radio" name="color" value="blue" /> Blue
</div>
</form>
```

The following figure shows the output for the preceding code:

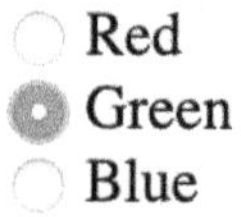

○ Red
◉ Green
○ Blue

Figure 4.5: Radio buttons shown in the browser

In contrast to checkboxes, with radio buttons, the user can select only one value. A common use case for radio buttons is when selecting a delivery option when ordering online.

The label Element

Now that we know how to create text inputs, checkboxes, and radio buttons, we need to look at the **label** element. In the previous examples, you might have noticed that we had text associated with the input fields either before or after an **input** element. The **label** element allows us to associate a piece of text with a **form** element and allows us to select the **form** element by clicking on the text. If we were to just include some text, as we did in *Figure 4.1*, we would lose this benefit and make our form less accessible for screen reader users since there would not be an associated **label** to call out when presenting a form element. The **label** element has an attribute called **for**, which we need to give the **id** for the element we wish to associate the label with. The following code snippet shows this in action:

```
<!-- text inputs with labels -->
<form action="url_to_send_form_data" method="post">
  <div>
    <label for="first_name">First name:</label><br />
    <input type="text" name="firstname" id="first_name" />
```

```
    </div>
    <div>
      <label for="last_name">Last name:</label><br />
      <input type="text" name="lastname" id="last_name" />
    </div>
  </form>
```

The following figure shows the output for the preceding code:

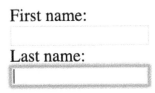

Figure 4.6: Text inputs with labels shown in the browser

The textarea Element

Imagine you are creating a "comments" section for a community web page. You might want a user to comment on a video or a blog post. However, using a text input is not ideal for long text messages. In such scenarios, when you want to allow the user to add more than one line of text, you can use the **textarea** element to capture larger amounts of text. You can specify the size of the **textarea** with the **rows** and **cols** attributes. The following code snippet shows how to include **textarea** within a form:

```
<!-- textarea -->
<form action="url_to_send_form_data" method="post">
  <div>
    <label for="first_name">First name:</label><br />
    <input type="text" name="firstname" id="first_name" />
  </div>
  <div>
    <label for="last_name">Last name:</label><br />
    <input type="text" name="lastname" id="last_name" />
  </div>
  <div>
    <label for="message">Message:</label><br />
    <textarea id="message" rows="5" cols="20"></textarea>
  </div>
</form>
```

The following figure shows the output for the preceding code:

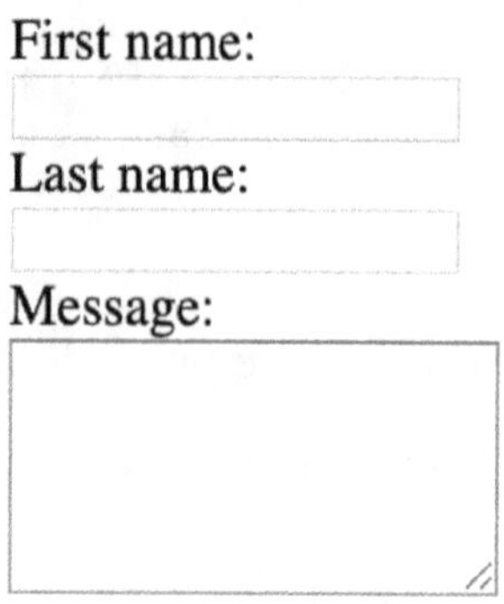

First name:

Last name:

Message:

Figure 4.7: Textarea shown in the browser

The fieldset Element

HTML provides us with a semantic tag to group related form elements and it is called the **fieldset** element. This element is mostly used with larger forms when you want to group related form elements together. You will probably have used online forms that make use of the **fieldset** element without realizing it. A common use case is when you have a large form with a section for personal details and a section for delivery details. Both of these sections of the form would be wrapped in a **fieldset** element. The following shows how we could include more than one form using **fieldset**:

```
<!-- fieldset -->
<form action="url_to_send_form_data" method="post">
  <fieldset>
    <div>
      <label for="first_name">First name:</label><br />
      <input type="text" name="firstname" id="first_name" />
    </div>
    <div>
      <label for="last_name">Last name:</label><br />
      <input type="text" name="lastname" id="last_name" />
    </div>
    <div>
      <label for="message">Message:</label><br />
      <textarea id="message" rows="5" cols="20"></textarea>
    </div>
  </fieldset>
  <p>Do you like HTML?</p>
  <fieldset>
    <div>
```

```
      <input type="radio" id="yes">
      <label for="yes">Yes</label>
    </div>
    <div>
      <input type="radio" id="no">
      <label for="no">No</label>
    </div>
  </fieldset>
</form>
```

The following figure shows the output for the preceding code:

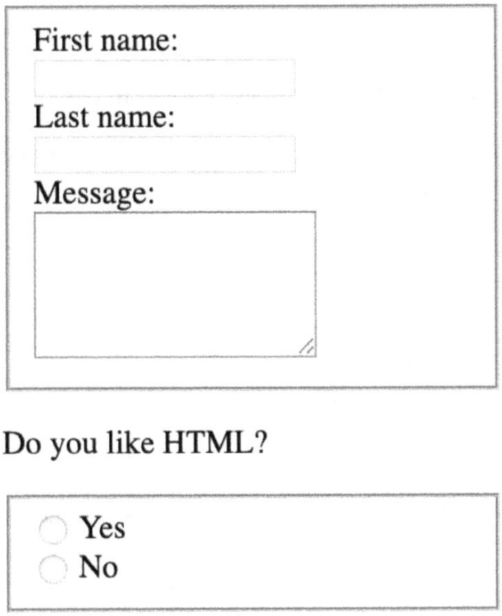

Do you like HTML?

Figure 4.8: Form with fieldset shown in the browser

The select Element

HTML provides us with the **select** element for creating select boxes. These are typically used when you have a long list of options and you want the user to select only one. Some common examples include lists of countries, addresses, and year of birth. Inside the **select** element, you provide a list of options inside of an **option** element. The following shows an example of how this looks in HTML:

```
<!-- select -->
<form action="url_to_send_form_data" method="post">
  <fieldset>
    <label for="countries">Country:</label><br />
    <select id="countries">
```

```
        <option value="england">England</option>
        <option value="scotland">Scotland</option>
        <option value="ireland">Ireland</option>
        <option value="wales">Wales</option>
      </select>
    </fieldset>
  </form>
```

The following figure shows the output for the preceding code:

Figure 4.9: Select box closed, shown in the browser

By clicking on the blue arrows on the right-hand side of the select box, we will get the options displayed in the following figure:

Figure 4.10: Select box open, shown in the browser

The button Element

Finally, now that we have a range of form elements we can use for building web forms, we now just need to know how to submit a form. The **button** element requires a **type** attribute that can have three different values. Firstly, the **button** value, which has no default behavior; the **"reset"** value, which, once clicked, will reset all form values; and finally, the **"submit"** value, which will submit the form once clicked. For this, we use the **button** element and give a value of **"submit"** in the **type** attribute:

```
<button type="submit">Submit</button>
```

Exercise 4.01: Creating a Simple Form

In this exercise, we will write the HTML to create a simple web form. Our aim will be to produce the following web form:

Figure 4.11: A simple web form shown in the browser

Let's complete the exercise with the following steps:

1. First, start by creating a new folder named **Chapter04** in VSCode. Within this folder, create a file called **Exercise 4.01.html** and use the following code as your starting point:

```
<!DOCTYPE html>
<html lang = "en">
<head>
  <title>Simple form</title>
</head>
<body>
  <form action="url_to_send_form_data" method="post">
    <fieldset>
      <!-- your code will go here -->
    </fieldset>
  </form>
</body>
</html>
```

2. First, we will add the HTML for the **title** and **first name** fields between the opening and closing **fieldset** elements:

```html
<fieldset>
  <div>
    <label for="title">Title:</label><br />
    <select id="title">
      <option value="Mr">Mr</option>
      <option value="Mrs">Mrs</option>
      <option value="Ms">Ms</option>
      <option value="Miss">Miss</option>
    </select>
  </div>
  <div>
    <label for="first_name">First name:</label><br />
    <input type="text" name="firstname" id="first_name" />
  </div>
</fieldset>
```

3. Next, we will add the HTML for the **last name** field. We will add this after the closing **div** element, which wraps the HTML for the **first name** field:

```html
<div>
  <label for="first_name">First name:</label><br />
  <input type="text" name="firstname" id="first_name" />
</div>
<div>
  <label for="last_name">Last name:</label><br />
  <input type="text" name="lastname" id="last_name" />
</div>
```

4. Now we will add the HTML for the **message** field. Notice how this will be a **textarea** element instead of an **input** element to allow the user to enter more than one line of text:

```html
<div>
  <label for="first_name">First name:</label><br />
  <input type="text" name="firstname" id="first_name" />
</div>
<div>
  <label for="last_name">Last name:</label><br />
  <input type="text" name="lastname" id="last_name" />
```

```
  </div>
  <div>
    <label for="message">Message:</label><br />
    <textarea id="message" rows="5" cols="20"></textarea>
  </div>
```

5. Finally, we need to add the **submit** button as follows:

```
<div>
  <label for="first_name">First name:</label><br />
  <input type="text" name="firstname" id="first_name" />
</div>
<div>
  <label for="last_name">Last name:</label><br />
  <input type="text" name="lastname" id="last_name" />
</div>
<div>
  <label for="message">Message:</label><br />
  <textarea id="message" rows="5" cols="20"></textarea>
</div>
<button type="submit">Submit</button>
```

If you now right-click on the filename in VSCode, on the left-hand side of the screen, and select **Open In Default Browser**, you will see the form in your browser.

You should now have a form that looks like the following figure:

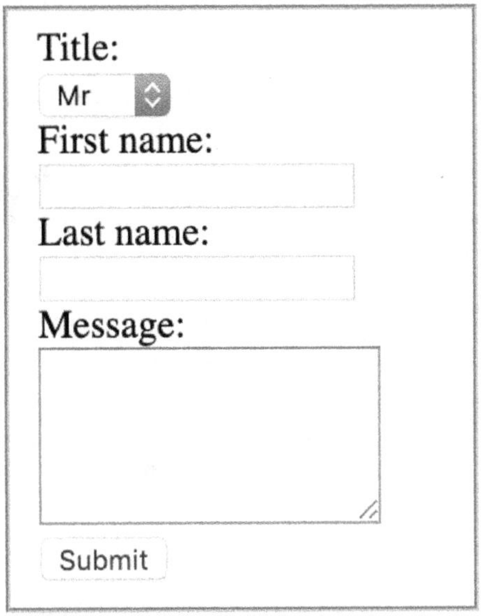

Figure 4.12: Submit button shown in the browser

Now that we have become acquainted with the most commonly used HTML form elements, we will now look at how to style them.

Styling Form Elements

In the examples in the previous section, the forms do not look very visually appealing by default, but luckily, we have the ability to improve the look and feel of our forms using CSS.

In this section, we will look at styling the following:

- Textbox
- Textarea
- Label
- Button
- Select box
- Validation styling

Label, Textbox, and Textarea

The first form elements we will look at styling are the **label**, textboxes, and **textarea** elements. These are probably the most common form elements and it is very straightforward to improve the look and feel of these elements with minimal code.

To style labels, you will typically just adjust the font you use and the size of the text. It is common to have the **label** element sit either on top of its associated form element or to the left.

For textboxes and textareas, typically, you will be interested in changing the size of these elements. It is common to remove the default border around these elements. Another common stylistic addition to textboxes and textareas is to add a placeholder attribute that provides the user with some text that helps them decide what needs to be typed into the textbox or textarea.

To illustrate an example of how to style these elements, we will start with the following markup, noting the addition of placeholder attributes:

```html
<!-- HTML -->
<form action="url_to_send_form_data" method="post">
  <div>
    <label for="first_name">First name:</label><br />
```

```
      <input type="text" name="firstname" id="first_name" placeholder="Your
        first name" />
    </div>
    <div>
      <label for="last_name">Last name:</label><br />
      <input type="text" name="lastname" id="last_name" placeholder=
        "Your last name" />
    </div>
    <div>
      <label for="message">Message:</label><br />
      <textarea id="message" rows="5" cols="20" placeholder="Your
        message"></textarea>
    </div>
</form>
/* CSS */
* {
  font-family: arial,sans-serif;
}
label {
  font-size: 20px;
}
div {
  margin-bottom: 30px;
}
input,
textarea {
  border: 0;
  border-bottom: 1px solid gray;
  padding: 10px 0;
  width: 200px;
}
```

In the preceding CSS, you will notice that we have applied a font family to all text elements. We have set the label text size to 20px and added a bottom margin to the **div** elements so that the form elements are nicely spaced, vertically. Finally, we have removed the default border applied to the **input** and **textarea** elements, replacing it with just a **border** on the bottom.

With just minimal CSS, we have improved the look and feel of our form drastically, as can be seen in the following screenshot:

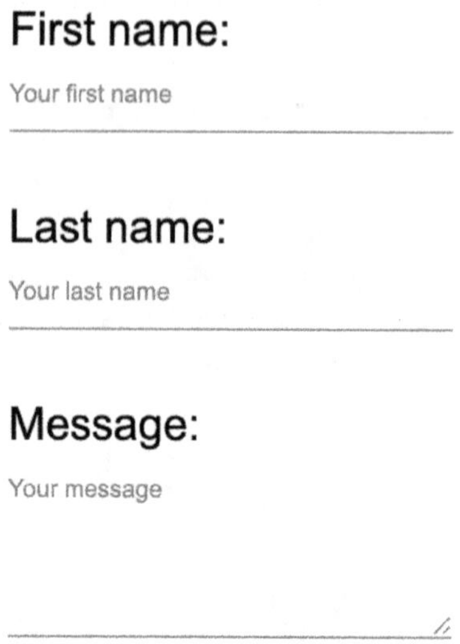

First name:

Your first name

Last name:

Your last name

Message:

Your message

Figure 4.13: Styled labels, textboxes, and textarea elements

Buttons

We will now look into styling the buttons that are used to submit a web form. Typically, you will see buttons with various different background colors and with different sizes applied when viewing websites with forms. Out of the box, the **button** element looks pretty ugly and so you will rarely see buttons without some CSS applied to them. The following is an example of how you could style a **submit** button:

```
<!-- HTML -->
<button type="submit">Submit</button>
/* CSS */
button {
  background: #999;
  border: 0;
  color: white;
  font-size: 12px;
  height: 50px;
  width: 200px;
  text-transform: uppercase;
}
```

The preceding CSS sets a background color, removes the border that is added to buttons by default, applies some styling to the button text, and finally, sets a width and height:

Figure 4.14: A styled submit button

Select Boxes

The last form element we will look at for styling is the **select** box. Typically, these are styled with the intention of making the **select** box look similar to a **textbox** within a form. It is common for developers to add a custom styled downward-pointing arrow to the right-hand side of the **select** box. The following is an example of how you could style a **select** box:

```
<!-- HTML -->
<div class="select-wrapper">
  <select id="countries">
    <option value="england">England</option>
    <option value="scotland">Scotland</option>
    <option value="ireland">Ireland</option>
    <option value="wales">Wales</option>
  </select>
</div>
/* CSS */
select {
  background: transparent;
  border: 0;
  border-radius: 0;
  border-bottom: 1px solid gray;
  box-shadow: none;
  color: #666;
  padding: 10px 0;
  width: 200px;
  -webkit-appearance: none;
}
```

The preceding CSS contains styling that essentially overrides what a **select** box looks like in the browser by default. Firstly, we need to remove the default background color, cancel the border, and apply just the bottom border. We also remove the custom **box-shadow** property, which is also applied to select boxes by default. Finally, to add a custom select box icon, we use the **after** pseudo selector to add the '< >' characters:

```
.select-wrapper {
  position: relative;
  width: 200px;
}
.select-wrapper:after {
  content: '< >';
  color: #666;
  font-size: 14px;
  top: 8px;
  right: 0;
  transform: rotate(90deg);
  position: absolute;
  z-index: -1;
}
```

The following figure shows the output:

Figure 4.15: Styled select box

Validation Styling

In real-world scenarios, simply formatting and styling the form appropriately is not enough. As a web user, you may encounter cases where form validation is performed before submitting a form. For example, while registering on a website, a user may accidentally submit a form before it is filled in completely or submit an incorrectly filled in form. Validation styling comes into play when you want to highlight the fact that a form is incomplete or incorrectly filled in.

You will probably have experienced form validation on web forms you have used in the past. HTML provides us with a **required** attribute, which we can apply to any form elements that we require input for. The **required** attribute plays an important role in contact forms; for example, on the Packt website's contact form (https://packt. live/35n6tvJ), you will notice that the name and email fields are required and the user cannot submit the form until a value for each is added.

This is in contrast with some form elements where the input is optional. With CSS, we can use the `:valid` and `:invalid` pseudo `selectors` in order to style elements based on valid or invalid form values. We will now do an exercise that will walk us through an example of validation styles in action.

Exercise 4.02: Creating a Form with Validation Styling

In this exercise, we will develop a simple web form that contains some **validation** styling. Our aim will be to produce a web form like the one shown in the following figure:

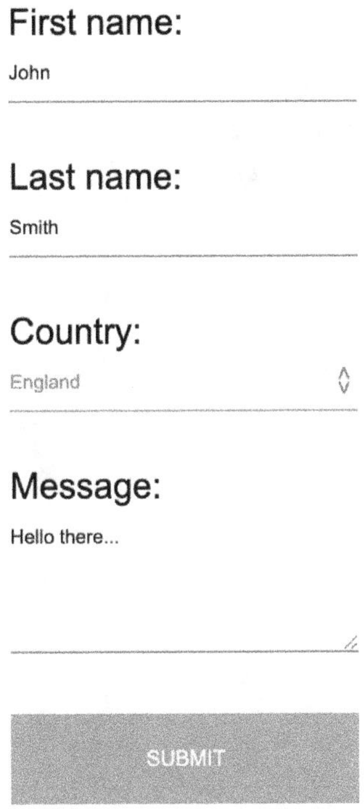

First name:

John

Last name:

Smith

Country:

England

Message:

Hello there...

SUBMIT

Figure 4.16: Expected output

Let's complete the exercise with the following steps:

1. First, start by creating a new file called **Exercise 4.02.html** inside the **Chapter04** folder and use the following code as your starting point:

```
<!DOCTYPE html>
<html lang = "en">
<head>
<title>Validation form</title>
<style>
 body {
```

```
      font-family: arial, sans-serif;
    }
  </style>
</head>
<body>
  <form action="url_to_send_form_data" method="post">
    <fieldset>
      <!-- your code will go here -->
    </fieldset>
  </form>
</body>
</html>
```

2. We will now add the HTML for the **first name** and **last name** form fields between the opening and closing **fieldset** tags. Notice how we have added **required** attributes to both of the **input** elements:

```
<fieldset>
  <div>
    <label for="first_name">First name:</label><br />
    <input type="text" name="firstname" id="first_name"
      placeholder="Your first name" required />
  </div>
  <div>
    <label for="last_name">Last name:</label><br />
    <input type="text" name="lastname" id="last_name"
      placeholder="Your last name" required />
  </div>
</fieldset>
```

3. Add the HTML for the remaining form elements. Notice that it is only the **textarea** element where we require the **required** attribute:

```
<label for="country">Country:</label>
<div class="select-wrapper">
  <select id="country">
    <option value="england">England</option>
    <option value="scotland">Scotland</option>
    <option value="ireland">Ireland</option>
    <option value="wales">Wales</option>
  </select>
</div>
<label for="message">Message:</label><br/>
<textarea id="message" rows="5" cols="20" placeholder=
```

```
    "Your message" required></textarea>
<br/><br/>
<button type="submit">Submit</button>
```

4. Now we will turn to the CSS. We will first add some styling, which will deal with **spacing** the **div** and **fieldset** elements:

```
div {
    margin-bottom: 30px;
}
fieldset {
    border: 0;
    padding: 30px;
}
```

5. Next, we will style the individual form elements one by one. The label's font size is set to 20px and the styling for the **input** and **textarea** elements is same as shown in the following code snippet:

```
label {
    font-size: 20px;
}
input,
textarea {
    border: 0;
    border-bottom: 1px solid gray;
    padding: 10px 0;
    width: 200px;
}
```

With respect to the expected output as shown in the *Figure 4.16*, we style **select** as shown in the following code snippet:

```
select {
    background: transparent;
    border: 0;
    border-radius: 0;
    border-bottom: 1px solid gray;
    box-shadow: none;
    color: #666;
    -webkit-appearance: none;
    padding: 10px 0;
    width: 200px;
}
```

We will use the following snippet of code to complete styling the **select**:

```css
.select-wrapper {
  position: relative;
  width: 200px;
}
.select-wrapper:after {
  content: '<>';
  color: #666;
  font-size: 14px;
  top: 8px;
  right: 0;
  transform: rotate(90deg);
  position: absolute;
  z-index: -1;
}
```

For styling the button, we will use the styling as shown in the following code snippet:

```css
button {
  background: #999;
  border: 0;
  color: white;
  font-size: 12px;
  height: 50px;
  width: 200px;
  text-transform: uppercase;
}
```

6. Finally, we will add the styles to validate the form elements that have a **required** attribute:

```css
input:valid,
textarea:valid {
  border-bottom-color: green;
}
input:invalid,
textarea:invalid {
  border-bottom-color: red;
}
```

If you now right-click on the filename in VSCode, on the left-hand side of the screen, and select **Open In Default Browser**, you will see the form in your browser. When you try to submit an incomplete form that has validation in it, the form will not submit and you will see something like the following screenshot:

Figure 4.17: Screenshot of the resultant form

Video Store Forms

From the video store product page examples from the previous chapter, where we built a whole web page, component by component, we will now build two complex forms for the video store project in order to put our new knowledge into practice.

Exercise 4.03: New Account Signup Form

In this exercise, we will write the HTML and CSS to create an account signup form. Our aim will be to produce a web form like the following wireframe:

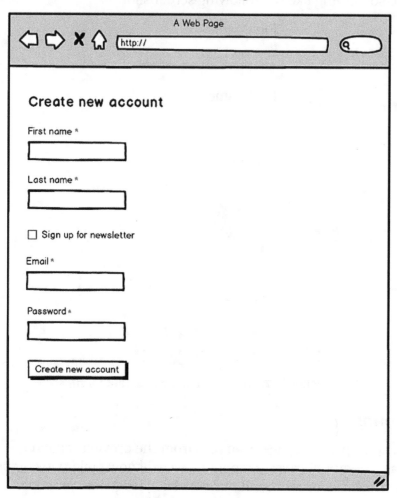

Figure 4.18: Wireframe of a new account form

Let's complete the exercise with the following steps:

1. First, start by creating a new file called **Exercise 4.03.html** within the **Chapter04** folder and use the following code as your starting point:

```html
<!DOCTYPE html>
<html lang = "en">
<head>
  <title>Signup form</title>
  <style>
    body {
      font-family: arial, sans-serif;
    }
  </style>
</head>
<body>
  <h1>Create new account</h1>
  <form action="url_to_send_form_data" method="post">
    <fieldset>
    </fieldset>
  </form>
</body>
</html>
```

2. Now, we will add the HTML for the form up until just before the checkbox. Notice how we have added the **required** attribute to the elements that we require the user to provide a value for:

```html
<fieldset>
  <div>
    <label for="first_name">First name: <span>*</span></label>
    <input id="first_name" type="text" name="firstname" required />
  </div>
  <div>
    <label for="last_name">Last name: <span>*</span></label>
    <input id="last_name" type="text" name="lastname" required />
  </div>
</fieldset>
```

3. Next, we will add the checkbox for signing up to the newsletter. In order to have the checkbox first and the description later, **label** comes after **input**:

```
<fieldset>
  <div>
    <label for="first_name">First name: <span>*</span></label>
    <input id="first_name" type="text" name="firstname" required />
  </div>
  <div>
    <label for="last_name">Last name: <span>*</span></label>
    <input id="last_name" type="text" name="lastname" required />
  </div>
  <div class="checkbox">
    <input id="newsletter" type="checkbox" name="newsletter"
      value="yes" />
    <label for="newsletter">Sign up for newsletter</label>
  </div>
</fieldset>
```

4. Now, we will add the remaining fields and the **submit** button below the **checkbox** shown in the **wireframe**. Again, pay attention to adding the **required** attribute to elements that need a value to be provided:

```
<div class="checkbox">
  <input id="newsletter" type="checkbox" name="newsletter"
    value="yes" />
  <label for="newsletter">Sign up for newsletter</label>
</div>
<div>
  <label for="email">Email: <span>*</span></label>
  <input id="email" type="email" name="email" required />
</div>
<div>
  <label for="password">Password: <span>*</span></label>
  <input id="password" type="password" name="password" required />
</div>
<button type="submit">Create new account</button>
```

5. If you look at the HTML file in your web browser, you should now have a web page that looks like the following screenshot:

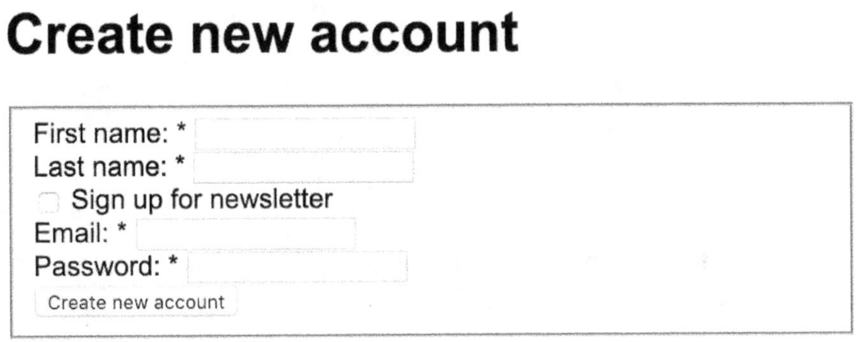

Create new account

First name: *
Last name: *
Sign up for newsletter
Email: *
Password: *
Create new account

Figure 4.19: Unstyled signup form in the browser

6. We can now add the CSS to style the form so that it more closely resembles the form shown in the preceding wireframe. We will start by applying some styles that deal with the spacing of elements and styling the red asterisks shown in the wireframe. The following code snippet will be inside the **style** tag following the font style applied to the body element:

```
fieldset {
  border: 0;
}
fieldset>div {
  margin-bottom: 30px;
}
label {
  display: block;
  margin-bottom: 10px;
}
label span {
  color: red;
}
input {
  padding: 10px;
  width: 200px;
}
```

7. Next, we will apply some styles for the **valid** and **invalid** states for the text inputs. Feel free to change the colors of these styles to suit your own taste:

```css
input:valid {
    border: 2px solid green;
}
input:invalid {
    border: 2px solid red;
}
```

8. Next, we will add some styling for the **checkbox** and the **submit** button, as shown in the following code. Again, feel free to adjust the colors to suit your needs:

```css
.checkbox input {
    float: left;
    margin-right: 10px;
    width: auto;
}
button {
    background: green;
    border: 0;
    color: white;
    width: 224px;
    padding: 10px;
    text-transform: uppercase;
}
```

If you now right-click on the filename in VSCode, on the left-hand side of the screen, and select **Open In Default Browser**, you will see the form in your browser.

You should now have a form that looks like the following figure:

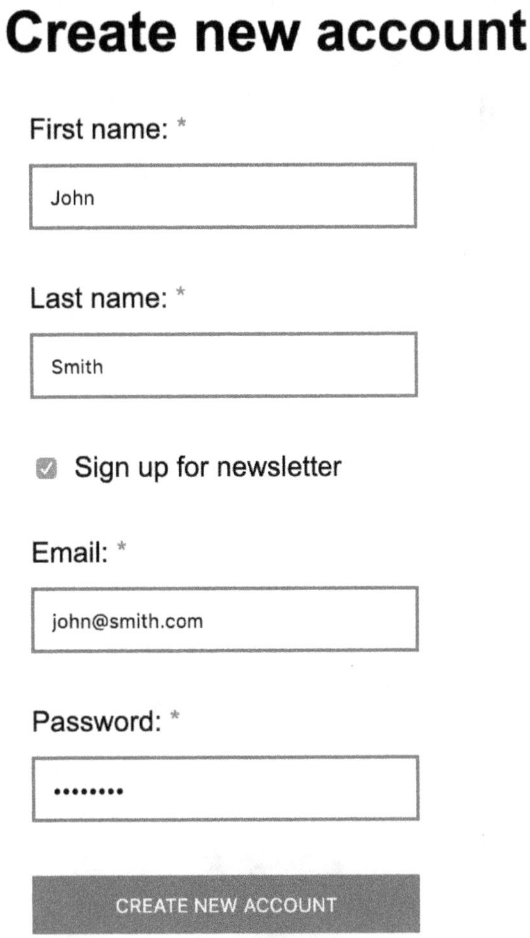

Create new account

First name: *

    John

Last name: *

    Smith

☑ Sign up for newsletter

Email: *

    john@smith.com

Password: *

    ••••••••

    CREATE NEW ACCOUNT

Figure 4.20: Styled signup form in the browser

9. We will now check that the email validation works correctly for the form by adding some text in the email field that isn't a valid email address:

First name: *

John

Last name: *

Smith

☑ Sign up for newsletter

Email: *

john

Password: *

••••••••

CREATE NEW ACCOUNT

Figure 4.21: Form showing an invalid email address

10. Next, we will check all of the required field validations by trying to submit the form without adding input for any of the required fields:

First name: *

Last name: *

☑ Sign up for newsletter

Email: *

Password: *

CREATE NEW ACCOUNT

Figure 4.22: A form highlighting fields that are required to be filled in to submit the form

We have now completed the signup form for our video store. We will also need to create a checkout page to allow the user to complete their online purchases. The next exercise will show how you can create a checkout form.

Exercise 4.04: Checkout Form

In this exercise, we will write the HTML and CSS to create a checkout form. We will make use of all the concepts and form elements we have learned about in the chapter. Our aim will be to produce a form similar to the following figure:

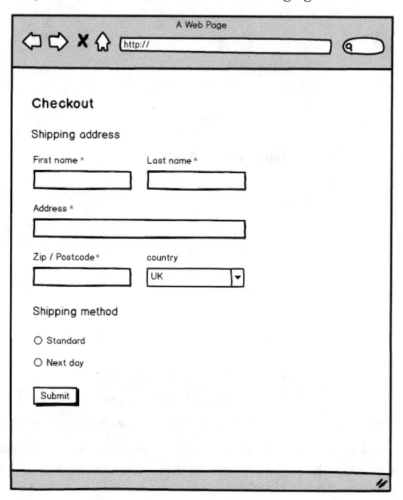

Figure 4.23: Checkout form wireframe

Let's complete the exercise with the following steps:

1. First, start by creating a new file called **Exercise 4.04.html** inside the **Chapter04** folder and use the following code as a starting point. This gives us the skeletal HTML and CSS we will need to get started writing the checkout form:

```html
<!DOCTYPE html>
<html lang = "en">
<head>
    <title>Checkout form</title>
    <style>
        body {
            font-family: arial, sans-serif;
        }
    </style>
</head>
<body>
    <h1>Checkout</h1>
    <form action="url_to_send_form_data" method="post">
        <fieldset>
        </fieldset>
    </form>
</body>
</html>
```

2. Now, we will add the HTML for the form inside the **fieldset** up until the **address** field. Notice how we have added the **required** attribute to the elements that we require the user to provide a value for:

```html
<h2>Shipping address</h2>
<div class="double-input">
  <div>
    <label for="first_name">First name: <span>*</span></label>
    <input id="first_name" type="text" name="firstname" required />
  </div>
  <div>
    <label for="last_name">Last name: <span>*</span></label>
    <input id="last_name" type="text" name="lastname" required />
  </div>
</div>
<div class="single-input">
  <label for="address">Address: <span>*</span></label>
  <input id="address" type="text" name="address" required />
</div>
```

3. Now, below the **address** field we write the HTML and CSS for the **postcode** with **input id** and country using **select** as shown in the wireframe:

```html
<div class="double-input">
  <div>
    <label for="postcode">Postcode: <span>*</span></label>
    <input id="postcode" type="text" name="postcode" required />
  </div>
  <div>
    <label for="country">Country:</label>
    <div class="select-wrapper">
      <select id="country">
        <option value="england">England</option>
        <option value="scotland">Scotland</option>
        <option value="ireland">Ireland</option>
        <option value="wales">Wales</option>
      </select>
    </div>
  </div>
</div>
```

4. With respect to the wireframe we will now add a level 2 header followed by the **radio** type checkboxes as shown in the following code:

```html
<h2>Shipping method</h2>
<div class="checkbox">
  <input id="standard" type="radio" name="shipping-method"
    value="standard" />
  <label for="standard">Standard</label>
</div>
<div class="checkbox">
  <input id="nextday" type="radio" name="shipping-method"
    value="nextday" />
  <label for="nextday">Next day</label>
</div>
<button type="submit">Submit</button>
```

5. If you look at the HTML file in your web browser, you should now have a web page that looks like the following screenshot:

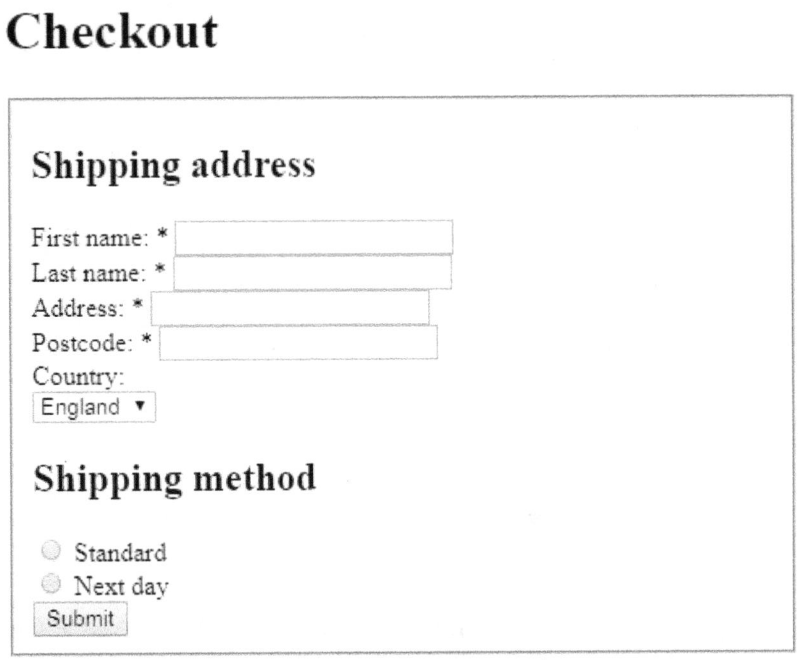

Checkout

Shipping address

First name: *

Last name: *

Address: *

Postcode: *

Country:

England ▼

Shipping method

○ Standard

○ Next day

Submit

Figure 4.24: Unstyled checkout form in the browser

6. We can now add some CSS to style the form so that it more closely resembles the form shown in the preceding wireframe. We will start by applying some styles that deal with the spacing of elements and styling the red asterisks shown in the wireframe:

```css
fieldset {
  border: 0;
  padding: 0;
}
fieldset > div {
  margin-bottom: 30px;
}
label {
  display: block;
  margin-bottom: 10px;
}
label span {
  color: red;
}
```

Notice that the **input** and **select** have same style as shown in the following code:

```
input,
select {
   border: 1px solid gray;
   padding: 10px;
   width: 200px;
}
```

7. Now let's add the CSS for the **select** boxes. We style **select** as shown in the following code snippet:

```
select {
   background: transparent;
   border-radius: 0;
   box-shadow: none;
   color: #666;
   -webkit-appearance: none;
   width: 100%;
}
```

8. Complete styling **select** using the following code:

```
.select-wrapper {
   position: relative;
   width: 222px;
}
.select-wrapper:after {
   content: '< >';
   color: #666;
   font-size: 14px;
   top: 8px;
   right: 0;
   transform: rotate(90deg);
   position: absolute;
   z-index: -1;
}
```

9. Finally, we will finish off the styling by adding some CSS for the inputs and the **button** styles, as shown in the following code:

```
.single-input input {
   width: 439px;
}
```

```
.double-input {
  display: flex;
}


.double-input > div {
  margin-right: 15px;
}
.checkbox input {
  float: left;
  width: auto;
  margin-right: 10px;
}
```

10. Following code is added to style the button:

```
button {
  background: green;
  border: 0;
  color: white;
  width: 224px;
  padding: 10px;
  text-transform: uppercase;
}
```

11. To style the inputs, we add the following code:

```
input:valid {
  border: 2px solid green;
}
input:invalid {
  border: 2px solid red;
}
```

If you now right-click on the filename in VSCode, on the left-hand side of the screen, and select **Open In Default Browser**, you will see the form in your browser.

You should now have a form that looks like the following figure:

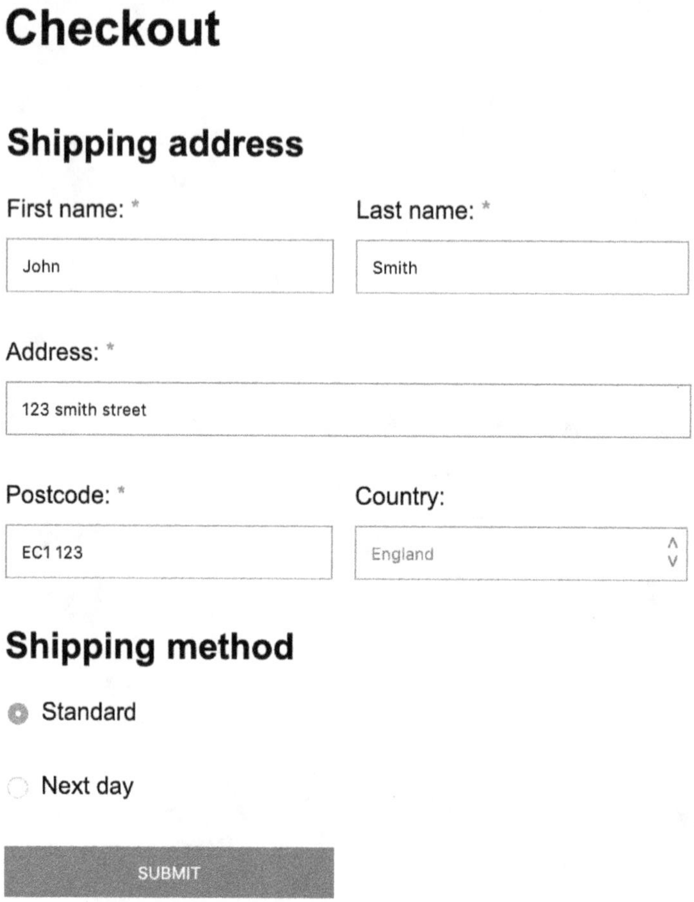

Checkout

Shipping address

First name: *

John

Last name: *

Smith

Address: *

123 smith street

Postcode: *

EC1 123

Country:

England

Shipping method

⦿ Standard

○ Next day

SUBMIT

Figure 4.25: Styled checkout form in the browser

12. Now, we will check whether the validation works for our required fields by submitting the form with empty fields. We will see something like the following screenshot:

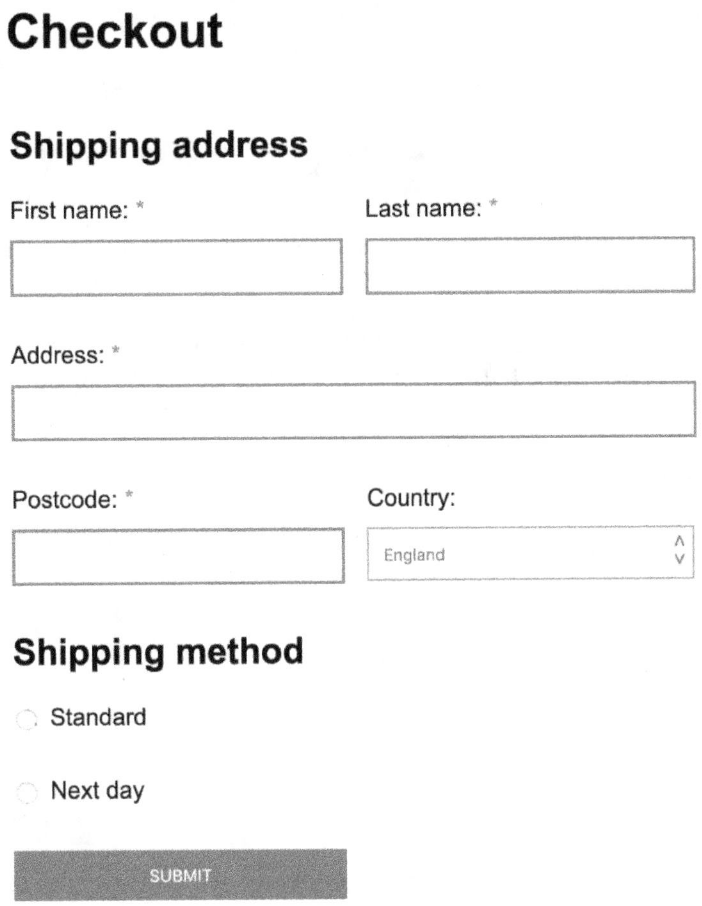

Checkout

Shipping address

First name: *

Last name: *

Address: *

Postcode: *

Country:

England

Shipping method

○ Standard

○ Next day

SUBMIT

Figure 4.26: Form highlighting the required fields before submitting

Activity 4.01: Building an Online Property Portal Website Form

An international online property website has approached you to design a search form for their listings page. This form should have the following fields: **search radius**, **price range**, **bedrooms**, **property type**, and **added to the site**, and an option to include sold properties in the user's search. Create your own solution using the skills you have learned in this chapter:

1. Start by creating a new file named **Activity 4.01.html** inside the **Chapter04** folder in VSCode.

2. Then, start writing the HTML for the form using the description of the fields required for this form.

3. Once you are happy with the HTML you have written, you can let your creativity run wild and style the form using CSS. Make sure that you include styling for validation:

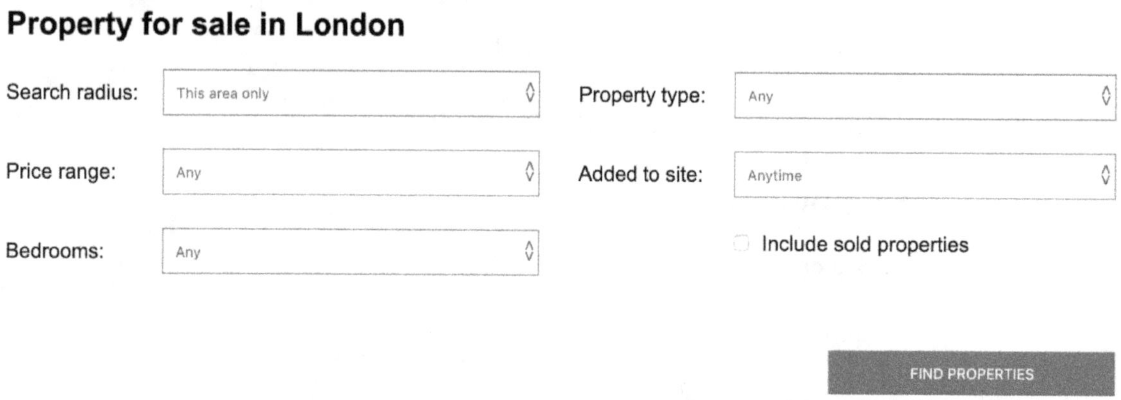

Property for sale in London

Search radius:	This area only	Property type:	Any
Price range:	Any	Added to site:	Anytime
Bedrooms:	Any		☐ Include sold properties

FIND PROPERTIES

Figure 4.27: Expected output of the activity

Note

The solution for this activity can be found in page 600.

Summary

In this chapter, we continued our journey into building web pages by exploring web forms. We first studied the most common form-based HTML elements, including inputs, select boxes, textareas, and buttons. We then looked at the most common styling methods for styling forms. To put this new knowledge into practice, we then built different forms.

We took some time to understand when you should use checkboxes and when to use radio buttons. We also spent some time looking at how you can add validation styles for web forms.

In the next chapter of this book, we will learn how to take our web pages to the next level. We will learn how to make our web pages even more appealing by creating and adding themes to our web pages.

5

Themes, Colors, and Polish

Overview

By the end of this chapter, you will be able to create a CSS theme for a website using different CSS properties that control color, borders, and backgrounds; use different CSS color value types (Hex, RGB, RGBA, and HSL) to define colors; create a dark theme using hsl() and the CSS invert filter; and customize a theme using CSS hooks. This chapter introduces some of the fundamental real-world uses of CSS. If you work for any period of time as a web developer, you will work with CSS to change the look and feel of a website or application. This chapter will show you how to do just that.

Introduction

In the previous chapter, you learned about forms, which allow you to work with user data in different ways. Data is important, of course – the modern world runs on it. It's not the only thing, however. For most people, all the data in the world is useless unless there's also a beautiful interface to access it with. **Cascading Style Sheets (CSS)** is the technology that will allow you to bring that interface to life on the web.

The power of CSS is its ability to change, sometimes radically, the appearance of a web page without updating the markup. This allows you to update the look and feel of a web page without you having any control at all over the content. This power expresses itself most commonly in themes, that is, CSS files that change the look and feel of a site or application and that can be applied easily. While some theming systems, like the ones in WordPress, allow and encourage dynamic changes to the markup, there's so much you can do with CSS alone. The classic example is the CSS Zen Garden site (https://packt. live/36HYha7). Founded in 2003, CSS Zen Garden has continued to showcase the wide range of possibilities that CSS offers to change the look and feel of a site:

Figure 5.1: Three examples of CSS Zen Garden

We won't be doing anything as dramatic as the designs on the CSS Zen Garden site, but the same lessons will apply here. We're going to take one HTML page, a simplified version of this author's WordPress blog, and apply multiple different themes to it to change its look and feel in different ways. In addition to applying different looks, we're going to also use different techniques to manipulate colors using different color **value types**.

The base that we'll be working with looks as follows. As you can see, it's a wholly text-based site. All of the design is provided by CSS, which makes it perfect for our purposes. We don't need to worry about creating new images – we can simply work with CSS files to update the look and feel of the site:

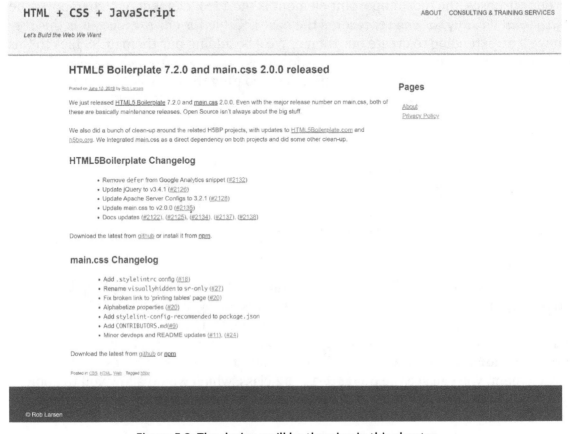

Figure 5.2: The design we'll be theming in this chapter

The folder structure is very simple. There is a file called **baseline.html** in the root of the folder and there is a folder called **assets** with one file in it: **style.css**. We will add new HTML files and new CSS files to this directory in order to create our themes:

assets	10/7/2019 4:22 PM	File folder	
baseline.html	10/7/2019 4:39 PM	Chrome HTML Do...	8 KB

Figure 5.3: The file structure

Let's get started.

The Markup

Before we get started with themes, let's familiarize ourselves with the markup we're going to be working with. Open up **baseline.html** in a text editor (preferably VSCode) to see what we're working with. The **head** of the document is straightforward. For our purposes, the most important element is the **link** element, which imports the **style.css** file. **style.css** represents the base CSS file for our exercise. It is this file that we'll be extending to create our theme. We'll be adding our theme CSS files to this section. With CSS, later rules override earlier rules, so updating styles is as simple as appending a new style sheet after the first and updating the rules with your new styles:

```
<!DOCTYPE html>
<html lang="en-US">
<head>
  <meta http-equiv="Content-Type" content="text/html; charset=UTF-8">
  <meta name="viewport" content="width=device-width, initial-scale=1">
  <title>HTML5 Boilerplate 7.2.0 and main.css 2.0.0 released -
    HTML + CSS + JavaScript</title>
  <link rel="stylesheet" id="hcj2-0-style-css" href="./assets/style.css"
    type="text/css" media="all">
</head>
```

If you're familiar with WordPress, you'll be familiar with some of the markup landmarks in the **body**, as well as many of the CSS classes present. WordPress themes take advantage of classes to customize pages from the template level, all the way down to the individual post level. In this case, on the **body** element, you can see the **post-template-default** class, which would allow you to style all post pages. In the same body element, you can see the **postid-11002** class, which would allow you to apply styles to an individual post and no other posts. Such is the power of a well-defined theme system:

```
<body class="post-template-default single single-post postid-11002
  single-format-standard">
<div id="page" class="site">
```

The header contains two areas that we'll manipulate: the **h1.site-title** element and **nav.site-navigation**. The **h1** is the title of the site and is therefore one of the most visible elements. The initial design is quite simple. We'll be updating it with something a little more eye-catching in our theme. A site navigation menu is a common element across most sites, so paying some attention to that in our example will be important. It is constructed as a **nav** element with a child **ul** element containing the menu items.

Notice again that there are classes that represent unique IDs attached to the `li` elements. If you wanted to style one of those individually, you could:

```html
<header id="masthead" class="site-header" role="banner">
  <div class="site-branding">
    <h1 class="site-title"><a href="https://htmlcssjavascript.com/"
      rel="home">HTML + CSS + JavaScript</a></h1>
    <p class="site-description">Let's Build the Web We Want</p>
  </div>
  <nav id="site-navigation" class="main-navigation"
    role="navigation">
    <div class="menu-menu-1-container">
      <ul id="primary-menu" class="menu">
        <li id="menu-item-10815" class="menu-item
          menu-item-type-post_type menu-item-object-page
          menu-item-10815">
          <a href=
          "https://htmlcssjavascript.com/about/">About</a></li>

        <li id="menu-item-10816" class=
          "menu-item menu-item-type-post_type menu-item-object-page
          menu-item-10816"><a href=
          "https://htmlcssjavascript.com/consulting-and-training-
          services/">Consulting &amp; Training Services</a></li>

      </ul>
    </div>
  </nav>
</header>
```

The primary and secondary sections of page content are contained in **div#content**:

```html
<div id="content" class="site-content">
```

div#primary is where the primary site content lives. It wraps a **main#main** element, which contains the **article** element, which is where the actual post lives. The **article** element has many classes that would allow you to target the post in many ways. We will actually use the **tag-h5bp** class in one of our themes to show that this particular post is about the HTML5 Boilerplate project:

```html
<div id="primary" class="content-area">
  <main id="main" class="site-main" role="main">


    <article id="post-11002" class="post-11002 post type-post
      status-publish format-standard hentry category-css
      category-html category-web tag-h5bp">
```

Inside the **article** is the article **header** element, which contains the article **header**, an **h1**, and some metadata about the post in **div.entry-meta**. Note that there are two versions of marked up metadata in this section: the **time** element with its associated **datetime** attribute and the vcard, which indicates the relationship of the author of the post with a specific URL:

```
<header class="entry-header">
  <h1 class="entry-title">HTML5 Boilerplate 7.2.0 and
    main.css 2.0.0 released</h1>
  <div class="entry-meta">
    <span class="posted-on">Posted on <a href="https://
      htmlcssjavascript.com/web/html5-boilerplate-7-2-0-and-
      main-css-2-0-0-released/" rel="bookmark"><time
      class="entry-date published updated" datetime=
      "2019-06-10T13:51:01-04:00">June 10, 2019</time></a>
    </span>
    <span class="byline"> by <span class="author vcard"><a
      class="url fn n" href="https://htmlcssjavascript.com/
      author/robreact/">Rob Larsen</a></span></span></div>
  </header>
```

Inside **div.entry-content** is the article itself. In this case, it is just a series of **p** elements, links, and one list, but in the world of WordPress, it could be pretty much anything. For our purposes, some simple text is enough to theme, so we'll work with what we've got:

```
<div class="entry-content">
  <p>We just released <a href="https://github.com/h5bp/html5-
    boilerplate">HTML5 Boilerplate</a> 7.2.0 and <a
    href="https://github.com/h5bp/main.css">main.css</a>
    2.0.0. Even with the major release number on main.css,
    both of these are basically maintenance releases. Open
    Source isn't always about the big stuff. </p>
  <p>We also did a bunch of clean-up around the related H5BP
    projects, with updates to <a href="https://github.com/
    h5bp/html5boilerplate.com">HTML5Boilerplate.com</a> and
    <a href="https://github.com/h5bp/h5bp.github.io">h5bp.
    org</a>. We integrated main.css as a direct dependency
    on both projects and did some other clean-up. </p>
```

Below the level 2 heading comes the unordered list as represented in the following code:

```
<h2>HTML5Boilerplate Changelog</h2>
<ul>
  <li>Remove <code>defer</code> from Google Analytics
      snippet (<a href="https://github.com/h5bp/html5-
      boilerplate/pull/2132">#2132</a>)</li>
  <li>Update jQuery to v3.4.1 (<a ref=
      "https://github.com/h5bp/
      html5-boilerplate/pull/2126">#2126</a>)</li>
  <li>Update Apache Server Configs to 3.2.1 (<a
      href="https://github.com/h5bp/html5-boilerplate/
      pull/2128">#2128</a>)</li>
  <li>Update main.css to v2.0.0 (<a href="https://github.
      com/h5bp/html5-boilerplate/pull/2135">#2135</a>)</li>
</ul>
<p>Download the latest from <a href="https://github.com/
   h5bp/html5-boilerplate/releases/download/v7.2.0/html5-
   boilerplate_v7.2.0.zip">github</a>or install it from <a
   href="https://www.npmjs.com/package/html5-
   boilerplate">npm</a>.</p>
<!--trimmed, see full content in downloaded files -->
   </div>
```

Following the body of the article, there's a small **footer** element for links to the different categories and tags for the content:

```
<footer class="entry-footer">
  <span class="cat-links">Posted in <a href="https://
     htmlcssjavascript.com/category/css/"
     rel="category tag">CSS</a>, <a href="https://
       htmlcssjavascript.com/category/html/"
     rel="category tag">HTML</a>, <a href="https://
       htmlcssjavascript.com/category/web/"
     rel="category tag">Web</a></span><span class="tags-
       links">Tagged <a
     href="https://htmlcssjavascript.com/tag/h5bp/"
       rel="tag">h5bp</a></span> </footer>
```

That ends the primary content section:

```
    </article>
    </main>
  </div>
```

The secondary content section is contained in an **aside**. It normally contains a number of section elements with different content. In this case, there is one **section** with an **h2** and a small list of links:

```
    <aside id="secondary" class="widget-area" role="complementary">
      <section id="pages-3" class="widget widget_pages">
        <h2 class="widget-title">Pages</h2>
        <ul>
          <li class="page_item page-item-2"><a href="https://
            htmlcssjavascript.com/about/">About</a></li>
          <li class="page_item page-item-5"><a href="https://
            htmlcssjavascript.com/privacy-policy/">Privacy Policy</a>
          </li>
        </ul>
      </section>
    </aside>
  </div>
```

And finally, there is a **footer** element with copyright information:

```
    <footer id="colophon" class="site-footer" role="contentinfo">
      <div class="site-info">
        <p>© Rob Larsen</p>
      </div>
    </footer>
  </div>
</body>
```

Now that we've taken a look at the markup, let's work at creating a "dark" theme for the site.

Inverting Colors

Many of the examples we'll be going through will use complementary inverted colors to create dark themes. To get the complementary or "opposite" color, you can use something like the invert tool in Photoshop or the one from https://packt. live/2WOOY3K.

Using the Opposite Color Tool (https://packt.live/2pGhUyS) is as simple as pasting the color you want to invert into the left text box and copying the output from the right text box:

You are here: Color Tools » **Opposite Color Tool**

Opposite Color Tool

Two colors are considered complimentary if they produce a neutral color — black, white, or grey — when mixed evenly. But, that is a mouthful. Simply put, a complimentary color has the opposite hue of a given color.

This tool will display the opposite color from a base color, and gives the hexadecimal code for your css/html:

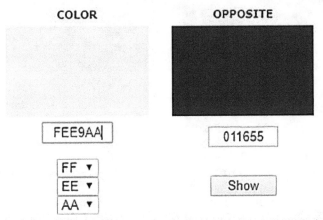

Alternatively, you could use the "invert" option in photoshop to get the same results...

Figure 5.4: The Opposite Color Tool

To invert colors in Photoshop, open the file and run the **invert** command. From the menu, go to **Image** > **Adjustments** > **Invert**. On the keyboard, use *Ctrl* + *I* on Windows or *Apple* + *I* on mac. This produces the following output:

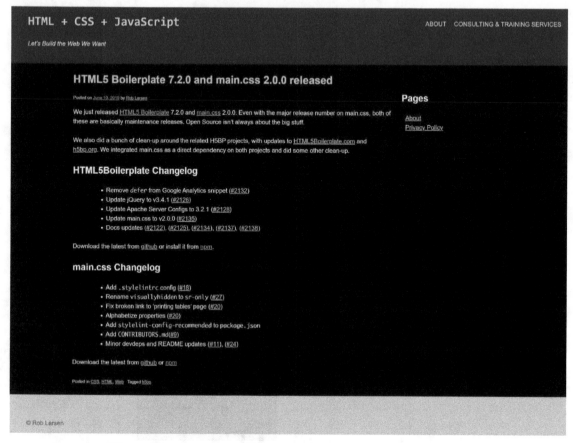

Figure 5.5: The output of the invert command in Photoshop

While this is a good baseline, we'll be adjusting some of the colors by hand in order to maintain some web design conventions with regard to traditional link colors.

We'll learn about another, much easier, way to invert colors in a later exercise.

New HTML Elements in the Theme

Theme files are, by nature, supposed to be inclusive of any markup you may use on a page. To that end, there are a few new HTML elements present in these files. This section will briefly introduce them:

- The **pre** (preformatted) element presents text that matches the text exactly as it was written in the HTML file, including all whitespace. The text is typically rendered in a monospace font, where each character takes up the exact same space in a line of text (like in a code editor).

- The **abbr** (abbreviation) element represents an acronym or abbreviation.

The deprecated **acronym** element serves the same purpose as **abbr**. While deprecated, it's included because themes have to handle a wide range of markup, including older, obsolete elements.

New CSS Background Properties

The following exercise uses three new CSS properties that all relate to background images. Let's introduce them before we go onto this new exercise.

background-image controls the background image of an element. Pass in the URL of your image to **url()** so that image appears as the background of your element.

background-repeat controls how the background image tiles or whether it tiles at all. The possible values are as follows:

- **repeat** indicates that the image should repeat as much as is needed to cover the available space. The last image will be clipped if it doesn't fit into the available space.

- **space** indicates that the image should repeat as much as possible without clipping. The original aspect ratio of the image will be maintained and the spacing between the first and last images, which are pinned to the edges, will change to fill the area.

- **round** indicates that the repeated images should stretch or compress (depending on whether greater than one half or less than one half of the image will fit the next slot).

- **no-repeat** indicates that the image is not repeated.

Let's use all the CSS knowledge we have gained so far to complete the next exercise.

> **Note**
>
> At this point, we are assuming you have already downloaded the necessary files (**baseline.html** and **style.css**) and set up your folders (**Chapter05** and **assets**) as outlined previously in this chapter. If not, you will need to go back and complete this part of the chapter to have your files ready before you perform the exercises and activity in this chapter.
>
> The link to the code and supporting files of this chapter can be found here: https://packt.live/35xhPOb

Exercise 5.01: Creating a Dark Theme

The first exercise we're going to go through will be creating a dark theme. Since many of the colors were originally generated using Hex values, we'll continue to use them in this first example. As you'll see, there are other, easier, ways to invert colors. Hex values are very common, however, so doing this example with Hex values is going to match itself to your real-world experience.

The finished theme will look like what's shown in the following screenshot:

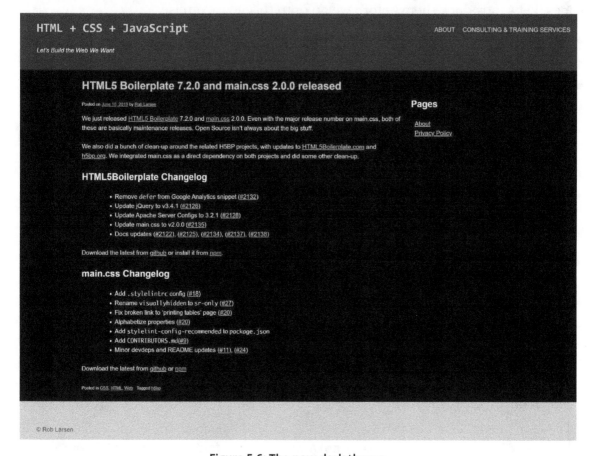

Figure 5.6: The new dark theme

Here are the steps to complete this exercise:

1. Create a file called **dark-theme.css** in the **Chapter05/assets** folder of your sample project.

2. Copy the contents of the **baseline.html** file and paste it into a new file called **Exercise 5.01.html**, which should be created inside the **Chapter05** folder.

3. Add the following line (see highlighted code) to the **head** of your document (**Exercise 5.01.html**) in order to include the new file. You do this by including **dark-theme.css** in your document in a new **link** element:

```
<head>
    <meta http-equiv="Content-Type" content="text/html;
        charset=UTF-8">
    <meta name="viewport" content="width=device-width,
        initial-scale=1">
    <title>HTML5 Boilerplate 7.2.0 and main.css 2.0.0 released -
        HTML + CSS + JavaScript</title>
    <link rel="stylesheet" id="hcj2-0-style-css"
        href="./assets/style.css" type="text/css" media="all">
    <link rel="stylesheet" id="dark-theme" href=
        "./assets/dark-theme.css" type="text/css" media="all">
</head>
```

4. Now, open **dark-theme.css** and start adding new rules to invert the colors that are present on the site. The existing style sheet is broken up into sections. We'll follow the same pattern in our new style sheet. We'll start with the *Typography* section, setting the text color for **body**, **button**, **input**, **select**, and **textarea** to be **#ffffff** (pure white):

```
body,
button,
input,
select,
textarea {
    color: #ffffff;
}
```

5. Next, we'll change some colors that don't actually show up in our demo but are text elements that may show up sometime in the future. We'll set the background of **pre** (preformatted) elements to **#111111** (a very dark gray) and the bottom border of the **abbr** (abbreviation) and **acronym** elements to be **#999999** (a lighter gray):

```
pre {
    background: #111111;
}
abbr,
acronym {
    border-bottom: 1px dotted #999999;
}
```

6. The next few changes will be in the **Elements** section, where we redefine certain generic HTML elements. This first edit will certainly be visible in the demo. We're going to change the background of the **body** element from white to black, **#000000**:

```
/*----------------------------------------------------------
# Elements
--------------------------------------------------------*/
body {
    background: #000000;
}
```

7. Next, we'll change the color of horizontal rules to be a dark gray:

```
hr {
    background-color: #333333;
}
```

8. In the Navigation section, we will change the color of links:

```
/*----------------------------------------------------------
# Navigation
--------------------------------------------------------*/
/*----------------------------------------------------------
## Links
--------------------------------------------------------*/
a {
    color: #add8e6;
}
a:visited {
    color: #b19cd9;
}
a:hover,
a:focus,
a:active {
    color: #4169e1;
}
```

Overall, these are not inverted or complementary colors. Since there are traditional colors for the different states of links, we will simply use the same blue and purple colors people expect for links and visited links – we will just use a lighter shade of each so that they show up better when placed on a black background. The first change will be to the generic **a** element, which we will set to **#add8e6**, a light blue. Next, change the color of visited links to be **#b19cd9**, a light purple. Finally, change the color of the **hover**, **focus**, and **active** states to be **#4169e1**, royal blue.

9. This final section of changes is going to be targeted at specific markup present in this specific WordPress theme. First, change the background of our generic **.content-area** element to be pure black:

```
.content-area {
    background: #000000;
}
```

10. Next, we need to change the color of the site header's **h1** to be **#EDC8AD**, which is a sandy yellow:

```
header.site-header h1 {
    color: #EDC8AD;
}
header.site-header h1 a {
    color: #EDC8AD;
}
```

Note that we are changing the color for both the **h1** element and the **a** element inside of it. This is to ensure that the generic **a** color we defined earlier doesn't override the sandy color we've generated for the **h1** element.

11. Continuing with the header, the next change adjusts the color of the navigation menu items and links to **#AAE8FF**, a pale blue, to contrast with the dark blue background we will set in the next step:

```
header.site-header nav .menu,
header.site-header nav li a {
    color: #AAE8FF;
}
```

12. The dark blue we just mentioned is added here, where we set **.site-header** to be **#001655**, which is, in fact, a dark blue:

```
.blog .site-header,
.single-post .site-header{
    background: #001655;
}
```

13. Next, we need to adjust the **h1** element inside of the main **article** element so that it's readable against the dark background. The complementary color of the original color, **#53276e**, is **#ACD891**. This is a light grayish green:

```
article h1 {
    color: #ACD891;
}
```

14. Next, we need to make sure that the body text is legible. The original content text defaulted to **#666666**, which is a dark gray. Set it to **#999999**, which is a lighter gray that will show up perfectly well against the black of the background:

```
article .content {
    color: #999999;
}
article picture img {
    border-top: 2px solid #6F66F6;
    border-bottom: 4px solid #6F66F6
}
```

15. Finally, we will update colors in the footer:

```
footer.site-footer {
    color: #123752;
    background: #AAC0FF;
}
footer.site-footer a {
    color: #123752;
}
```

First, we set the color of the default text and also the color of the elements to be **#123752**, which is a very dark blue. That's important because we set the background of **footer.site-footer** to **#AAC0FF**, which is a pale blue/magenta color.

16. Save **dark-theme.css**, and if you now right-click on the filename in VSCode on the left-hand side of the screen and select **Open In Default Browser**, it will show your dark theme:

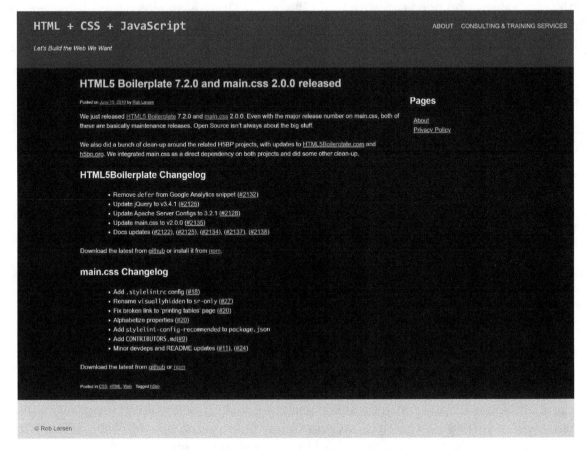

Figure 5.7: Successfully implemented theme

> **Note**
>
> Make sure that the **assets** folder contains the right file such as **style.css** to get the above output.

While the CSS involved here is very basic, the entirety of the work is simply changing the color values of several individual properties – calculating the complementary colors of the different Hex values isn't straightforward. Unless you're very good at working with color values, you need software to do the work. There is an alternative, however, that makes getting the complementary value of a color much easier. Using the `hsl()` (**Hue/Saturation/Lightness**) function for color values is going to make it a breeze. Let's look at how this works.

Creating a Dark Theme with the HSL Function

The HSL function allows you to update the color value of a property by using one of three arguments: **Hue**, **Saturation**, or **Lightness**:

- H represents the hue as an angle on the color wheel. You can specify this, using degrees (or, programmatically, radians). When provided as a unitless number, it is interpreted as degrees, with **0** as pure red, **120** as pure green, and **240** as pure blue.

- S represents the saturation, with **100%** saturation being completely saturated, while **0%** is completely unsaturated (gray). **50%** is a "normal" color.

- L represents the saturation, with **100%** saturation being completely saturated, while **0%** is completely unsaturated (gray).

This color system allows you to manipulate colors by hand in a way that doesn't really happen very often when working with colors in CSS. Assuming you have an encyclopedic knowledge of named colors, you could do this work off the top of your head, but otherwise, it's very difficult to manipulate Hex values or even RGB values without software. Working with the color wheel is much more intuitive, especially when dealing with inverting colors. To do so, you simply point to the opposite side of the color value. If you think of the color wheel in terms of **360** degrees, then the color **180** degrees away from your target color is on the opposite side of the color wheel.

The following image shows the color wheel and an illustration of **180** degrees being on the opposite side of the color wheel:

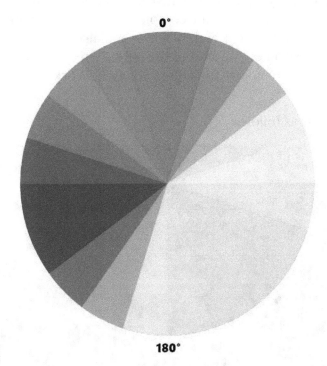

Figure 5.8: Color wheel

When we want to have grays inverting the color, this means we need to flip the value of the lightness value. The Hue, in that case, can be anything since the "grayness" of an HSL color is managed by the Saturation. Set the Saturation to 0 and you have a gray. From there, you adjust the Lightness from **0%** (black) to **100%** white to move along the grayscale.

Exercise 5.02: Creating a Dark Theme Using hsl()

For this exercise, we're not going to simply invert every color, since we've already seen how that turns out. Instead, we'll look for complementary colors where appropriate, look for entirely new colors where it makes sense, and also adjust the saturation and lightness of colors to make a new version of the dark theme. As you're following along with this exercise, feel free to adjust the colors yourself. Working with HSL makes it easy to experiment.

The new theme will look as follows:

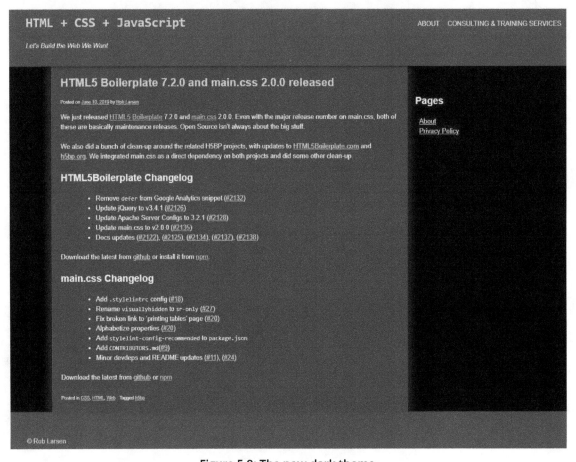

Figure 5.9: The new dark theme

Here are the steps to complete this exercise:

1. Create a file called **dark-theme-hsl.css** in the **Chapter05/assets** folder of your sample project.

2. Copy the contents of the **baseline.html** file and paste it into a new file called **Exercise 5.02.html**, which should be created inside the **Chapter05** folder.

3. Add the following lines (see highlighted code) to the **head** of your document (**Exercise 5.02.html**) in order to include the new file (**dark-theme-hsl.css**):

```
<head>
   <meta http-equiv="Content-Type" content="text/html;
      charset=UTF-8">
   <meta name="viewport" content="width=device-width,
      initial-scale=1">
   <title>HTML5 Boilerplate 7.2.0 and main.css 2.0.0
      released - HTML + CSS + JavaScript</title>
   <link rel="stylesheet" id="hcj2-0-style-css"
      href="./assets/style.css" type="text/css" media="all">
   <link rel="stylesheet" id="dark-theme-hsl" href=
      "./assets/dark-theme-hsl.css" type="text/css" media="all">
</head>
```

4. Now, open **dark-theme-hsl.css** and start adding new rules to create a new dark theme. The existing style sheet is broken up into sections:

```
/*-----------------------------------------------------------
# Typography
-----------------------------------------------------------*/
body,
button,
input,
select,
textarea {
    color: hsl(0, 0%, 100%);
}
```

Next we'll once again change some colors that don't actually show up in our demo. We'll set the background of **pre** elements to **hsl(0, 0%, 7%)** This is the opposite of the very light gray in the original design as is indicated by the Lightness value of the original color which was written in HSL as **hsl(0, 0%, 93%)** We then set the bottom border of **abbr** and **acronym** elements to be **hsl(0, 0%, 60%)**, the opposite of the original **hsl(0, 0%, 40%)**:

```
pre {
    background: hsl(0, 0%, 7%);
}
abbr,
acronym {
    border-bottom: 1px dotted hsl(0, 0%, 60%);
}
```

We'll follow the same pattern in our new style sheet. We'll start with the **Typography** section, setting the text color for **body**, **button**, **input**, **select**, and **textarea** to **hsl(0,0,100%)** (pure white). Remember that the hue here (pure red) doesn't matter since the saturation is 0% (pure gray). All that matters is the lightness, which is set to 100%.

5. Next, change the background of the **body** element from white to black, **hsl(0, 0%, 0%)**; like the example with white earlier, the **hue** here is still set to pure red, but since the lightness is set to 0%, it doesn't matter:

```
/*-------------------------------------------------------------
# Elements
-------------------------------------------------------------*/
body {
    background: hsl(0, 0%, 0%);
}
```

6. Next, update the **hr** definition with a color that inverts the gray from a light gray to something very, very dark, with the lightness set to 20%:

```
hr {
    background-color: hsl(0, 0%, 20%);
}
```

7. In the Navigation section, we will change the color of the links. Once again, these are not inverted. They're similar to the colors in the original dark theme, but using HSL allows us to align them a little bit. Set the color of the **a** elements to **hsl(195, 75%, 80%)**. Finally, set the **a:hover**, **a:focus**, and **a:active** colors to **hsl (225, 75%, 60%)**:

```
/*-------------------------------------------------------------
# Navigation
-------------------------------------------------------------*/
/*-------------------------------------------------------------
## Links
-------------------------------------------------------------*/
a {
    color: hsl(195, 75%, 80%);
}
a:visited {
    color: hsl(260, 75%, 80%);
}
```

```
a:hover,
a:focus,
a:active {
    color: hsl(225, 75%, 60%);
}
```

The color for basic links is a blue with a bit of green in it (imagine where 195 would be on the color wheel). It's highly saturated (**75%**) and very light (**80%**). Set **a:visited** to **hsl(260, 75%, 80%)**. This is a light purple (follow the color wheel three quarters of the way around to see where the purples live on the color wheel). We set the same lightness and saturation to keep the colors feeling similar on the page. In this case, the blue is nearly a "true" blue (blue being **240**) and then it's got the same saturation and a slightly darker lightness. These states are temporary, so they have some more weight to them than the links that need to be part of the text.

8. Next, set the background color of **.content-area** to **hsl(205, 20%, 20%)**. This is the biggest change from the original dark theme, where we've gone from pure black to a very dark blue:

```
.content-area {
    background: hsl(205, 20%, 20%);
}
```

Because of the previous change, you should add some padding to the child nodes of **.widget-area** in order to ensure that the text doesn't bump up against our dark blue **content-area**.

9. Set **padding-left** of the **h2** elements in the widget area to **20px** and set the padding of **ul** to **30px**. This will give you a comfortable amount of space:

```
.widget-area h2 {
    padding-left: 20px;
}
.widget-area ul {
    padding-left: 30px;
}
```

This isn't a radical change to the layout but illustrates how easy it is to work with the look and feel of a site without needing to get into the weeds with the markup.

10. Next, we need to change the color of the site header's **h1** to **hsl(205, 50%, 80%)** which, unlike the sandy yellow of the previous dark theme, is a blue:

```
header.site-header h1 {
    color: hsl(205, 50%, 80%);
}
```

You'll note that this blue is the same Hue as the background of the content area – it's just had the saturation upped to **50%** and the lightness upped to a very bright **80%**.

11. Once again, make the same change to the **header.site-header**, h1, that is, a color definition so that the **a:visited** color we defined earlier doesn't mess up the site header:

```
header.site-header h1 a {
    color: hsl(205, 50%, 80%);
}
```

12. Continuing with the header, the next change adjusts the color of the navigation menu items and links to **hsl(195, 100%, 80%)**:

```
header.site-header nav .menu,
header.site-header nav li a {
    color: hsl(195, 100%, 83%);
}
```

This is a more saturated version of the color we're using for links elsewhere on the page. It's the same Hue and Lightness. It will look like a link in the context of the page but will be different enough to signify that's it's not the same as a content link.

13. Set the background of **.site-header** to **hsl(205, 100%, 20%)**. You'll notice that this is a more saturated version of the same dark blue we used for the background of the content area. Using the same hues and changing the saturation and lightness makes adding coherence to the theme easy:

```
.blog .site-header,
.single-post .site-header{
    background: hsl(205, 100%, 20%);
}
```

14. Next, set the **h1** in the article elements to **hsl(205, 50%, 80%)**, which is the exact same color as the main site header:

```
article h1 {
    color: hsl(205, 50%, 80%);
}
```

15. The next change to make is setting the body text to be a gray with 60% lightness. We do this by setting the base color of text in the **.content** element to **hsl(0, 0%, 60%);**:

```
article .content {
    color: hsl(0, 0%, 60%);
}
article picture img {
    border-top: 2px solid hsl(244, 89%, 68%);
    border-bottom: 4px solid hsl(244, 89%, 68%)
}
```

16. Finally, we will update the colors in the **footer**:

```
footer.site-footer {
    color: hsl(205, 60%, 80%);
    background: hsl(205, 100%, 20%);
}
footer.site-footer a {
    color: hsl(205, 60%, 80%);
}
```

First, we adjust the text color and the associated link color to **hsl(205, 60%, 80%)**. Unlike the previous version of the dark theme, this text color is a bright blue. It's based on the same **205** Hue we've been using throughout this theme. This has a saturation of **60%** and the same **80%** lightness that many of the other text elements have. Next, set the background color to **hsl(205, 100%, 20%)**, which is a dark blue.

17. Save **dark-theme-hsl.css** and right-click on the filename in VSCode on the left-hand side of the screen and select **Open In Default Browser**, to see your dark theme:

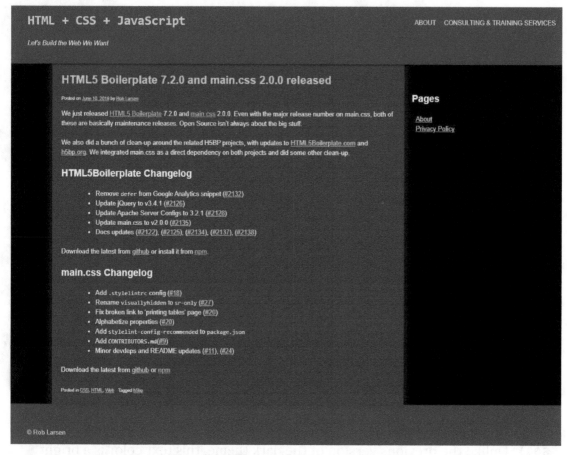

Figure 5.10: The new dark theme executed

Before we exit the topic of dark themes, there's one more CSS property and value that can provide a shortcut to a dark theme. Let's take a quick look at the CSS invert filter.

CSS Invert Filter

While you wouldn't use this particular CSS property and value in this way to create a full-on theme, it does present an interesting CSS-based shortcut for inverting the color scheme of an HTML element. CSS provides a number of filters that apply graphical effects to an element. There are many available. The one we're going to focus on is **filter: invert()**. This filter does what you would expect if you've been following along in this chapter – it inverts the color of the element.

A basic example of using invert is to invert the colors of an image, as shown in the following example. The following is the original image:

Figure 5.11: Image before using the invert filter

The following is the image after the invert filter has been applied:

Figure 5.12: Image after using the invert filter

Figure 5.11 shows the original image, while *Figure* 5.12 shows the inverted image after applying the invert filter. This is produced by the following markup. The **.invert** class that's defined in the **style** element in the head uses the invert filter and inverts the colors of the image. It's applied to the second of the two images:

```
<!DOCTYPE html>
<html lang="en-US">
<head>
   <meta http-equiv="Content-Type" content="text/html; charset=UTF-8">
   <title>Invert Filter</title>
   <style type="text/css">
     .invert {
        filter: invert(100%)
     }
   </style>
</head>
<body>
   <p><img src="./assets/react-2019-7.jpg" width="100%"
      alt="a mural by the author" ></p>
   <p><img src="./assets/react-2019-7.jpg" width="100%"
      alt="a mural by the author" class="invert"></p>
</body>
</html>
```

While it takes away flexibility and doesn't make for the best possible design, you can, with just two lines of CSS, create a rough dark theme using this CSS property.

Exercise 5.03: Creating a Dark Theme with the CSS Invert Filter

In this exercise, we will use the CSS invert filter to achieve the results that we obtained in the previous exercises. Let's take a look at how this works. The finished product will look as follows:

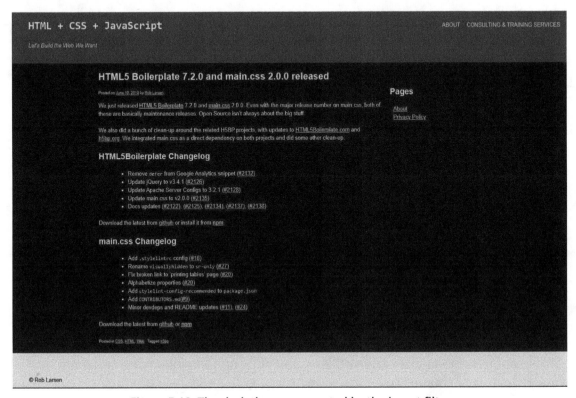

Figure 5.13: The dark theme generated by the invert filter

You'll notice that it's not quite the same as the original dark theme we created since this wholesale approach means we don't have access to individual elements such as the links we hand-tweaked previously. Still, it's pretty great to be able to change things this drastically with just a couple of lines of CSS.

Let's get started:

1. Create a file called **dark-theme-invert.css** in the **Chapter05/assets** folder of your sample project.

2. Copy the contents of the **baseline.html** file and paste it into a new file called **Exercise 5.03.html**, which should be created under the **Chapter05** folder.

3. Add the following line (see highlighted code) to the **head** of your document (**Exercise 5.03.html**) in order to include the new file (**dark-theme-invert css**):

```
<head>
  <meta http-equiv="Content-Type" content="text/html; charset=UTF-8">
  <meta name="viewport" content="width=device-width, initial-scale=1">
  <title>HTML5 Boilerplate 7.2.0 and main.css 2.0.0 released - HTML +
    CSS + JavaScript</title>
  <link rel="stylesheet" id="hcj2-0-style-css" href="./assets/style.
    css" type="text/css" media="all">
  <link rel="stylesheet" id="dark-theme-invert" href="./assets/dark-
    theme-invert.css" type="text/css" media="all">
</head>
```

4. Now, open **dark-theme-invert.css** and add the following rules to create our quick and dirty theme. Apply the invert filter to the **body** element:

```
body {
  filter: invert(100%);
  background: #000000;
}
```

The **100%** argument indicates that the inversion should be complete. You could apply a smaller percentage if you wanted, but the full version is what we're looking for here. Next, you need to apply a black background color because the invert doesn't work on the background color of the **body** element.

5. Save **dark-theme-invert.css** and if you now right-click on the filename in VSCode on the left-hand side of the screen and select **Open In Default Browser**, to see the following output:

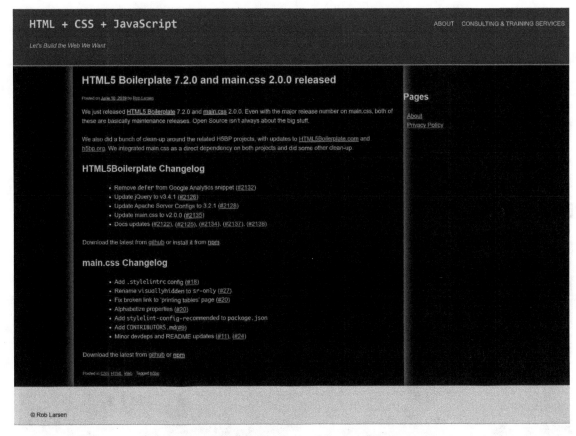

Figure 5.14: White shadows in the web page

You'll notice a couple of drop shadows that have been inverted into white shadows. Our previous dark theme work didn't show them because they were dark shadows on dark backgrounds. Here, they're distracting. Let's get rid of them.

6. First, remove **box-shadow** from the **.content-area** element by setting it to the **none** keyword value:

```
.content-area {
  box-shadow: none;
}
```

7. Next, remove **box-shadow** from the **.site-header** element by setting it to the **none** keyword value:

```
.blog .site-header, .single-post .site-header {
  box-shadow: none;
}
```

8. Save **dark-theme-invert.css** and refresh **Exercise 5.03.html**. You will see that your cleaned up dark theme has been generated with just a couple of lines of CSS:

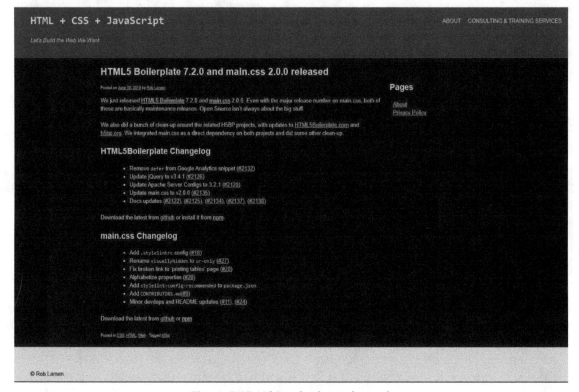

Figure 5.15: White shadows cleaned

It's not perfect, with the lime green links and yellow visited links, but we've avoided a lot of work. If you wanted to use this as part of a more complete theme, you could couple the invert technique with one of the previous techniques, that is, you could use the invert function in some places and then hex values or HSL functions where the application of individual styles is more important.

Now that we've covered the basics of theming, let's look at a few of the ways we can alter the look and feel of a site using the various hooks that WordPress (or another theme) can provide.

CSS Hooks

As I mentioned earlier, WordPress adds a lot of CSS hooks to the output markup to allow for a lot of customization regarding the look and feel. This starts with classes that are unique to the page and post and includes classes for the type of page, any tags or categories that were added to the post, and a number of other attributes. We're going to take advantage of those classes to make a few changes to the theme that would be specific to this particular post, as well as any posts that are tagged in a specific way.

Imagine you were a blogger and you wrote about web technology. You could tag posts with **"css"**, **"html"**, or **"javascript"** and then change, for example, the background color of the page:

```html
<!DOCTYPE html>
<html lang="en-US">
<head>
  <meta http-equiv="Content-Type" content="text/html; charset=UTF-8">
  <title>Using CSS hooks</title>
  <style type="text/css">
    .tag-css {
      background: #003366;
    }
    .tag-html {
      background: #006600;
    }
    .tag-javascript {
      background: #660000;
    }
  </style>
</head>
<body class="tag-css">

</body>
</html>
```

Two classes have been applied to this page in two places. We're going to use them to change the appearance of *just this page*. These are the unique **postid-11002** class on the **body** (that class references the internal post ID in WordPress) and the **tag-h5bp** class on the article element (which indicates that it's been tagged as **h5bp**).

Exercise 5.04: Customizing a Theme with CSS Hooks

In this exercise, we're going to build a theme on the same foundation as the rest of the themes in this chapter. The finished product should look as follows:

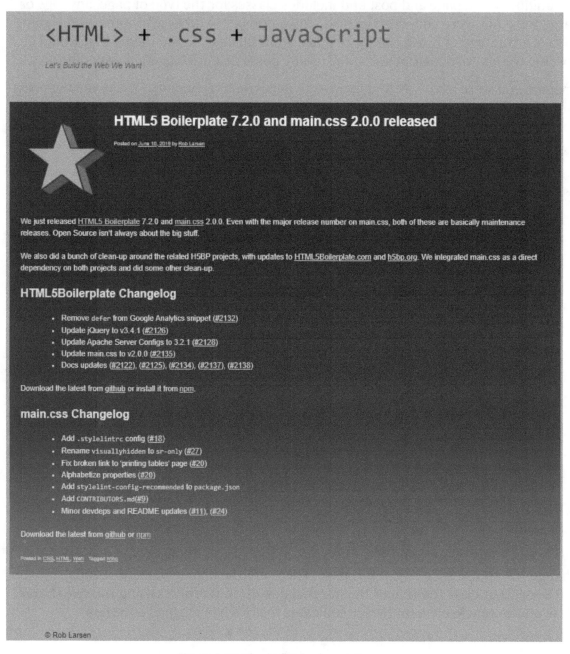

Figure 5.16: The individual post theme

As before, no changes have been made to the markup. All of the differences are down to CSS and if you applied this style sheet to any other page in that site without the **postid-11002** class, nothing would happen. Let's get started:

1. Create a file called **post-theme.css** in the **Chapter05/assets** folder of your sample project.

2. Copy the contents of the **baseline.html** file and paste it into a new file called **Exercise 5.04.html**, which should be created inside the **Chapter05** folder.

3. Add the following line (see highlighted code) to the **head** of your document (**Exercise 5.04.html**) in order to include the new file (**post-theme.css**):

```html
<head>
    <meta http-equiv="Content-Type" content="text/html;
      charset=UTF-8">
    <meta name="viewport" content="width=device-width,
      initial-scale=1">
    <title>HTML5 Boilerplate 7.2.0 and main.css 2.0.0 released -
      HTML + CSS + JavaScript</title>
    <link rel="stylesheet" id="hcj2-0-style-css" href=
      "./assets/style.css" type="text/css" media="all">
    <link rel="stylesheet" id="theme" href=
      "./assets/post-theme.css" type="text/css" media="all">
</head>
```

4. Now, open **post-theme.css** and start adding new rules to create your single post theme. Since this is going to be a darker theme, let's start by redefining the link colors. First, using the **postid-11002** class to differentiate it from other pages on the site, change the color of the **a** elements to be **#a5dff2**, a light blue. Then, once again, using the **postid-11002** class to differentiate it from other pages on the site, change the color of the **a:visited** elements to be **#c0a5f2**, a light purple. Finally, in the same manner, change the **hover**, **focus**, and **active** links to **#4c72e5**, which is a bright blue:

```css
.postid-11002 a {
  color: #a5dff2;
}
.postid-11002 a:visited {
  color: #c0a5f2;
}
.postid-11002 a:hover,
.postid-11002 a:focus,
.postid-11002 a:active {
  color: #4c72e5;
}
```

5. Next, we'll change `.content-area` so that it fits our new style:

```
.postid-11002 .content-area {
  background: linear-gradient(to bottom, #222222 30%, #333333 70%,
    #666666 100%);
  width: 100%;
  float: none;
  padding: 0;
  box-shadow: none;
  border: 10px solid #999;
}
```

In the preceding snippet, first, we add a **linear-gradient** from top to bottom, flowing from a very dark gray, **#222222**, to a dark gray, **#333333**, to a medium gray at the bottom. Next, we increase the width to be **100%** with **width: 100%**. Remember that, previously, this content area was floated to the left with the **widget-area** element floated to the right. So, we need to also remove the float with **float: none**. We also need to remove the padding from this element because we want **site-header** to fit nicely inside this containing element. We do that with **padding: 0**. Finally, we remove the **box-shadow** with **box-shadow: none** and replace it with a chunky light gray border with **border: 10px** solid **#999**. Note that the shorthand border definition is equivalent to setting **border-width**, **border-style**, and **border-color** individually.

6. Next, we hide the `.widget-area` and `.main-navigation` elements in order to make a slightly cleaner look for this specific post. This is done by setting the **display** property to **none** for both elements:

```
.postid-11002 .main-navigation,
.postid-11002 .widget-area {
  display: none;
}
```

7. Next, we're going to adjust the header. In the original design, it was the width of the page. In this version, we're going to cap the **width** at **1200px**, change the color of the background, add a logo, and remove the **box-shadow**:

```
.postid-11002 header.site-header {
  max-width: 1200px;
  margin: auto;
  background: #999;
  box-shadow: none;
}
```

In the preceding code, first, we set **max-width: 1200px** and add **margin: auto** to center the header element on the page. Then, we set the background to **#999** (which you'll notice is the same color as the border of the .**content-area** element – this is intentional) and the **box-shadow** to **none**. We still have to add the logo, but we'll do that in the next section.

This is the downside of pure theming. If we could change the markup, this would be done in another way – maybe by using one of the many available images replacement classes or by just inserting the image directly with proper accessible markup. We don't have any option to do anything like that, so instead, we're going to replace the text of the link inside the **h1** element with an image.

8. Let's do this by setting the background shorthand property to use **url(logo. png)** as the source of the background and to not repeat the background image with **no-repeat**. The rest of the rules are designed to allow the entire area to be clickable. To that end, give the **a** element a **height** of **75px**, a **width** of **750px**, and set the display to **inline-block** so that the height and width will stick. Finally, we use **text-indent: -9999px** to move the text off screen:

```
.postid-11002 header.site-header h1 a {
    background: url(logo.png) no-repeat;
    height: 75px;
    text-indent: -9999px;
    width: 750px;
    display: inline-block;
}
```

9. Next up, we target the **article.tag-h5bp** element. First, change the text color to #ddd, a light gray. Remember the padding we removed from the .**content-area**? We've moved the padding here with **padding: 0 2%**:

```
.postid-11002 article.tag-h5bp {
    color: #ddd;
    padding: 0 2%;
}
```

A new wrinkle to this design is targeting the .**entry-header** specifically. We do this in order to add a logo specific to the **tag-h5bp** class. This image is **200px** square and is a star on a dark gray (**#222**) background.

10. First, we set the background color to match **background-color: #222**. Next, we add the background image itself by adding **background-image: url(icon.png)** and suppress tiling of the image with **background-repeat: no-repeat**. Adding **background-size: contain** scales the background image so that it's as large as possible without cropping or stretching the image. Finally, set a **height** of **200px** on the **entry-header** and a **padding-left** of **200px** on the left to ensure that the text doesn't cover the icon:

```
.postid-11002 article.tag-h5bp .entry-header {
    background-color: #222;
    background-image: url(icon.png);
    background-repeat: no-repeat;
    background-size: contain;
    height: 200px;
    padding-left: 200px;
}
```

11. Earlier, we set the color of this article element to **#ddd**. Here, change the color of the article **h1** to be **#fff** (white) to make it pop a little bit compared to the body text:

```
.postid-11002 article.tag-h5bp .entry-header h1 {
    color: #fff;
}
```

12. The final update is going to be made to the **.site-footer** element. The background, like the header, is set to be **#999**, which matches the border on the article. This connects the three sections of the page. Finally, set the color of the text to **#222**, which is a dark gray:

```
.postid-11002 .site-footer {
    max-width: 1200px;
    margin: auto;
    background: #999;
    color: #222;
}
```

Like the header, the footer in the original design fits the **full-width** of the screen. This version has a **max-width** of **1200px**. **margin: auto** is added to ensure that the element is centered on the page.

13. Save **post-theme.css** and right-click on the filename in VSCode on the left-hand side of the screen and select **Open In Default Browser**. Now, you will be able to see your changes, as shown in the following screenshot:

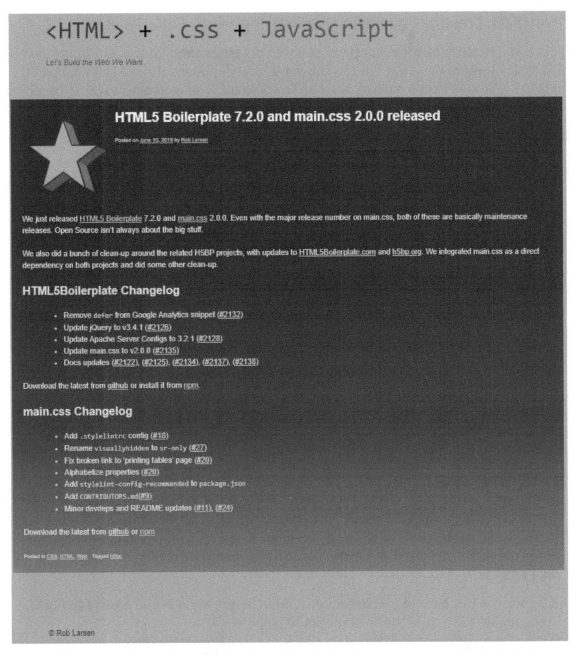

Figure 5.17: Final output

We began this chapter by creating a dark theme using the standard way, **hsl()**, and the invert filter. Then, we learned about CSS hooks. We will use all of these newly developed skills in the upcoming activity.

Activity 5.01: Creating Your Own Theme Using a New Color Palette

Suppose you're working for a company as a web developer and your boss comes to you with a task – it seems that the folks in the marketing department have just changed the company's colors or brand palette. Your boss would like you to create a theme based on those colors. You get the following image, which shows the brand colors:

Figure 5.18: Brand colors

This activity will allow you to create your own theme based on the brand colors. Assuming that **#4C72E5** is the main brand color and that **#FFE13F** (yellow) and **#FF9B3F** (orange) are secondary and tertiary colors, you will create a theme based on our existing markup that embraces the new brand's color scheme.

Here's how you'll do it:

1. Create file named **Activity 5.01.html** inside the **Chapter05** folder.

2. Change the background of the **body** element to use the lightest blue (**#9DB3F4**) and change the default text for the page to be something dark blue (something like **#333355**).

3. Change the header and footer background to be the main brand blue (**#4c72e5**) and change the text in those elements to be pure white.

4. Change the links to match the brand colors and choose a purple of the same value as the darker brand blue (**#0e3ece**) for the visited link color (something like **#bf0ece**).

5. Make the background of the whole **site-content** area white and add a brand blue border to the right and left-hand sides.

6. Change the color of the content area's h1s to brand blue.

7. Add a definition for the content area's h2s so that they use the secondary orange (**#FF7a00**).

8. Remove the **box-shadow** definitions from **site-header** and **content-area**.

It will look something like this when you're done:

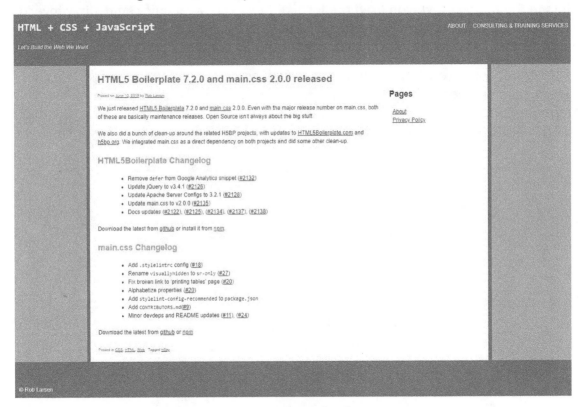

Figure 5.19: Expected web page

> **Note**
>
> The solution to this activity can be found on page 606.

Summary

In this chapter, you learned about using CSS to theme a web page with new colors, backgrounds, and borders. By adding a small CSS file on top of an existing CSS design, you were able to easily create four different versions of a web page using nothing but CSS properties and values that you applied to your existing markup.

First, you created an inverted dark theme using Hex values for complementary colors of the colors in the original theme. Next, you used HSL colors to create a more polished theme, where design considerations trumped pure complementary values for the colors in the theme. Next, you used the invert filter to create a quick-and-dirty inverted theme. Finally, you created a post-specific theme by taking advantage of the hooks that a good CMS such as WordPress provides to be able to style individual pages.

In the next chapter, you'll learn about some very important technologies for today's modern, many-device web – Media Queries and Responsive Web Design. With Media Queries and Responsive Web Design, you'll be able to make sites and applications that elegantly scale to work with whatever browser/screen/device your user happens to be using at that moment.

Responsive Web Design and Media Queries

Overview

By the end of this chapter, you will be able to implement the mobile-first principle when designing web pages; explain the fundamentals of responsive web design; use a variety of CSS media queries; create a mobile-friendly web page; and implement printer-friendly web page styles. This chapter introduces you to the world of responsive web design and aims to teach you how to develop web pages using media queries to build mobile-first websites that are suitable for a whole variety of devices, regardless of the device's screen size or hardware. In addition to this, the chapters introduce printer-friendly web page styling.

Introduction

In the previous chapters, we learned about the basics of HTML5 and CSS3 in terms of structure, layout, and typography. We have also explored HTML forms, and how to go about theming websites.

Now, we're going to look into adapting our web pages for the world of mobile-first responsive web design. A key element of a successful website starts with a good user experience design, providing the user with a seamless browsing experience on the website regardless of their device. Many users access websites through their mobile devices; it's second nature for them. From finding a location on a map to gathering information over the internet, users are finding that everything can be done using their phone, which is convenient since, when on the go, users are hardly likely to have a computer with them. With the rise of the mobile internet, it's realistic to expect to be able to get online in many parts of the world, even the more remote places, these days. Given this, it's not uncommon for users to use their phones more than their computers. As such, it would be wise to assume that users expect websites to just work on all their devices, regardless of screen size, and so we'll be looking at the mobile-first principle and how we can implement it for a website.

The majority of clients these days will ask for a mobile version of their website. It's become an expectation in the industry for a user to be able to browse and/or shop online from their phone, tablet, or computer; and if they can't, then it's generally considered a missed opportunity for custom. With every year that passes, many more new devices with different screen sizes are released and it would be incredibly difficult for a web developer to keep up with them all. This is where the responsive web design approach comes into play, which you'll learn about in this chapter.

Mobile-First

With the rapid uptake of the mobile internet over the past decade, it's now very common to see websites that have more mobile web traffic than desktop web traffic. Considering new (and existing) websites from a mobile point of view is now seen as an essential business case, so it's an essential web design principle to put into practice. It's vital, for instance, to consider the mobile user's experience when using the website, right down to the basics of navigation and the experience of buying items if it's an e-commerce site.

In the following figure, we can see three wireframes illustrating an unresponsive website. You can see how the layout of the web page doesn't change despite being displayed on different-sized devices. This figure also shows the output of *Chapter 3, Text and Typography Exercise 3.06, Putting It All Together*, which is also unresponsive, as it maintains its layout when viewed on a mobile device:

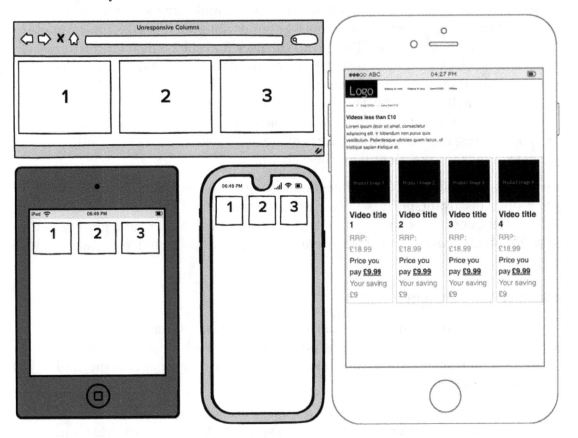

Figure 6.1: Wireframes of unresponsive websites, which maintain their layout regardless of device

The layout remains the same on each device, even though the narrower the device is, the more zoomed out the display of the website has to be in order to maintain the layout. This may force the user to zoom in to read website content, which creates a bad user experience and will likely cause users to leave the website when using smaller devices. As a web developer, you would want to optimize the website user's experience to increase your client's potential website conversion. The mobile-first principle plays a key part in this.

The mobile-first principle consists of two main points:

- **Responsive web design**: This refers to a website adapting its design as the web browser is resized, to fit to whatever device the user is choosing to use at the time.

- **Feature enhancements and graceful degradation**: This means making the most of the device you are accessing a website on. For example, on mobile websites, we can take full advantage of mobile features where possible, such as **GPS** (for example, a mobile website could use GPS to help the user find the nearest store), touch gestures (such as swiping across the screen to see more content or using pinch gestures to zoom), and the gyroscope (device movement can be used to help build a more interactive website). For the desktop website, you can add new website features that utilize the increase of available screen space and gracefully degrade other features; mouse-click gestures can be used instead of touch gestures to achieve similar interactions with the website.

As a web designer following the mobile-first principle, the lack of space on a mobile device is not a disadvantage. It is, in fact, an advantage of the mobile-first design, as the most important content is delivered first, often cutting the "waffle" and getting to the point so that mobile users don't lose interest and instead have a much better mobile user experience.

Responsive Web Design

With a multitude of different devices currently on the market, responsive web design is an approach where a website is designed to respond to work smoothly irrespective of the device's screen size, software (for instance, iOS Safari, Android Chrome, and so on), or orientation (portrait or landscape).

A website built using the responsive web design approach should be built to automatically adjust to the user's device and respond to the user's behavior, such as changing the device from portrait to landscape orientation, resizing the browser window, and scrolling.

The following figure shows a wireframe of a responsive website stacking content, along with an example of stacking from https://packt.live/31e3aUv:

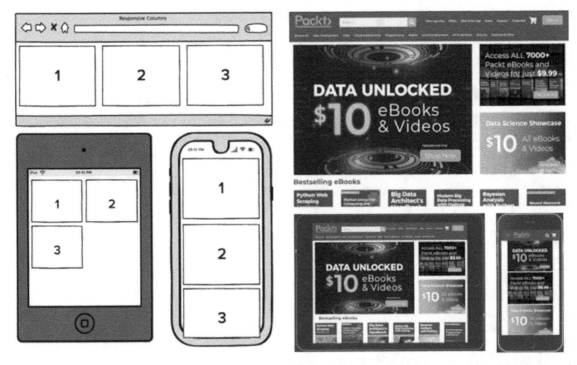

Figure 6.2: A wireframe of a responsive website stacking content, and an example of stacking in practice with the Packt website

The preceding figures are a good demonstration of putting responsive web design into practice. Content that is currently displayed over three columns on a desktop computer can be stacked vertically across two columns on a tablet and stacked vertically in a single column on a smaller mobile device. Using HTML5 and CSS3, this layout could be achieved with a combination of the **flex**-based grid layouts discussed in *Chapter 2, Structure and Layout*, and media queries, which we'll learn about in this chapter.

Responsive Viewport

In order to be able to build a responsive website, it is necessary to understand media queries, but first, we need to ensure that we have the viewport meta tag in the **<head>** of our HTML document; this was previously introduced in *Chapter 1, Introduction to HTML and CSS*. The viewport meta tag is widely used in modern web browsers to help ensure cross-browser compatibility in responsive web design. The following code snippet shows where you would place this within your HTML document:

```
<!DOCTYPE html>
<html>
<head>
<meta name="viewport" content="width=device-width, initial-scale=1" />
</head>
<body>
    <!-- Website content here -->
</body>
</html>
```

By setting the viewport **width** property to **device-width**, we are instructing the web page to display with the device's natural resolution, as opposed to the default zoomed-out display of the website at desktop resolution on the device, like the example in *Figure 6.1*. The **initial-scale** viewport controls the initial zoom on the web page; you should keep this at 1 unless you need to enlarge the page for any reason. Remember that the user can zoom in on the web page themselves using the normal gestures, unless this is disabled by setting another additional viewport property, **maximum-scale**, to the value of 1 as well. However, this is not recommended for accessibility reasons as some users may find it beneficial to zoom in when browsing, especially if they have a visual impairment.

Understanding Basic Media Queries

Media queries are a core feature of CSS3 and they allow the developer to change the display of a page based on the device screen width and orientation (portrait/landscape), or even to display the page differently for printers if a physical copy of the web page is to be created.

First, let's understand what a breakpoint is. A breakpoint is the point at which the web page responds to the device or web browser's width (or height). When the breakpoint threshold has been reached, different styles can be applied in CSS. Take the Packt Publishing website, for example (https://packt.live/34BkPHM). If you load this on a computer and then resize the width of the web browser window from full width to narrow width, you will notice that changes in styling are triggered at certain points: breakpoints.

This is made possible using CSS3 media queries. We can have breakpoints with the height of a device or browser screen, too. See *Figure 6.3*:

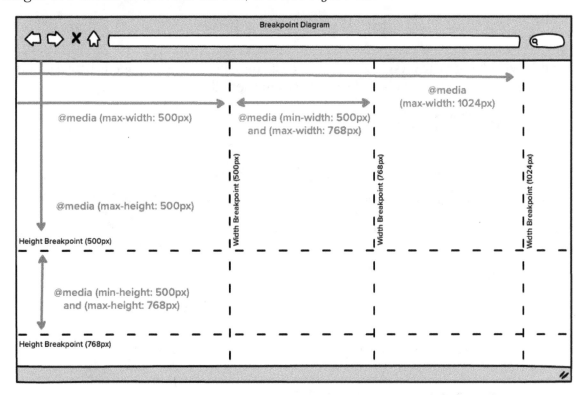

Figure 6.3: Breakpoint diagram with media query examples

A media query is represented in CSS3 with the `@media` syntax, followed by a condition or breakpoint. For example, if we wanted styles to be applied to a screen with a maximum width of **768** pixels, then we could write this as `@media only screen and (max-width: 768px)`, followed by curly braces `{ }` containing the code to apply when this media query condition is met.

For example, take the following CSS code:

```
@media only screen and (max-width: 768px) {
    p {
        color: blue;
    }
}
```

This code will change the **p** tag's text **color** to **blue** when the web page is browsed on a device that is up to **768** pixels wide, as shown in the following figure:

HTML and CSS

How to create a modern, responsive website with HTML and CSS

Figure 6.4: <p> tag displayed in blue on a screen that is up to 768 pixels wide

The following code will change the **p** tag's text color to red when the web page is browsed on a device wider than **769** pixels, as shown in the figure that follows the code:

```
@media only screen and (min-width: 769px) {
    p {
        color: red;
    }
}
```

HTML and CSS

How to create a modern, responsive website with HTML and CSS

Figure 6.5: <p> tag displayed in red on a screen wider than 769 pixels

So, as you can see, CSS media queries allow you to apply styling based upon the device/browser width. You can combine the screen minimum and maximum width properties of **min-width** and **max-width** to target certain breakpoint windows, meaning certain styles can be applied according to the screen width.

For example, this CSS code combines the **min-width** and **max-width** properties:

```
@media only screen and (min-width: 480px) and (max-width: 768px) {
    p {
        color: purple;
    }
}
```

This code will change the **p** tag's text to **purple** when the **screen** width is between **480** pixels and **768** pixels, as shown in the following figure:

HTML and CSS

How to create a modern, responsive website with HTML and CSS

Figure 6.6: <p> tag displayed in purple on a screen whose width is between 480 and 768 pixels inclusive

It's also possible to use media queries to style things based upon the device/browser screen height, as in this example:

```
@media only screen and (min-height: 1000px) {
    body {
        background-color: pink;
    }
}
```

The preceding code snippet will change the web page's background color to pink, using the **body** HTML tag, when the web browser's height is 1,000 pixels or higher, as shown in the following figure:

HTML and CSS

How to create a modern, responsive website with HTML and CSS

Figure 6.7: <body> tag displayed with a pink background for a screen height of 1,000 pixels or above

These @media code examples would be written within a CSS file and work well if you have one stylesheet for your web page. However, if you wanted to have separate CSS stylesheets for different device width breakpoints, this can also be achieved. For example, we can include the stylesheets with this HTML code:

```
<link rel="stylesheet" media="screen and (max-width: 480px)"
  href="mobile.css" />
<link rel="stylesheet" media="screen and (min-width: 481px) and
  (max-width: 768px)" href="tablet.css" />
```

This example code would be placed within the **<head>** of the HTML document and would use the **mobile.css** file on screen widths of **480** pixels or lower, and use the **tablet.css** file on screen widths between **481** pixels and **768** pixels.

This can be a very useful feature, from a code maintenance point of view, for helping to keep device-specific styles together in separate files, particularly if they vary significantly from breakpoint to breakpoint. But to debunk any performance optimizations myths here, sadly, most modern browsers will still load all **<link />** files when loading a web page, regardless of screen size – even though the styles are only applied when the media query specified becomes true.

Now, cast your minds back to *Chapter 3, Text and Typography Exercise 3.06, Putting It All Together*, in which you created a video store web page (see *Figure 6.8*). We'll be developing this exercise further to introduce media queries to the web page in our first exercise of this chapter:

Figure 6.8: Video store web page

Exercise 6.01: Using Media Queries to Change the Page Layout

In this exercise, we'll make the video store navigation work responsively based upon browser screen width.

Let's complete the exercise with the following steps:

1. Create a folder named **Chapter06**. And then create a file named **Exercise 6.01. html** under **Chapter06** folder. Start with copying the final code of **Exercise 3.02.html** from *Chapter 3, Text and Typography Exercise 3.02, Navigation* . Second, we'll change the CSS styles on the navigation menu to make the code styled for mobile devices. Then, we'll remove **display: flex** from the **nav ul** element to make the navigation list items collapse underneath one another, instead of sitting horizontally in the same row. We'll also change the styling on the **a** tag element further to make it more suitable for mobiles by giving the links a background color and changing this on its hovered/tapped state.

It's worth noting the importance of the * { margin: 0; padding: 0; } part of the code, as this resets the browser's default margins and the padding of all elements on the web page to zero. By default, many web browsers apply a margin and/or padding to list styles (ul/li), which would cause us issues when trying to style the navigation. For complete control and cross-browser fluency, it's better to reset, as we have done in the following example code. The following code snippet shows the example code of the completed CSS changes:

```
<style>
    * {
      margin: 0;
      padding: 0;
    }
    nav ul {
      list-style: none;
    }
    nav a {
      color: white;
      background: black;
      font-weight: bold;
      display: block;
      padding: 15px;
      font-size: 15px;
      text-decoration: none;
      text-align: center;
    }
    nav a:hover {
      background: darkgray;
    }
</style>
```

If you now open your exercise web page in the web browser, you'll be able to see how the navigation has changed to be vertically stacked, as opposed to being displayed in a row as it was before. You can see this in the following figure:

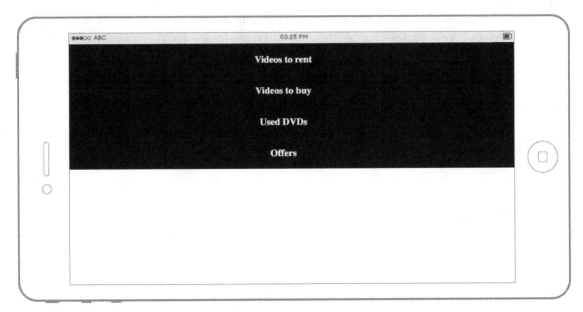

Figure 6.9: Navigation display after removing flex

2. Next, we will add our first media query to apply the **display: flex** style to only screen sizes of **480** pixels or above (which commonly means landscape mobile and wider):

```
@media (min-width: 480px) {
    nav ul {
        display: flex;
    }
    nav li {
        flex: 1 1 auto;
    }
}
```

We were introduced to flexbox in *Chapter 2, Structure and Layout*, but to recap briefly, adding **display: flex** to the **ul** parent element ensures that the child **li** elements are displayed in the same row. Then, by providing the **flex: 1 1 auto;** directive, we tell the **li** elements that 'the basis of growing and shrinking in width changes, this **flex** property combines the **flex-grow**, **flex-shrink**, and **flex-basis** properties into one shorthand property. This will create a single row of navigation elements, with the spacing and element widths being automatically calculated as the web page screen width increases or decreases.

Now, the navigation system will comprise flexible blocks in a row, each with an on-screen width of **480** pixels or above, as shown in the following figure:

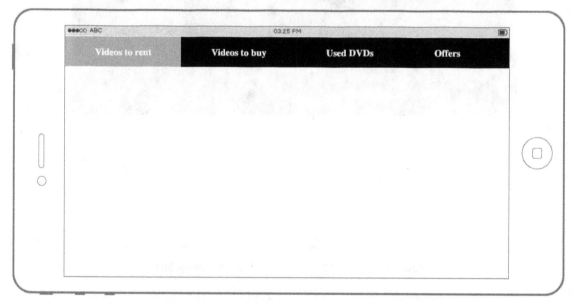

Figure 6.10: Navigation display after adding flex

3. Now, to cater for even wider screens, we're going to increase the font size for screens of **768** pixels or higher (which commonly would be screens for tablets and other bigger devices):

```
@media only screen and (min-width: 768px) {
    nav a {
        font-size: 20px;
    }
}
```

4. Save your HTML file and right-click on the filename in **VSCode** on the left-hand side of the screen and select **open in default browser**. In the web browser, if you resize the browser window (by placing your cursor in the bottom-right corner of the browser window and clicking and dragging the corner of the web browser to make the window narrower and wider), you should notice the styling change as different breakpoints are reached, as in the following screenshots.

The following figure shows what the navigation blocks at **479** pixels (or less) look like with a stacked layout; this is most commonly seen on mobiles in portrait orientation:

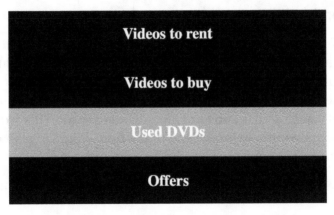

Figure 6.11: Navigation at 479 pixels or below

The following figure shows the navigation blocks at the breakpoint of **480** pixels or above, using the flex column layout and with the font size fixed at **15px**:

Figure 6.12: Navigation at 480 pixels or above

The following figure shows the navigation blocks displayed at **767** pixels or above, maintaining the previously set flex column layout, but with the font size increased to **20px**:

Figure 6.13: Navigation at 767 pixels or above

As a recap, in this exercise, you have learned how to take an existing web page that was not responsive or mobile-friendly and change the styling structure to make it a mobile-first web page, and you have used media queries to apply responsive CSS3 styling as the screen size increases to change the styles/layout to suit the screen better. Next, we will be looking at device orientation in media queries to help design responsive web pages.

Device Orientation Media Queries

Mobile and tablet devices have two orientations: portrait (when the device is upright) and landscape (when the device is on its side). When coding websites for such devices, it can be useful to target the device's orientation with media queries to apply different styles to give the user a better experience. The syntax of the media query for checking device orientation is **@media (orientation: landscape)**. This would make the rules apply to devices currently in landscape mode only. You can also have **@media (orientation: portrait)** for portrait devices only instead.

The following figure shows a screenshot of what this could look like in action on a tablet device:

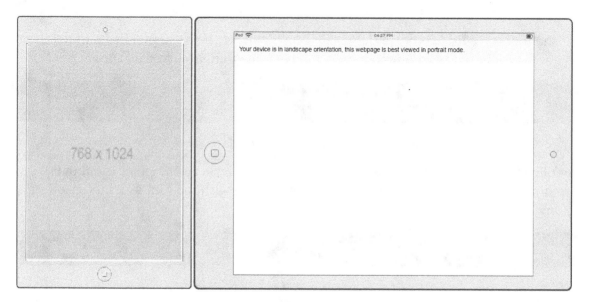

Figure 6.14: Orientation warning demo on a tablet device

Here is the full source code for the preceding figure, demonstrating the media query in action:

Example 6.01.html

```
13    @media (orientation: landscape) {
14      p.warning {
15        display: block;
16      }
17    }
```

The complete code for this example is available at: https://packt.live/2JXLsyM

You can combine orientation and width media queries together to form more advanced rules. Take this, for example:

```
@media (orientation: landscape) and (min-width: 400px) and (max-width:
    768px)
```

This would display the **p** tag with a warning only on screens that are at least **400** pixels wide, but that are no more than **768** pixels wide if the screen is also in landscape orientation mode. You can see this demonstrated in the following figure:

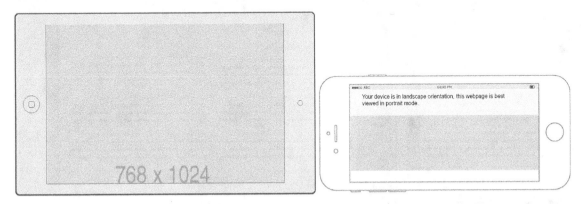

Figure 6.15: Orientation warning demo on a tablet and mobile device

Here is the full source code for the preceding figure, demonstrating the combination of orientation and width media queries:

Example 6.02.html

```
13   @media (orientation: landscape) and (min-width: 400px)
        and (max-width: 768px) {
14     p.warning {
15       display: block;
16     }
17   }
```

The complete code for this example is available at: https://packt.live/33s5gBY

As you can see, we can build very effective media queries to change the styling and display of a web page based on many environmental conditions.

Exercise 6.02: Using Media Queries to Detect Device Orientation

In this exercise, we will create a web page to detect device orientation using media queries, and then we will change the layout of the page within these media queries to create portrait and landscape layouts for the website. In the following figure, you can see a wireframe demonstrating what we are trying to achieve in this exercise:

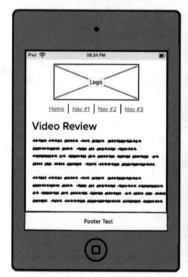
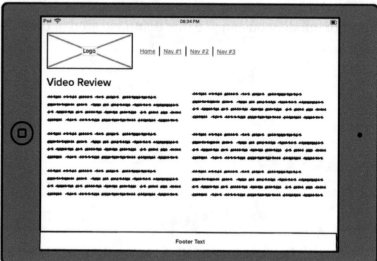

Figure 6.16: Sketch of the expected output of the exercise with portrait and landscape device orientation layouts

Let's complete the exercise with the following steps:

1. Create a file named **Exercise 6.02.html** within the **Chapter06** folder. Start with the following code as our template. We've got a **<header>** tag containing a logo image and navigation links, followed by an **<article>** tag containing some text, and finally, a **<footer>** tag containing the author credit:

```
<!DOCTYPE html>
<html lang = "en">
<head>
  <title>Exercise 6.02: Device Orientation</title>
</head>
<body>
  <header>
      <img src="https://dummyimage.com/200x100/000/fff&text=Logo"
        alt="" />
      <nav>
          <ul>
```

```
              <li><a href="">Home</a></li>
              <li><a href="">Nav #1</a></li>
              <li><a href="">Nav #2</a></li>
              <li><a href="">Nav #3</a></li>
          </ul>
      </nav>
  </header>
  <article>
      <h1>Video Review</h1>
      <p>

      Lorem ipsum dolor sit amet, consectetur adipiscing elit.
      Phasellus scelerisque, sapien at tincidunt finibus, mauris
      purus tincidunt orci, non cursus lorem lectus ac urna. Ut
      non porttitor nisl. Morbi id nisi eros.
      </p>

      <p>

      Donec et purus sit amet odio interdum accumsan eleifend ut
      sapien. Praesent bibendum turpis non nisl elementum, sed
      ornare purus semper. In vel sagittis felis.
      </p>

      <p>

      Suspendisse vitae scelerisque est. Aenean tempus congue
      lacus vehicula congue. Vivamus congue ligula nec purus
      malesuada volutpat.
      </p>

      <p>

      Vestibulum ante ipsum primis in faucibus orci luctus et
      ultrices posuere cubilia Curae; Curabitur sit amet mattis
      urna. Morbi id luctus purus, sodales facilisis odio. Orci
      varius natoque penatibus et magnis dis parturient montes,
      nascetur ridiculus mus.
      </p>

  </article>
  <footer>
      <p>Website by Author Name</p>
  </footer>
</body>
</html>
```

2. Add the responsive viewport meta tag to the **<head>** of the HTML code:

```
<meta name="viewport" content="width=device-width, initial-scale=1" />
```

3. Now, add some CSS styling to the **<head>**, coding it with the portrait mobile display in mind. The following code adds some navigation styles to put the links in a row with some spacing between them:

```
<style>
  * {
    margin: 0;
    padding: 0;
  }
  header {
      text-align: center;
  }
  nav ul {
      list-style: none;
  }
  nav li {
      display: inline-block;
      margin-right: 5px;
      padding-right: 5px;
      border-right: 1px solid black;
  }
  nav li a {
      color: black;
      text-decoration: none;
  }
  nav li a:hover {
      text-decoration: underline;
  }
  nav li:last-child {
      border-right: 0;
      padding-right: 0;
      margin-right: 0;
  }
  article, footer, nav {
      padding: 10px;
  }
  article h1, article p {
      margin-bottom: 10px;
  }
```

```
    footer {
        background: black;
        text-align: center;
        color: white;
    }
    </style>
```

You'll notice **:last-child**, which means that the final element in the navigation, in this case, has slightly different rules to finish off the styling. In addition to styling the navigation, the CSS also adds a basic style to the article content and footer of the web page.

4. Add a **@media** query onto the end of the existing CSS to apply the appropriate layout change to devices in landscape orientation, as per *Figure* 6.16. In the following code, when in landscape orientation, we are applying the flex layout and center alignment in the header tag. This enables the logo and navigation to sit horizontally next to one another. The navigation gets some margin space to separate itself from the logo in the header, and the article uses a CSS3 column to achieve a two-column layout, utilizing the extra space on a screen viewed in landscape orientation:

```
<style>
    @media (orientation: landscape) {
        header {
            display: flex;
            align-items: center;
        }
        nav {
            margin-left: 20px;
        }
        article {
            columns: 100px 2;
        }
    }
</style>
```

5. You should now have the outcome in your web browser (*Figure 6.17*). You can use Developer Tools in Chrome, for example, to simulate devices and their current display orientation. The easiest way to do this in Chrome would be to right-click anywhere on the web page and click **Inspect**. This will open Developer Tools in Chrome. Once opened, you can click the mobile and tablet device icon to toggle the device toolbar, which allows you to simulate mobile devices and rotate the device's current orientation. You can see a screenshot of this in *Figure 6.18*.

The following diagram shows the output of the exercise in a mobile device, both in portrait and landscape orientation:

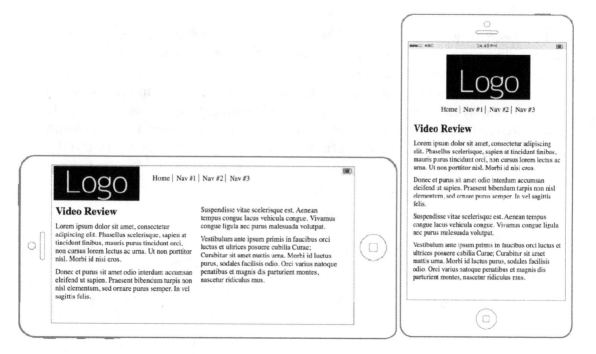

Figure 6.17: Completed solution in mobile portrait and landscape orientation

The following figure shows a screenshot of the device simulator in Chrome's Developer Tools:

Figure 6.18: Chrome Developer Tools Device Simulator

We've extended our knowledge of media queries to now be able to change the styling of a web page based on the device's orientation. We've also utilized the display of the copy within the landscape orientation 'display, by making use of the CSS3 columns property to achieve a two-column text layout. Next, we'll look at combining multiple media queries with the OR operator for media queries.

Combining Multiple Media Queries

If you want to apply a certain style change to more than one breakpoint, then you can combine media queries with comma-separated lists. The comma acts as an OR condition in the media query string. Refer to the following code snippet, for example:

```
@media screen and (max-width: 480px) and (min-width: 320px),
   (min-width: 1280px) {
      body {
         background: red;
      }
}
```

This would apply a red background to the web page's **body** tag when the screen is between **320** pixels and **480** pixels wide, and 1,280 pixels or wider. You can see this in the following figure:

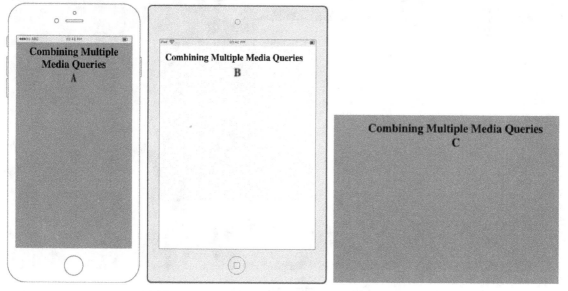

Figure 6.19: Demo of combining multiple media queries on mobile (A), tablet (B), and desktop (C)

As you can see in the preceding figure, the red background was only visible within the media queries specified, even though it was combined into a single query. This can really help a developer when there is a need to share a certain style between multiple breakpoints in CSS3.

Print Stylesheets

CSS3 media queries allow the developer to specify print-only styles to help control how a web page is displayed when printed out as a physical copy. This can be very helpful on websites where the user may want to print off information; for example, the user may want to print a cooking recipe or resources for an educational activity for later offline use.

There are two ways in which we can tell the browser that these styles are print only:

- Using a new HTML **<link />** tag to import a completely separate stylesheet for print styles

- Within an existing stylesheet using the **@media** query syntax to adjust some styles for print

To implement a print-only stylesheet, you can use the following code:

```
<link rel="stylesheet" media="print" href="print.css" />
```

This code allows the **print.css** stylesheet to be used when printing mode is in use within the browser.

The other method is using **@media** queries to enable your print mode styles to become active. Take a look at the following example:

```
@media only print {
    /* style changes here */
}
```

This can be a very useful feature, particularly if you would like the user to be able to print something off, such as a returns label or application form, that's displayed within a web page. We'll now put this into practice in the following exercise.

Exercise 6.03: Generating a Printable Version of a Web Page Using CSS Media Queries

The aim of this exercise is to update some supplied HTML and CSS code and then apply a CSS layout change to the printable version of the page. The overall aim is to make a printable version of this web page by removing unnecessary items from the printing view and tidying up the styling elsewhere in order to make the page ready for printing.

The following figure demonstrates the layout changes that are expected to be visible within print preview mode:

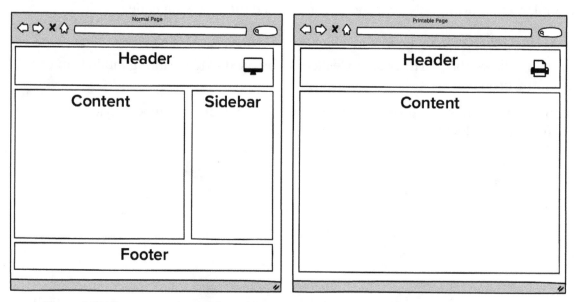

Figure 6.20: Demonstration of the print media query in action, changing the page layout

Let's complete the exercise with the following steps:

6. Let's start with the **Exercise 6.03_Raw.html** from https://packt.live/2q5mS8f. In the HTML we've got a **<header>** block where a logo would normally go, a **<main>** tag section containing a **<section>** for the main content, and **<aside>** for the sidebar content. Finally, there is a **<footer>** tag to complete the web page.

7. You can see the result of this code displayed in the following figure:

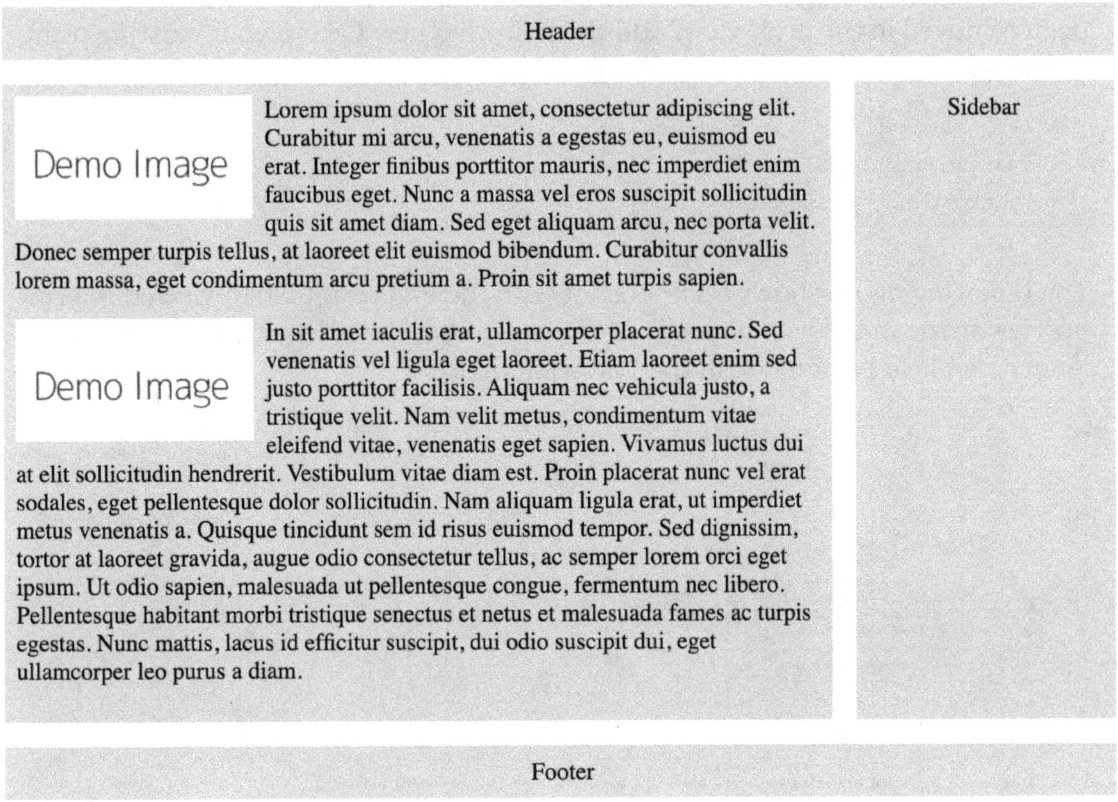

Figure 6.21: Screenshot of the layout HTML

8. Create a file named **Exercise 6.03.html** within the **Chapter06** folder and ensure that the code available in **Exercise 6.03_Raw.html** is available in the new file as well. Let's remove the advert sidebar and footer navigation from the print view. Add the following new CSS media query:

```
@media only print {
    footer, aside {
        display: none;
    }
}
```

This code will hide the display of the **footer** and **aside** tags when the web page is printed off. You should be able to view your print media style changes in the print preview mode of your device.

9. Let's tidy up the final display of the web page when in print mode by removing the unnecessary margin now that the sidebar and the background colors have been removed in the print view. Add the following new CSS to your existing print media query:

```
header, section {
    margin-right: 0;
    background: transparent;
}
```

The following figure shows a screenshot with the end result when viewed in print preview mode in the web browser. You can see that the sidebar and footer are no longer visible in print mode in the following figure:

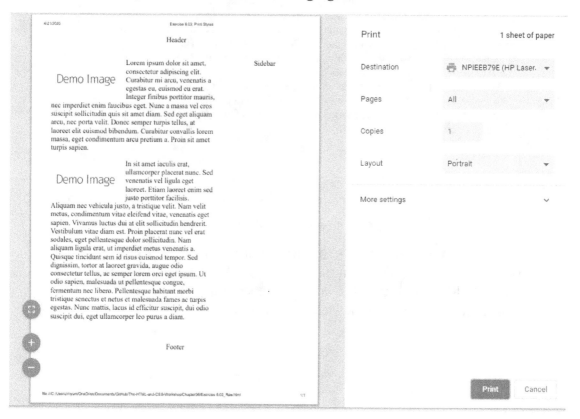

Figure 6.22: Output in a web browser (on the left) and print preview (on the right)

Activity 6.01: Refactoring the Video Store Product Cards into a Responsive Web Page

For this activity, we're going to turn the product page you've already created for your video store in *Chapter 3, Text and Typography*, into a responsive web page, coding it while following mobile-first principles, ensuring that it's cross-browser compatible, and making it a robust responsive web page for modern devices. The following diagram shows the video store product page wireframe from *Chapter 3, Text and Typography*:

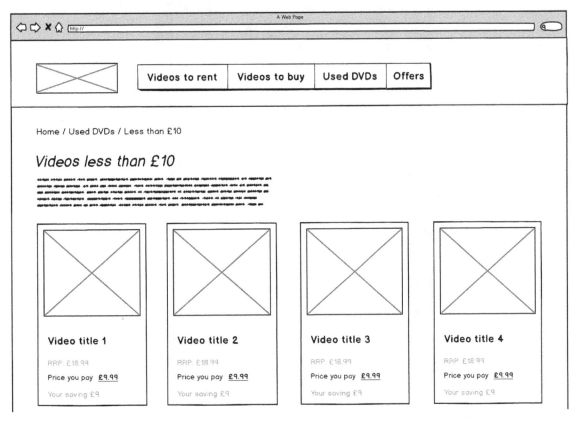

Figure 6.23: The video store product page from Chapter 3

The following figure is a mock-up design of the mobile view:

Figure 6.24: Mock-up design of the mobile view

In the preceding figure, we are showing a mock-up of the existing desktop layout for the video store product page. The preceding figure displays the proposed mobile design of the page that you will create in this activity. The instructions for this activity are as follows:

1. Get a copy of the code for your existing product page from *Chapter 3, Text and Typography Exercise 3.06, Putting It All Together*, and save it as a new HTML file named `Activity 6.01.html` under `Chapter06` folder

2. Apply the responsive viewport meta tag to the `<head>` of the new HTML file.

3. Keeping the rest of the HTML on the web page the same, amend the CSS to make the website suitable for mobiles. Don't apply any media queries just yet, as we'll start with making the web page for mobiles first. You can improve the styling/ appearance of the page elements at this point also, to make it more mobile-friendly where possible. Make a note of anything you remove to make the web page mobile-first, though, as this will help you in the next step.

4. Now, amend your CSS further to include media queries to adapt the web page to utilize the additional screen space on tablet and desktop devices.

5. Finally, apply a print-only media query to hide the navigation of the website if the web page is to be printed.

> **Note**
> The solution to this activity can be found on page 610.

Summary

In this chapter, we have explored the topic of mobile-first and looked at what building a responsive website entails. You should have a good understanding of what media queries are in CSS3 and what you can do with them to aid your web development for multiple screen sizes. You should also know how to modify the styles of the printed version of a website, and, as part of this chapter's activity, you have refactored some existing website code into responsive website code using media queries to help create a mobile-friendly web page.

In the next chapter, we will learn about using the HTML5 media elements, including the audio, video and canvas elements.

Media – Audio, Video, and Canvas

Overview

By the end of this chapter, you will be able to append audio and video elements to a web page; create custom controls for media playback; implement different ways of adding images to a web page and evaluate the pros and cons of each; and create and animate graphics using the canvas element. This chapter aims to get you acquainted with the visual and audio elements that you can add to a web page. By adding audio, video, and image elements to a web page, we can improve the visual interest of it and add content that will enrich our users' experience in a way text alone cannot.

Introduction

In the previous chapter, we laid the foundations of our web pages by learning about structural elements, how to style text, and how to add interactive forms.

In this chapter, we will learn about elements that we can use to greatly enrich the experience a user has on a web page. We will look at how we can showcase video, music, images, and programmatic art on our web pages.

With the advent of HTML5, powerful tools for rich interaction have become available in the browser. We can now embed video and audio directly in a web page. We can use the canvas to draw interactive graphics and make games. Previously, you would have needed a plugin such as Flash or Silverlight to be able to show a video or play a song in the browser.

There are so many reasons to add these elements to a web page. We can use images as content in a blog or a news article. We can use images and video to showcase a product in a catalog or on an e-commerce site. Logos and images can be used to add a visual style to a web page or brand. Video is often used for adverts on web pages and also to showcase content on sites such as Vimeo and YouTube. With audio embedded in a web page, you can provide your own radio channel online or release a podcast that is easily accessible to users of your web page.

Animation can further enrich the style of – and even give personality to – a web page. You will learn more about that topic in the next chapter.

By adding a media player to the case study website, we will see how we can use the `video` and `audio` elements to playback media, and how we can style the controls for a media player. We will then see how the `canvas` element can be used to add graphical effects for video effects, visualizing audio, and animation.

Audio

If we want a user of our web page to listen to our weekly podcast without leaving the page or we want them to be able to hear our latest song recording or mixtape, we can do so in HTML5 with the `audio` element. The `audio` element allows us to embed audio in a web page.

The **audio** element can have a single audio source as an **src** attribute or as a **source** element, or it can have multiple audio sources (in which case, these are added as children of the **audio** element as a list of **source** elements). We will look at how and why we would want to provide multiple sources for an audio file later in this chapter. Until then, we will only consider the MP3 format, which is one of the most commonly used formats for music and audio tracks.

Let's look at a basic example:

```
<audio
      controls
      src="media/track1.mp3">
It looks like your browser does not support <code>audio</code>.
</audio>
```

This simple example will load a single audio source and provide your browser's default interactive controls for you to play the audio.

We have added content between the opening and closing **<audio></audio>** tags. This content acts as a fallback on browsers that do not support the **audio** element. Most modern browsers do support the **audio** element, but it is still a good idea to provide a fallback message to let users of older browsers know that they aren't able to play the content back.

We added the **controls** attribute to allow you to play the audio. The audio will not automatically play in the browser. We will look at the various attributes that are available for the **audio** element in the next section.

If an **audio** element has no controls and is supported by the browser, that element will not appear on the web page; instead, it will load the audio referenced in the **src** attribute.

Having looked at an example of an **audio** element being embedded in a web page, let's try it out.

Exercise 7.01: Adding Audio to a Web Page

In this exercise, we will add an **audio** element to a web page. We will use the browser's default controls and a provided audio track. The steps are as follows:

1. First, we will create a web page. Create a folder named **Chapter07**, and under that create a new HTML file and name it **Exercise 7.01.html**. Then, copy and paste the following code to create a minimal web page with a container (**audio-container**) for our audio player:

```
<!DOCTYPE html>
<html lang="en">
    <head>
        <meta charset="utf-8">
        <title>Exercise 7.01: Audio</title>
    </head>
    <body>
        <div class="audio-container">
        </div>
    </body>
</html>
```

2. Create a directory called media and download the audio track, **track1.mp3**, into it. We will use this audio track as the source for our **audio** element.

3. Next, we need to add the **audio** element to embed the audio track source in our web page. We will add an **audio** element to the **div.audio-container** element and set the **controls** attribute so that we can control the audio playback from our web page:

```
<div class="audio-container">
    <audio id="audio-1" src="media/track1.mp3" controls>
        Your browser does not support the HTML5 <code>audio</code>
        element.
    </audio>
</div>
```

4. Save the page. The full code listing is as follows:

```html
<!DOCTYPE html>
<html lang="en">
    <head>
        <meta charset="utf-8">
        <title>Exercise 7.01: Audio</title>
    </head>
    <body>
        <div class="audio-container">
            <audio id="audio-1" controls src="media/track1.mp3">
            Your browser does not support the HTML5 <code>audio
            </code> element.
            </audio>
        </div>
    </body>
</html>
```

5. Now right-click on the filename in VSCode on the left-hand side of the screen and select **Open In Default Browser**, shows the result of this code that is, we can now play the audio track from our web page.

The following screenshot shows the media controls provided by Google Chrome:

Figure 7.1: Chrome media controls playing audio

In this exercise, we have learned how to add an audio track to our web page. We have learned how to create an **audio** element and how to add the **controls** attribute to the **audio** element, allowing us to play and control the audio playback with minimal effort.

Next, we will look at the **audio** element in a bit more detail by considering the attributes we can use to modify the audio playback before we look at our options for customizing the controls a user has for audio playback.

Attributes

The following attributes can be used to modify the **audio** element:

- autoplay
- preload
- loop
- controls

The autoplay Attribute

The **autoplay** attribute is a Boolean that lets the browser know to start the audio track as soon as it has enough information to do so. The whole file does not need to download before playback begins.

If you decide to use the **autoplay** attribute for any reason, be very careful with it. Not many users appreciate music or sound blaring out from a web page without them having any control over it. It is much better to give the user control over starting and stopping audio and this can be done with the **controls** attribute or with your own custom control UI.

While there are some valid use cases for **autoplay**, including working with switching audio sources in a web app or game, the different **autoplay** policies of various browsers can be a bit difficult for developers to work with. We'll look into **autoplay** in a bit more detail later, in the *Limitations* section.

The preload Attribute

The **preload** attribute lets you tell the browser how you would like it to handle the loading of media sources. The attribute has a few options:

- **none**: The audio will not be preloaded.
- **metadata**: The browser will load audio metadata but not the whole file.
- **auto**: This is the same as an empty string; that is, **preload=""** – the whole audio file will be downloaded.

Different browsers handle the default preload behavior in different ways. The W3C HTML5 specification recommends that the default setting is **metadata**. In most use cases, preloading metadata is the preferred method because this gives us enough information to tell the user about the audio but does not load the rest of the audio file until the user has clicked play. This prevents unnecessary loading, which can be quite costly for those on mobile plans with limited data.

You can see the difference in behavior between a preload value of **none** and **metadata** in the following screenshots. In Firefox (version 66.0), in the Network data section, at the bottom of each image, you can see that when preload is set to none, there is no network traffic and for preload set to metadata, 560 KB of an audio file has been loaded. The loaded metadata means the browser can include the duration of the audio in the user interface (UI) and will have to load less data before playback can commence.

The following screenshot shows the network activity of the browser when an **audio** element has the **preload** attribute set to none. It shows that no part of the audio file has been downloaded:

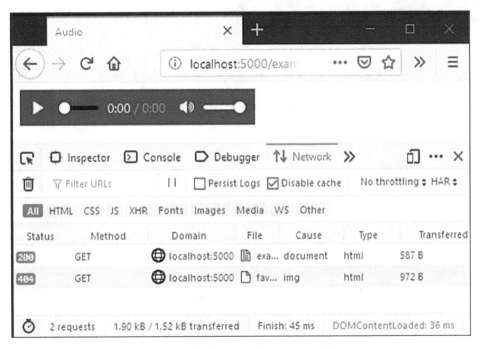

Figure 7.2: Preload attribute set to None

In contrast to the preceding screenshot, the following screenshot shows the network activity of the browser when an **audio** element has the **preload** attribute set to metadata. It shows that part of the audio file has been downloaded, allowing the user to get information about its duration:

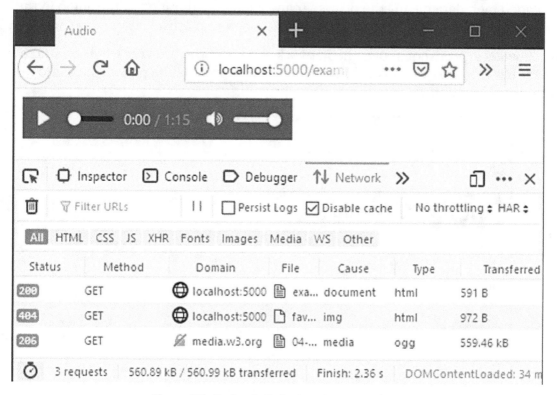

Figure 7.3: Preload attribute set to metadata

The loop Attribute

The **loop** attribute is a **Boolean** attribute that loops the audio back to the beginning once the end of the audio track has been reached.

An example use case is in an HTML5 game where, in addition to sound effects that play once, we may want a background score to play in a loop during a level.

The controls Attribute

The **controls** attribute adds the default media controls for the browser. This allows a user to control the playback of audio. While the browser's media playback UI shares many common features (volume, mute toggle, and scrub bar, to name a few), the style of these controls varies greatly. The following screenshot shows more examples:

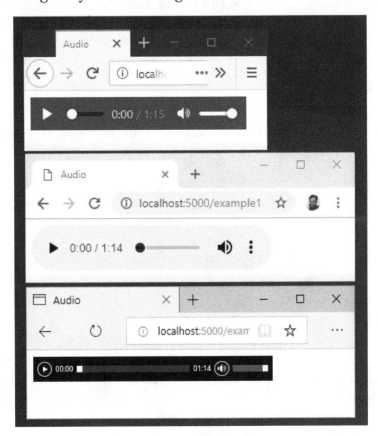

Figure 7.4: Default audio controls in the Firefox, Chrome, and Edge browsers

Styling Audio Controls

The default controls provide a lot of functionality for free. They may well be all that you need. However, as you've seen, the styles are different for each browser and may not fit with your own design or requirements. The best option for customizing the controls is to not set the **controls** attribute and to create your own custom controls with HTML, CSS, and a little bit of JavaScript.

In the next exercise, we will look at how to style our own audio controls. This means we can create custom media players and keep the style in sync with our web page's look and feel.

Exercise 7.02: Styling Controls

In this exercise, we will add a minimal set of custom controls to our audio player. We will then style the controls. The behavior will be handled by JavaScript, which we will add with a **script** element. The design of the controls will be handled with CSS.

We will use a minimal amount of JavaScript in this exercise to handle some events from the **audio** element and to trigger playback. A thorough explanation of JavaScript is beyond the scope of this book, but we will provide an explanation of the small snippets that are used in the exercise.

The steps are as follows:

1. We start by creating a new file, which we will name **Exercise 7.02.html** within the **Chapter07** folder. We will copy this basic web page structure into the file and save it:

```
<!DOCTYPE html>
<html lang="en">
    <head>
        <meta charset="utf-8">
        <title>Exercise 7.02: Styling Controls</title>
    </head>
    <body>

    </body>
</html>
```

2. In the **body** element, we want to add a wrapper **div** with the **audio-player** class and add our **audio** element in it. We deliberately do not want to set the **controls** attribute on the **audio** element:

```
<div id="audio-player" class="audio-player">
    <audio id="audio-1" src="media/track1.ogg">
        Your browser does not support the HTML5 <code>audio</code>
        element.
    </audio>
</div>
```

3. Next, we will add another wrapper, **div.audio-controls**, after the **audio** element. In there, we will add buttons for playing and pausing the audio track and information for the duration of the track and the time played so far. The markup looks like this:

```html
<div id="audio-player" class="audio-player">
    <audio id="audio-1" src="media/track1.ogg">
      Your browser does not support the HTML5 <code>audio</code>
      element.
    </audio>
<div class="audio-controls">
    <button id="play-btn" class="button">Play</button>
    <button id="pause-btn" class="button">Pause</button>
    <div class="audio-status">
        <span id="played">00:00</span> /
          <span id="duration">00:00</span>
    </div>
</div>
</div>
```

4. Next, we can add some initial styles to the audio player. We'll add a **style** element just beneath the **title** element in the head of the HTML document. We will also set the font and text color for the .**audio-player** and .**audio-controls** class selectors, respectively:

```html
<style>
  .audio-player,
  .audio-controls {
      color: white;
      font-family: Arial, Helvetica, sans-serif;
      font-size: 16px;
      font-weight: 200;
  }
</style>
```

5. We want to style the `.audio-controls` container as a media player. We'll give it a dark background, light border, and rounded rectangle shape (with **border-radius**). We'll also use the flex box to control the layout of the children of `.audio-controls`:

```
.audio-controls {
    align-items: center;
    background-color: #0E0B16;
    border: 3px solid #E7DFDD;
    border-radius: 30px;
    display: flex;
    justify-content: space-between;
    max-width: 240px;
    padding: 8px;
}
```

The result of our styles, so far, can be seen in the following screenshot:

Figure 7.5: .audio-controls styled

6. We want to make the **Play** and **Pause** buttons stand out more. By default, button elements have an **appearance** property with a value of **button**, which sets the appearance to that of the browser's native button. By setting this property to **none**, we can easily style the buttons. Add the following at the bottom of the **style** element:

```
.audio-controls .button {
    -webkit-appearance: none;
    appearance: none;
    background-color: #4717F6;
    border: 1px solid #E7DFDD;
    border-radius: 16px;
    cursor: pointer;
    outline: none;
    padding: 4px 16px;
    text-transform: uppercase;
}
```

7. We will also add the **button** class selector to the CSS ruleset for handling the text color and font size at the top of our **style** element:

```
.audio-player,
.audio-controls,
.audio-controls .button {
```

8. We want to control what appears, depending on the state of the audio player. A **data** element data state will be added to the **.audio-player** element via our JavaScript. We can use this to control styles with the attribute selector:

```
[data-state="playing"] #play-btn,
[data-state="paused"] #pause-btn {
    display: none;
}
```

9. For the audio player to control our **audio** element, we need to use JavaScript. We will add a **<script>** element beneath the **div.audio-player** element. We'll use **document.getElementById** to select the elements that are used by the audio player using their ID attributes:

```
<script>
    const audioPlayer = document.getElementById('audio-player');
    const audio1 = document.getElementById('audio-1');
    const playButton = document.getElementById('play-btn');
    const pauseButton = document.getElementById('pause-btn');
    const duration = document.getElementById('duration');
    const played = document.getElementById('played');
```

> **Note**
>
> In JavaScript, the **const** keyword is used with a name to create a constant value. We can store anything (**DOM** element, number, **string**) in it, but trying to replace it later will not work. If you need to replace the value, you can use the let keyword.

10. We will set the **data-state** attribute for the audio player to the **"paused"** value initially:

```
audioPlayer.dataset.state = "paused";
```

11. Next, we want to add a click **event handler** – that is, a function that will be triggered when the user clicks the button – to the **Play** and **Pause** button. This will change the state of the audio player and set the audio track to play or pause:

```
playButton.addEventListener("click", function() {
    if (audioPlayer.dataset.state === "paused") {
        audio1.play();
        audioPlayer.dataset.state = "playing";
    }
});

pauseButton.addEventListener("click", function() {
    if (audioPlayer.dataset.state === "playing") {
        audio1.pause();
        audioPlayer.dataset.state = "paused";
    }
});
```

12. Finally, we will set the time played and duration text based on the **duration** and **currentTime** values of the audio. We will use event listeners for **loadedmetadata** and **timeupdate** to set these values when the time has updated and the duration is known. We've added a **formatTime** function to format the time value as minutes and seconds (mm:ss):

```
audio1.addEventListener("loadedmetadata", function(event) {
    duration.textContent = formatTime(audio1.duration);
});

audio1.addEventListener("timeupdate", function(event) {
    played.textContent = formatTime(audio1.currentTime);
});

function formatTime(time) {
    const minutes = Math.floor(time / 60);
    const seconds = Math.round(time) % 60;
    return `${
      minutes < 10 ? "0" + minutes : minutes
    }:${
      seconds < 10 ? "0" + seconds : seconds
    }`;
}
</script>
```

If you now right-click on the filename in VSCode on the left-hand side of the screen and select **Open In Default Browser**, you will see a simple but usable HTML audio player:

Figure 7.6: The play state of our styled audio player

The paused state of our player will look as follows:

Figure 7.7: The pause state of our styled audio player

In this exercise, we have learned how to style the controls for an audio player. Using a small amount of JavaScript and data attributes to maintain the state of our audio player, we are able to create our own custom audio controls for a web page.

Next, we will look at adding more than one source to an **audio** element. This can be useful because browsers have varying support for audio formats.

Multiple Sources

While it is easy to get started with the **audio** element, browser support for different audio formats adds some complexity. Due to licensing issues and other historical reasons, not all codecs (mp3, ogg, and so on) are supported by all browsers. Because of this, it is recommended to support several formats when you publish audio on the web.

The webm format has good browser support and is optimized for loading on the web. However, there are a few browsers that don't support WebM (such as IE 11). The MP3 format is well supported and is used a lot for music. A good open source option is ogg, which is also well supported.

> **Note**
>
> You can learn more about media formats and browser compatibility from MDN at https://packt.live/33s2I6V.

Happily, the **audio** element provides nice support for multiple sources with the **source** element. The following is an example of this:

```
<audio controls>
    <source src="media/track1.webm" type="audio/webm">
    <source src="media/track1.mp3" type="audio/mpeg">
    <source src="media/track1.ogg" type="audio/ogg">
    It looks like your browser does not support <code>audio</code>.
</audio>
```

Rather than setting the **src** attribute of the **audio** element, we have set three **source** elements as children of the **audio** element. Each **source** element refers to a different audio source. We have also provided a type attribute for each **source** element to let the browser know the MIME type of the referenced audio source file.

The order of the **source** elements is important. The browser will attempt to play the first element but, if that codec is not supported, the browser will try the next source and then the next, and so on until it runs out of options.

The video Element

The **video** element and the **audio** element are both HTML media elements and have a lot in common. It is even possible to play audio files with the **video** element and video files with the **audio** element. The main difference is that the **video** element provides a display area.

Attributes

The **video** element shares most of its attributes with the **audio** element. There are a few additional attributes to do with the display area:

- **height**
- **width**
- **poster**

The width and height Attributes

The **width** and **height** attributes set the width and height of the video display area, respectively. Both values are measured in absolute pixel values – in other words, the values must be non-negative integers and they cannot be percentages.

The poster Attribute

The **poster** attribute allows you to provide the source for an image that will be shown while the video is being downloaded. If we do not set the **poster** attribute, a blank square will appear until the first frame of the video has been downloaded, and then the first frame of the video will show in place of the poster image.

Exercise 7.03: Adding Video to a Web Page

In this exercise, we will add a **video** element to a web page to embed a video. We will use the default browser controls, just like we did with the **audio** element.

The steps are as follows:

1. Create a file, **Exercise 7.03.html** within the **Chapter07** folder. Copy the following code into that file and save it. This will create the outline for a simple web page:

```
<!DOCTYPE html>
<html lang="en">
    <head>
        <meta charset="utf-8">
        <title>Exercise 7.03: Video</title>
    </head>
    <body>

    </body>
</html>
```

2. In the **body** element, we will add a **div** element to contain the video player with the **video-container** class. In that container, we will add our **video** element:

```
<div class="video-container">
    <video id="video-1">Your browser does not support the HTML5
      <code>video</code> element.
    </video>
</div>
```

3. We will add a **poster** attribute to the **video** element to show an image while the video is loading, and a **controls** attribute to show the browser's default controls for media playback:

```
<video id="video-1" poster="media/poster.png" controls>
```

4. We will add some source elements in several formats to the video file. Here, we are adding a WebM and MP4 version of the same video. The WebM format is designed especially for the web but is not supported in all browsers:

```
<video id="video-1" poster="media/poster.png" controls>
    <source src="media/html_and_css.webm" type="video/webm">
    <source src="media/html_and_css.mp4" type="video/mp4">Your
        browser does not support the HTML5 <code>video</code>
        element.
</video>
```

5. We will add a style element in the head of the HTML document and add rules to resize the video according to the size of the web page and to control the maximum size of the video:

```
<style>
    .video-container {
        max-width: 1080px;
                margin: 16px auto;
    }

     .video-container video {
        width: 100%;
        }
</style>
```

If you now right-click on the filename in VSCode on the left-hand side of the screen and select **Open In Default Browser**, you will see a video player in the browser, similar to what's shown in the following screenshot:

Figure 7.8: HTML video with a poster frame

In this exercise, we have learned how to embed a video in a web page using the HTML5 **video** element. We have added a poster image to that video and used the default browser controls for playback.

Next, we will consider some limitations of the media elements in HTML5.

Limitations

There are some limitations to the HTML **audio** element that can add complexity during development. These behaviors are different depending on how browsers and different devices implement the **audio** and **video** elements.

On most mobile devices, limitations on downloading content without the user's approval have been put in place to prevent a browser using up data allowance. Audio and video files, especially high-quality and stereo formats, can be very large.

One common policy is to only allow audio content to play when user input is recorded. This may mean content can only play when the user presses the **Play** button, even if we set the **preload** and **autoplay** attributes on a **media** element. This can be particularly problematic if you are using media in a JavaScript application, such as a game, and you want to have full control of media playback and loading.

On an iPhone, the volume at which audio plays in the browser is tied into the device's volume controls and cannot be controlled with JavaScript.

Another common policy is that you can preload and autoplay video content but only with the **muted** attribute set on the **video** element.

> **Note**
>
> You can learn more about various restrictions and policies from the browser vendors from a variety of sources. For example, good information about Safari and iOS can be found here: https://packt.live/2CivkDQ.
>
> Information on autoplay in Chrome can be found here: https://packt.live/2qrGTWH.

These policies are not defined by a standard and are subject to change as the browser vendors try their best to serve a diverse audience.

The track Element

The **track** element lets us specify a time-based text track that has some relationship to a **media** element. This means we can show text at time cues that are synchronized with media playback.

Using the **track** element, we can make our video and audio accessible to more users. This can be used to provide subtitles, closed captions, or a description of media content that may not otherwise be accessible to users for various reasons such as blind users and users with hearing difficulties or technical or contextual restrictions that mean the user cannot play sound.

Similar to how we can use the **source** element with video and audio, we can add multiple **track** elements to one **media** element and these can have different relationships to the media or provide different language translations of the content.

The following attributes let you modify a **track** element:

- **src**: The location of the external file with the text track.

- **default**: One **track** element per **media** element can be set as the default track. This attribute acts as an indication and it may be overridden by user settings (such as language).

- **kind**: Specifies how the text track is supposed to be used. There are several options, including **subtitles**, **captions**, **descriptions**, **chapters**, and **metadata**. The default value is **subtitles**. The difference between subtitles and captions may not be entirely obvious; the former is meant for the translation of content that a user may not understand, for example, a non-English dialogue in an English film, while the latter is more suited to users who are deaf or for when content is muted.

- **srclang**: The language of the track text; for instance, **en** for English, **fr** for French. If the track is a subtitle track, then **srclang** is required.

- **label**: A human-readable title that's used to differentiate between captions and subtitles.

Take a look at the following example code to see how the following **track** labels would appear in Chrome's video controls as caption options:

```
<track src="media/track1-en.vtt" kind="subtitles" label="English
  subtitles">
<track src="media/track1-sparse.vtt" kind="captions" label="Sparse
  captions">
<track src="media/track1-full.vtt" kind="captions" label="Full
  captions">
```

The following screenshot shows the output for the preceding code:

Figure 7.9: Caption options with multiple tracks in Chrome

Adding Subtitles

The most common format for text tracks used on the web is **Web Video Text Tracks Format (WebVTT)**. The format is a plain text means of describing timestamps and captions. A full explanation is beyond the scope of this book, but here's a simple example:

```
WEBVTT

00:00:01.000 --> 00:00:05.000
First subtitle text

00:00:10.000 --> 00:00:25.000
Second subtitle text
```

The first line declares that the file is in the WebVTT format. The rest of the file consists of cues. Each cue is separated by a blank line and has a starting time and ending time separated by the **-->** string. On the next line, the text content of the subtitle or caption is presented.

Exercise 7.04: Adding Subtitles

In this exercise, we will add two WebVTT files to a video player to provide subtitles for a video. When we turn captions on, we can see how this is presented in the browser.

> **Note**
>
> This exercise requires the use of the web server. If you have not already followed the steps to install and set up your web server, you will need to go back to *Preface* and follow those steps before you begin the exercise.
>
> To test this exercise, you will need to place your files into the XAMPP **htdocs** folder on your computer and access them through your browser using **localhost**. Review the web server instructions given in the *Preface* if you need help with these steps.

The steps for this exercise are as follows:

1. Create a file named **Exercise 7.04.html** within the **Chapter07** folder.

2. We will start with the final source code from the previous exercise, which gives us a **video** element with default browser controls:

```
<!DOCTYPE html>
<html lang="en">
    <head>
        <meta charset="utf-8">
        <title>Exercise 7.04: Adding Captions</title>
        <style>
            .video-container {
                max-width: 1080px;
                    margin: 16px auto;
            }
                .video-container video {
                    width: 100%;
                }
        </style>
    </head>
    <body>
     <div class="video-container">
            <video id="video-1" poster="media/poster.png" controls>
                <source src="media/html_and_css.webm"
                  type="video/webm">
                <source src="media/html_and_css.mp4"
                  type="video/mp4">
```

```
                    Your browser does not support the HTML5
                        <code>video</code> element.
                </video>
        </div>
        </body>
</html>
```

3. We will save a new WebVTT file to **media/subtitles-en.vtt**. The content of the file will be as follows:

```
WEBVTT


00:02.000 --> 00:04.000
HTML and CSS Video


00:06.000 --> 00:15.000
Hi, I'm Brett Jephson. I'm the author of this chapter on Video, Audio, and
Canvas in HTML5.


00:16.000 --> 00:23.000
In this part of the lesson, we've learned to add subtitles to a video element.
```

4. Next, we will save another new WebVTT file to **media/subtitles-es.vtt**. The content of that file is as follows:

```
WEBVTT


00:02.000 --> 00:04.000
HTML y CSS Video


00:06.000 --> 00:15.000
Hola, soy Brett Jephson. Soy el autor de este capítulo sobre Video, Audio e
Imágenes en HTML5.


00:16.000 --> 00:23.000
En esta parte de la lección, hemos aprendido a agregar subtítlos a un elemento
de video.
```

5. We want to add a **track** element to the **video** element, after the source elements, to provide the English subtitles as the default:

```
<track src="media/subtitles-en.vtt" kind="subtitles"
    label="English" srclang="en" default>
```

6. Next, we want to add a **track** element for the Spanish subtitles:

```
<track src="media/subtitles-es.vtt" kind="subtitles"
    label="Español" srclang="es">
```

7. Now save this file and place the **Chapter07** folder within **htdocs** folder and make sure you have started the web server.

> **Note**
>
> Refer to the instructions in the *Preface* to know how to start the web server.

8. Head to your browser, type **localhost/Chapter07/** in the URL and then hit *Enter*. You will obtain a list of files present in this folder.

9. Click on **Exercise 7.04.html** file.

    A video player would appear on your browser window. When you click on the play button, the following output will be visible on your browser window:

Figure 7.10: A video element with English subtitles

In the following still, the Spanish subtitles track is shown at the same point in the video playback:

En esta parte de la lección, hemos aprendido a agregar subtítulos a un elemento de video.

0:16 / 0:29

Figure 7.11: A video element with Spanish subtitles

In this exercise, we have learned about WebVTT and how we can add captions to a video or audio to make media playback more accessible to those who cannot hear the sound.

In the next part of this chapter, we will look at the ways we can embed images in a web page. We will look at several HTML elements that are available for the task and the different use cases they are for.

Images

When you want to add images to a web page, there are several options available. Each has its advantages and disadvantages.

The approaches we will look at for embedding images in a web page are:

- The `img` element
- The `picture` element
- Programmable graphics

We will consider each of these approaches one by one.

The img Element

We encountered the **img** element in *Chapter 1, Introduction to HTML and CSS*, but to recap, the simplest approach to embed an image into a web page is to add an external image file using the **img** element.

We specify an **src** attribute for the image file we want to embed. This could be a file location or a data URI.

For accessibility, it is important that you provide an **alt** attribute for any **img** element that is not pure decoration. This provides an alternative text description of the image for users who cannot see the image, for example, screen reader users and users with image loading blocked.

Here is an example **img** element:

```
<img src="media/poster.png" alt="HTML and CSS poster">
```

The picture Element

The **picture** element is similar to the **img** element but lets you set multiple image source elements and specify rules for why the image will be the one to appear. This is best illustrated with an example:

```
<picture>
    <source srcset="media/poster-small.png" media="(max-width: 639px)">
    <source srcset="media/poster-large.png" media="(min-width: 800px)">
    <img src="media/poster.png" alt="HTML and CSS poster">
</picture>
```

In this example, we have a **picture** element with two source elements and an **img** as a fallback. Both source elements have a **media** attribute.

The browser will pick the most appropriate image source. It will start at the first source element. It will test whether the condition set out by the **media** attribute is true. In this case, this condition is whether the width of the viewport is 639 pixels or less. If the condition is met, it will show that image and if not, it will try the next source element. If the viewport is 800 pixels or more, it will show the second source element. Finally, if neither condition is met or the **picture** element is not supported, the browser will show the fallback image element; that is, the last child of the **picture** element. The following diagram is a flow chart of this logic:

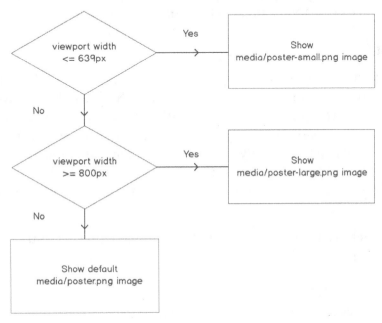

Figure 7.12: Flow chart of the responsive logic of the picture element

The condition can be a media query or it can be based on a type of image format; for example, if the **type** attribute was specified with a **mime** type that the browser can handle.

The **img** element fallback is required for browsers that don't support the **picture** element (such as Internet Explorer 11) as a default if no other source elements apply. Without it, no image will be shown, regardless of whether a source element's condition is met.

We will learn more about how we can handle the embedding of images for different display sizes and different device pixel densities in *Chapter 6, Responsive Web Design and Media Queries*.

Programmable Graphics

As well as loading and embedding rasterized image files, such as `.jpeg`, `.gif`, `.png`, and `.webp`, on a web page, there are some options for creating graphics programmatically:

- **Scalable vector graphics (SVG)** is an XML-based format for creating vector graphics on web pages.

- The **canvas** element gives you access to a JavaScript drawing API that you can use to create rasterized graphics on a web page.

The svg Element

An increasingly popular option and one that provides a lot of power for small file sizes is to use SVG to render vector graphics on a web page. SVG is a declarative XML-based format for defining vector graphics. Vector graphics are really useful for responsive design because they are mathematically defined and can be scaled without causing blurring. It is an extensive technology with shape and path drawing, filters, and animation capabilities. It is largely beyond the scope of this book, but we will look at a simple example to give you an idea of how it can be used to create graphics on web pages.

As an example, let's look at the SVG markup for a simple icon that could be used as a **Play** button icon for our media player:

```
<svg viewBox="0 0 100 100" xmlns="http://www.w3.org/2000/svg"
    width="100" height="100">
      <circle cx="50" cy="50" r="48" fill="#4717F6" />
          <path d="M 85,50 25,75 25,25 z" fill="white" />
</svg>
```

The XML format should be quite familiar. HTML5 is not a strict XML format but is syntactically similar.

The graphic declarations are nested in an **svg** element. In the **svg** element, we set a namespace value with the **xmlns** attribute. This lets the browser know the SVG version and the contents of the **svg** element.

We also set a **viewBox** attribute. Here, we have set the **viewBox** to have an x-coordinate of **0**, a y-coordinate of **0**, and dimensions of **100**. The **viewBox** does not represent pixels but a rectangular viewport in the userspace. We can scale this by setting the **width** and **height** attributes on the **svg** element. The coordinates on the children will always stay relative to their position in the **viewBox**.

Within the **svg** element, we draw a circle with its center at the x-coordinate of **50** and the y-coordinate of **50**. This means that, because our **viewBox** for the **svg** element has a **width** and **height** of **100**, the circle will be centered. It has a **radius** of **48**, which almost fills the view box. We have given the circle a fill color with an **RGB** hex-value **#4717F6**, which is the same as the background color of the button in our audio player styling exercise.

The second child of the **svg** element is a white path, which defines the three points of a triangle. This is given the fill color of white. This child will be drawn above the circle.

In the following screenshot, we can see the image as it would appear rendered in the browser:

Figure 7.13: The svg play button

SVG is often a really good choice if you are making graphics for a website. You can scale and change the graphics with CSS or attributes in markup. The graphics stay clean and sharp at different scales without **pixelation**. SVG is generally more accessible than the canvas element and doesn't require JavaScript to draw shapes.

It is not a good choice if you want direct control over pixels and, when there are a lot of shapes, it can create performance issues as the CPU is used to do a lot of mathematical calculations to work out what is being drawn. For these needs, the **canvas** element should be considered.

The canvas Element

You can programmatically render rasterized graphics with the **canvas** element. To do this, we use a drawing API in JavaScript. The API is quite low-level but allows you to create a great variety of graphics in the **canvas** element. We can also use the power of JavaScript to animate the **canvas**.

The **canvas** element has two commonly used rendering modes: 2D and WebGL. While WebGL provides a lot of power with shader support as well as GPU acceleration, it is beyond the scope of this book to look into this. However, we will be looking at some 2D examples.

SVG is a great solution for graphics such as icons that can appear in lots of sizes and where we may want to control the color with CSS. It has become more and more popular as a format with the rise of responsive design.

canvas is often used for programming art in creative coding and games. You can replace individual pixels with this tool, so it is good for producing effects on photos or videos. Browsers have better support for hardware acceleration in **canvas** than they do SVG and this means it has faster performance and consumes less processor, which can be useful on mobile devices or when you want to draw a lot of objects at once.

Checking for canvas Support

Before we can use the **canvas** element to render 2D graphics, we need to check that the element is supported by the browser. We do this by first creating a **canvas** element and then, with JavaScript, we check that the **canvas** element has a context of the type we request. Options for context include **'2d'** and **'webgl'**.

With the help of the succeeding code, we could check that our context is supported. If we can get a 2D context, we can begin to draw in it. It is also worth pointing out that any children nested in the **canvas** element will appear as a fallback in case the browser does not support the **canvas** element. The fallback can be used to provide an image or an appropriate message. Here is the code:

```
<canvas id="canvasArea" width="320" height="240">
    Your browser does not support the <code>canvas</code> element.
</canvas>
<script>
    const canvasElement = document.getElementById('canvasArea');
    if (canvasElement.getContext) {
            const context = canvasElement.getContext('2d');
          // we have a context so we can draw something
    }
</script>
```

Drawing in canvas

The **canvas** context provides us with a number of commands we can use to set styles and draw primitive shapes, text, and images in the **canvas** element. A complete guide to programming for the **canvas** element is outside the scope of this book, but let's look at some of the most commonly used commands before we try them out in an exercise.

The canvas Grid

When working with the **canvas**, it is important to understand how its coordinate system works.

The coordinate system for the **canvas** starts at the top-left corner with the x and y coordinates at 0. When moving to the right, the x value increases. When moving down, the y value increases. A unit in the **canvas** is **1** pixel as it relates to the natural size of the **canvas** (the size we set with the **width** and **height** attributes of the **canvas**).

The following diagram represents this canvas coordinate system diagrammatically, with a rectangle drawn on the canvas:

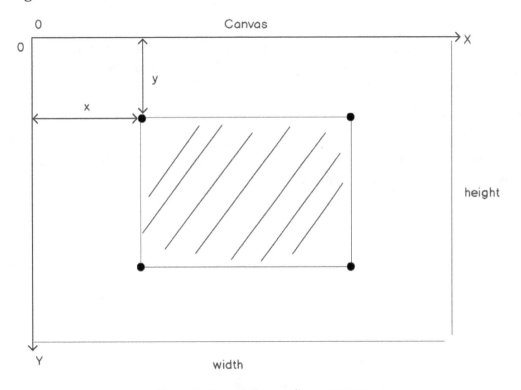

Figure 7.14: canvas coordinate system

Most of the commands in the **canvas** context require an x and y coordinate.

Drawing a Rectangle

To draw a square or rectangle with a fill color, we need to use the **fillRect** method in the canvas context. We give the **fillRect** method four numbers: x- and y-coordinates for the top-left corner and a width and height. To set the color that the rectangle will be filled with, we first need to set a **fillStyle** on the context. The **fillStyle** property can be any of the values that you could use in CSS.

When setting a color value, we can use all the same values as CSS will recognize the **color** or **backgroundColor** style properties, including color keywords such as **black**, **blue**, and **white**; hex values such as **#354555**; and rgba values such as **rgba(35, 45, 55, 0.5)**. You will learn more about the options that are available in *Chapter 5, Themes, Colors, and Polish.*

To draw a red rectangle with the top-left corner at the x-coordinate of **50** pixels and the y-coordinate of **50** pixels and with a width of **150** pixels and a height **100** pixels, we could use the following code:

```
context.fillStyle = "red";
context.fillRect(50, 50, 150, 100);
```

The following figure shows the red rectangle, that is, the output of this code, on a canvas with a width of **320** pixels and a height of **240** pixels:

Figure 7.15: fillRect used to draw a red rectangle on the canvas

To draw a rectangle with an outline around it, rather than a color filling it, we can set **strokeStyle** on the context and again give it a color value. We can set the width of the line with the **lineWidth** property on the context. We can then use the **strokeRect** method on the **canvas** context to draw the rectangle.

For example, if we wanted to draw a **red** border around a rectangle and give it a line width of **4** pixels, we could write the following set of commands:

```
context.strokeStyle = "red";
context.lineWidth = 4;
context.strokeRect(50, 50, 150, 100);
```

The following image shows a rectangle drawn with a **red** border with a line width of **4** pixels:

Figure 7.16: strokeRect used to draw a rectangle with a border on the canvas

To draw a rectangle with both a stroke and a fill color, we can set both the **fillStyle** and **strokeStyle** properties on the context. We can then use the **rect** method to create the rectangle path and then use separate **fill** and **stroke** methods on the context to fill the rectangle and draw a border stroke around it.

The following list of commands can be used to draw a rectangle that is filled blue and has an orange border with a 4-pixel line width around it:

```
context.fillStyle = "blue";
context.strokeStyle = "orange";
context.lineWidth = 4;
context.rect(50, 50, 150, 100);
context.fill();
context.stroke();
```

The following image shows the rectangle drawn with an **orange** border and with a line width of **4** pixels and a **blue** fill:

Figure 7.17: Fill and stroke used to draw a rectangle with a border and fill on the canvas

We've seen how to draw a rectangle with a border and a fill color. Next, we will look at how we can draw a circle.

Drawing a Circle

To draw a circle, we need to use the **arc** method. The **arc** method receives up to six arguments. The first two arguments for the **arc** method are the x and y position of the center of the circle. The third argument is the radius of the circle. The next two arguments are the starting angle and the ending angle. The last argument is a **true** or **false** value that decides whether the arc is drawn anti-clockwise (**true**) or clockwise (**false**). The default value is false.

To draw a circle with a **red** fill, the center of the circle is positioned **50** pixels right and **50** pixels down from the top-left corner, and the radius of the circle is **25** pixels, as shown in the following code:

```
context.fillStyle = 'red';
context.arc(50, 50, 25, 0, Math.PI * 2, false);
context.fill();
```

The following image shows the resulting red circle:

Figure 7.18: Fill used to draw a red circle on the canvas

> **Note**
>
> 2 π radians is equal to 360 degrees, so if we want to convert degrees into radians, we can use the following formula: degrees multiplied by π and divided by 180. To convert radians into degrees, we can do the opposite calculation with the following formula: radians multiplied by 180 and divided by π.

Clearing the canvas

Sometimes, we may want to clear the canvas, and we can do this with the **clearRect** command on the **canvas** context. You define a rectangle (x, y, width, and height) that you want to clear, which means you can clear all or part of a canvas.

When we want to clear the whole **canvas**, we can get the values of the canvas dimensions from the **canvas** element, where the width and height are stored. For example, if we wanted to clear the whole of a **canvas** stored as **canvasElement**, we would use the following **clearRect** command on the **canvas** context:

```
context.clearRect(0, 0, canvasElement.width, canvasElement.height);
```

We will make use of the **clearRect** command when we animate the **canvas** later in this chapter.

In the following exercise, we will try to draw some of these simple shapes with the **canvas** element. By combining a few shapes, we can start to draw pictures.

Exercise 7.05: Drawing Shapes

In this exercise, we will draw a picture of an igloo at night. It will be a simple picture made up of the two shapes we have already learned how to draw on a canvas – circles and rectangles. Through this exercise, we will learn how a picture can be composed on the canvas from very simple shapes.

The steps are as follows:

1. We will start by creating a new HTML file within the **Chapter07** folder. Name the file **Exercise 7.05.html** and save it. Copy the following code into the file. This will create a web page with a **canvas** element and a **script** tag with JavaScript to check for canvas support. If the canvas is supported, the script calls a function, **drawFrame**, which is where we will add our drawing code:

```
<!DOCTYPE html>
<html lang="en">
    <head>
        <title>Exercise 7.05</title>
        <style> canvas { border: 1px solid gray; } </style>
    </head>
    <body>
        <canvas width="320" height="240" id="canvasArea"></canvas>
        <script>
            function drawFrame(context) {
                context.clearRect(0, 0, canvasElement.width,
                    canvasElement.height);
                <!-- start drawing here -->
            }

            const canvasElement = document.getElementById
                ('canvasArea');
            if (canvasElement.getContext) {
```

```
                    const context = canvasElement.getContext('2d');
                    drawFrame(context);
                }
            </script>
        </body>
    </html>
```

2. To draw our picture, we need to do three things in the **drawFrame** function: draw the sky, draw the igloo, and draw the icy reflective floor. We will start with the background, which will be a dark blue rectangle representing the night sky. We'll create a function that we will call **drawSky**. We pass the function the context so that we have access to the canvas element's drawing commands. In the **drawSky** function, we set a fill style and draw a rectangle that's the size of the canvas:

```
function drawSky(context) {
    context.fillStyle = "#030339";
    context.fillRect(0, 0, canvasElement.width,
        canvasElement.height);
}
function drawFrame(context) {
    context.clearRect(0, 0, canvasElement.width,
        canvasElement.height);
    drawSky(context);

}
```

If you now right-click on the filename in VSCode on the left-hand side of the screen and select **Open In Default Browser**, you will see the following screenshot that shows the result of calling the **drawSky** function:

Figure 7.19: The drawSky function fills the canvas with a dark blue color

3. The next step is to draw the igloo. To do this, we will add another function to the **drawFrame** function, after the **drawSky** function. The order of the functions decides the order of drawing and so we must draw background elements before we draw foreground elements. The new function is **drawIgloo** and in it, we are going to start by drawing a large white circle:

```
function drawIgloo(context) {
    context.fillStyle = "white";
    context.arc(120, 160, 60, 0, Math.PI * 2, false);
    context.fill();
}
function drawFrame(context) {
    context.clearRect(0, 0, canvasElement.width,
        canvasElement.height);
    drawSky(context);
    drawIgloo(context);
}
```

The white circle that's drawn on the canvas is drawn over the sky. This will be the main dome of the igloo, as shown in the following image:

Figure 7.20: The drawIgloo function

4. So far, the **drawIgloo** function draws the main dome but it doesn't look much like an igloo. What it needs is a door. We are going to add the door using a white rectangle and another circle for the door's entrance. For the rectangle, we'll use **fillRect**. For the entrance, we will use a small circle the color of the night sky but with a light blue stroke around it. Here is the complete code for the **drawIgloo** function:

```
function drawIgloo(context) {
    context.fillStyle = "white";
    context.arc(120, 160, 60, 0, Math.PI * 2, false);
    context.fill();
    context.fillRect(160, 135, 30, 50); // igloo door
    context.strokeStyle = "#cffcff";
    context.lineWidth = 4;
    context.fillStyle = "#030339";
    context.beginPath();
    context.arc(194, 160, 23, 0, Math.PI * 2, false);
    context.fill();
    context.stroke();
}
```

The following image shows the result of our **drawIgloo** function. While it looks a bit like an igloo, it is floating in the sky. We will take care of this in the next step:

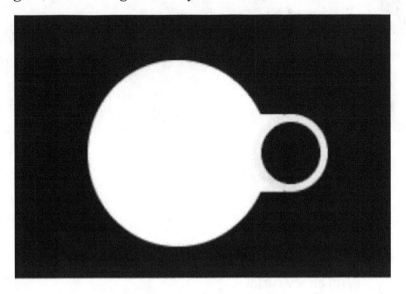

Figure 7.21: The drawIgloo function adds an igloo floating in the night sky

5. The last step is to add **drawFloor** to finish our picture by adding an icy floor. This is done with the **fillRect** function. We can use a light blue color for the fill (**#cffcff**) to make the floor look like ice. Here's the code for the **drawFloor** function, which is called at the end of the **drawFrame** function:

```
function drawFloor(context) {
    context.fillStyle = "#cffcff";
    context.fillRect(0, 160, canvasElement.width, 80);
}
function drawFrame(context) {
    context.clearRect(0, 0, canvasElement.width,
        canvasElement.height);
    drawSky(context);
    drawIgloo(context);
    drawFloor(context);
}
```

The following image shows the result of adding the light blue floor:

Figure 7.22: drawFloor grounds the igloo by adding an icy floor

The igloo now sits on an icy surface.

6. As an added touch, we can give the icy floor a reflective, mirror-like quality by reducing the alpha value of the rectangle's color. We can do this if we change **fillStyle** from **#cffcff** to a hex value with an alpha (**#cffcffef**). A standard hex value is 6 digits, with each pair representing red (**cf**), green (**fc**), and blue (**ff**) values. We can add an alpha value as an additional two digits (**ef**). This value decides the transparency of the fill.

The floor with a reflective surface is shown in the following image:

Figure 7.23: drawFloor with added reflection

In this exercise, we have learned how we can put multiple simple shapes (circles and rectangles) together in canvas to create a more complex picture. We have used these simple shapes to create an igloo on an icy surface, but we can take this method much further to create our pictures.

Next, we will look at how we can draw some other shapes using the **moveTo** and **lineTo** commands, both of which are available through the canvas context.

Drawing Other Shapes

To draw more complicated shapes, we can draw paths using the **moveTo** and **lineTo** methods. We can think of these commands as controlling a pen. When we want to lift the pen and position it on our canvas without drawing anything, we use **moveTo**, and when we want to move it across the canvas while drawing a line, we use **lineTo**.

First, we would start by drawing a path with the **beginPath** command. This tells our canvas context that we are ready to draw a path. When we set fill and stroke styles, they apply to a whole path, which means that if we want to change the styles, we have to start a new path with **beginPath**.

moveTo takes x and y arguments; these values represent the coordinates we want to move the pen to on the canvas without drawing anything.

lineTo also takes x and y coordinates but will draw a line to those coordinates. The line will start from where the pointer is currently positioned.

For example, we could use the following set of commands to draw a star, with a yellow fill:

```
context.fillStyle = "yellow";
context.beginPath();
context.moveTo(160, 60);
context.lineTo(220, 120);
context.lineTo(100, 120);
context.lineTo(160, 60);
context.fill();
context.beginPath();
context.moveTo(160, 140);
context.lineTo(220, 80);
context.lineTo(100, 80);
context.lineTo(160, 140);
context.fill();
```

The preceding code draws two paths. Each one draws a triangle and fills it with the fill style defined in the first line.

The following image shows the results of running this set of commands on a canvas:

Figure 7.24: A star shape created on the canvas with the moveTo and lineTo commands

As well as more complex shapes, we can define more complex patterns and fills for our shapes. Next, we will look at how we can define a gradient to use as a fill style.

Gradients

As well as solid fills, we can create radial and linear gradients to fill a shape. This can add greater texture and depth to a picture, as well as giving us access to more patterns and effects.

We can create two types of gradients:

- Linear
- Radial

For both, first, we create the **gradient** and then add color stops, which define the color at a certain point in the gradient.

To create a linear gradient with the canvas context, we would use the **createLinearGradient** command. This function takes two sets of x and y coordinates. These two points define the direction of the gradient. For example, if we wanted to create a linear gradient starting with white at the top and ending with black at the bottom, we could use the following commands:

```
const gradient = context.createLinearGradient(0, 0, 0, canvasElement.
    height);
gradient.addColorStop(0, 'white');
gradient.addColorStop(1, 'black');
```

This would create the gradient, which we could use via a fill command. For example, we could fill a rectangle the size of the canvas element with this gradient using the following code:

```
context.fillStyle = gradient;
context.fillRect(0, 0, canvasElement.width, canvasElement.height);
```

The result would be a canvas filled with a white to black gradient, as shown in the following image:

Figure 7.25: A top to bottom linear gradient from white to black

If we wanted the direction of the linear gradient to be from left to right, we could set the first point at the top left of the canvas and the second point at the top right of the canvas.

For example, let's say we make a change to the first line of the preceding code, as shown in the following code:

```
const gradient = context.createLinearGradient(0, 0, canvasElement.width,
    0);
```

The following image shows the result, with the white to black gradient from left to right:

Figure 7.26: A left to right linear gradient from white to black

To create a radial gradient, we would use the **createRadialGradient** command on the canvas context. This time, we pass the gradient two circles represented by the x and y coordinates of the center of the circle and a radius:

```
const gradient = context.createRadialGradient(
  canvasElement.width * 0.5,
  canvasElement.height * 0.5, 1,
  canvasElement.width * 0.5,
  canvasElement.height * 0.5,
  canvasElement.width * 0.5
);
gradient.addColorStop(0, 'white');
gradient.addColorStop(1, 'black');
```

Similarly, we could then use the **radial gradient** as a fill style. If we were to fill the whole canvas with this gradient, it would create the following image:

Figure 7.27: A radial gradient from white to black

In the next exercise, we will see how we can use gradients to enhance the canvas drawing we created in the previous exercise.

Exercise 7.06: Gradients

In this exercise, we will learn how to use gradients as the fill style for a canvas drawing. This will help us to enhance our drawings further and give us more tools with which to draw more complex and interesting graphics. We will use both linear and radial gradients to add a bit more interest to our igloo picture.

The steps are as follows:

1. This exercise will start from where *Exercise 7.05, Drawing Shapes*, left off. Make a copy of the **Exercise 7.05.html** file and save it as **Exercise 7.06.html** within the **Chapter07** folder. In *Exercise 7.05, Drawing Shapes*, we were able to create an icy floor with a reflective surface using a semi-transparent fill on a rectangle. Here, we will enhance the effect further by using a linear gradient to make the reflection fade. The gradient will have two-color stops starting at the top, with a semi-transparent light blue fading to an opaque version of the same light blue. Here is the updated **drawFloor** function:

```
function drawFloor(context) {
    const gradient = context.createLinearGradient
        (0, 160, 0, 240);
    gradient.addColorStop(0.05, "#cffcffcf");
    gradient.addColorStop(0.65, '#cffcff');
    context.fillStyle = gradient;
    context.fillRect(0, 160, canvasElement.width, 80);
}
```

The following image shows the result of the gradient fade being applied to the icy floor:

Figure 7.28: A fade effect on the icy floor created with a linear gradient

2. The sky of our igloo drawing is still a little bit dull. It is a flat dark blue fill. In this step, we will apply a radial gradient to create a green halo around the igloo. This will represent the natural phenomenon called aurora borealis, or the northern lights. Here is the updated **drawSky** function with a radial gradient fill emanating from the igloo:

```
function drawSky(context) {
    const gradient = context.createRadialGradient(120,
        160, 0, 120, 60, 160);
    gradient.addColorStop(0.3, "#030339");
    gradient.addColorStop(0.6, "#00F260");
    gradient.addColorStop(1, "#030339");
    context.fillStyle = gradient;
    context.fillRect(0, 0, canvasElement.width,
        canvasElement.height);
}
```

If you now right-click on the filename in VSCode on the left-hand side of the screen and select **Open In Default Browser**, you will see the following image that shows the result of adding this radial gradient to the sky:

Figure 7.29: Northern lights represented by a radial gradient

In this exercise, we have learned how to use gradients in canvas drawings to add interest and texture. In the next section, we will look at how we can animate a drawing in a **canvas** element.

Animating a Canvas

In the previous section, we learned how we can draw shapes in a canvas context. It is possible to animate the shapes we draw on the canvas by clearing the context on each frame. We can then draw a new frame with the shapes at slightly different positions and by modifying the shapes in some way.

We need to be able to clear the canvas and we need a way of timing frames in JavaScript so that we know when to clear and redraw the frame. Let's look at these two concepts before putting them into practice in the next exercise.

In the following diagram, we can see how we can create an animation on the canvas by drawing a frame, waiting some time, clearing a frame, and drawing another frame that is slightly different:

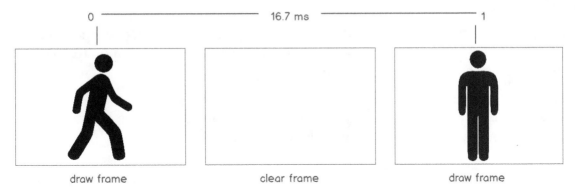

Figure 7.30: Drawing and clearing a canvas frame to create an animation

To clear the canvas context, we use the **clearRect** command and pass it x, y, **width**, and **height** values. The x and y values represent the top-leftmost point of the rectangle. The width and height represent the size of the rectangle. Any pixels within that rectangle's area will be set to a transparent black value, that is, **rgba(0, 0, 0, 0)**. In other words, those pixels will be cleared.

If we want to clear the whole canvas context for a **canvas** element with the **canvasArea** ID attribute, we could use the following code:

```
const canvasElement = document.getElementById('canvasArea');
const context = canvasElement.getContext('2d);
context.clearRect(0, 0, canvasElement.width, canvasElement.height);
```

Here, we get the width and height of the **canvas** element and clear a rectangle from the top-left corner of the canvas context to its bottom-right corner.

The next problem we have to solve to be able to animate our canvas is timing. If we draw on the canvas and then clear it, we will just see a blank canvas. If we draw something, wait for a bit, clear the canvas, and then draw on it again, we will see the illusion of animation.

There are several options for delaying actions with JavaScript. For example, we can set an interval or a timeout of a specified number of milliseconds using **setInterval** and **setTimeout**, or we can use **requestAnimationFrame** to run a function every time the browser renders a frame. We will focus on **requestAnimationFrame** as explaining all the options would be outside the scope of this book and **requestAnimationFrame** is the most appropriate way to handle animation in modern browsers.

The function we give to **requestAnimationFrame** will be triggered every time the browser paints a frame. The great benefit of **requestAnimationFrame** is that it responds to the browser's own paint events rather than a simple timer. This makes it perfect for animation.

On most desktop monitors, an animation frame happens 60 times per second. The number of frames per second is determined by the refresh rate of the monitor and in some cases may be more or less than 60.

Here is the code we could use to draw and clear a frame using **requestAnimationFrame**. To keep the code example simple, we have assumed we already have a reference to the **canvas** element and its context:

```
function drawFrame(context) {
  // first, we clear the canvas
  context.clearRect(0, 0, canvasElement.width, canvasElement.height);
  // ...draw something in our canvas
  // finally, request the next frame
  requestAnimationFrame(function() { drawFrame(context); });
}
drawFrame(context);
```

We have a function called **drawFrame**. The **drawFrame** function clears the canvas context and then draws the shapes that represent the current frame. The last line of the **drawFrame** function makes a new request to **requestAnimationFrame** to call the same **drawFrame** function again on the next animation frame.

On the last line, we are using the **requestAnimationFrame** method to call the **drawFrame** function, which will trigger the first frame of the animation.

Let's put these techniques for animating a canvas into practice with an exercise.

Exercise 7.07: Animated canvas

In this exercise, we will animate the igloo drawing we created in *Exercise 7.06, Gradients*. We are going to make the northern lights in the sky move. We will use the JavaScript modulo (%) operator to create a cyclical animation that loops.

Here are the steps we will follow:

1. Starting from where *Exercise 7.06, Gradients*, left off, we will make a copy of the **Exercise 7.06.html** file and save it as **Exercise 7.07.html** under **Chapter07** folder. In *Exercise 7.06, Gradients*, we added a radial gradient to represent the northern lights on our igloo drawing. In this exercise, we will animate this effect. To do so, we will use **requestAnimationFrame** to redraw the canvas frame 60 times per second. To do this, we add a single line to the end of the **drawFrame** function:

    ```
    requestAnimationFrame(function() { drawFrame(context) });
    ```

    We are now calling the **drawFrame** function at the rate of the screen updates. For most screens and devices, that is 1/60 of a second, or about once every 16 milliseconds.

2. If you look at the output here, you will see no obvious change. This is because our current image is the same for each frame that's drawn. To animate a drawing, we need to make changes to the image in each frame. To do this, we are going to store a global counter value that will start at 0 and increment by 1 each frame.

    At the top of the script, we add the following:

    ```
    let count = 0;
    ```

    At the beginning of the **drawFrame** function, we add the increment:

    ```
    count++
    ```

3. We can use this counter to update our sky. In the **drawSky** function, we will work out the size of our gradient with a calculation based on the count value. We use the modulo operator (%) here so that the value will be between count and the width of the **canvas** element. We then apply the size to the radial gradient:

    ```
    const countLoop = (count % canvasElement.width);
    const size = 1 + 3 * countLoop;
    const gradient = context.createRadialGradient(120,
        160, 0, 120, 160, size);
    ```

The result of this change is that we have an animated gradient. The size of the gradient expands and then loops so that the animation continues ad infinitum. If you now right-click on the filename in VSCode on the left-hand side of the screen and select **Open In Default Browser**, you will see this shown in the following image:

Figure 7.31: Canvas drawing with animated northern lights

In this section, we have learned how we can animate a canvas drawing by clearing a frame and applying changes to the drawing after a short delay.

Next, we will apply what we have learned throughout this chapter to an activity.

Activity 7.01: Media Page

In this activity, we have been asked to create a page to show a video on the *Films on Demand* website. This page will let a user play a video in the browser. For this activity, we are going to show a trailer for the website.

We want to add a **video** element to the page and styles for the play and pause buttons based on the custom controls we made earlier. We will add the provided icons to the buttons.

The steps are as follows:

1. Copy over the page template we created in *Chapter 1, Introduction to HTML and CSS*, and name it as **Activity 7.01.html**:

```
<!DOCTYPE html>
<html lang="en">
    <head>
        <meta charset="utf-8">
        <title>Films on Demand - <!-- Title for page goes here -->
        </title>
        <meta name="description" content="Buy films from our great
          selection. Watch movies on demand.">
        <meta name="viewport" content="width=device-width,
          initial-scale=1">
        <link href="styles/normalize.css" rel="stylesheet">
```

```
        <style>
            body {
                font-family: Arial, Helvetica, sans-serif;
                font-size: 16px;
            }
            #pageWrapper {
                background-color: #eeeae4;
            }
            .full-page {
                min-height: 100vh;
            }
        </style>
    </head>
    <body>
        <div id="pageWrapper" class="full-page">
         <!-- content goes here -->
        </div>
    </body>
</html>
```

2. Add a **styles** directory and copy over the **normalize.css** file.

3. Add a **scripts** directory and copy over the **media-controls.js** file.

4. Next, add a **media** directory and copy over the **html_and_css.mp4**, **html_and_css.webm**, and **poster.png** files.

5. We want to change the title of the page to say **Films on Demand - Video Promo**.

6. Replace the **<!-- content goes here -->** comment with a **video** element. We want to provide both video files using the **source** element. Also, provide a fallback message and an image during preloading with the **poster** attribute.

7. We want to add custom controls for playing and pausing the video and to show the current time and duration of the video. You can follow the design shown in the following screenshot, with the control bar positioned absolutely over the bottom of the **video** element . Now right-click on the filename in VSCode on the left-hand side of the screen and select **Open In Default Browser**:

Figure 7.32: Video controls design

8. Finally, we can use the **scripts/media-controls.js** file to add behavior for the controls to work.

> **Note**
>
> The solution to this activity can be found in page 614.

In this activity, we have put together a media player from what we have learned throughout this chapter. We have added a **video** element and markup for controlling media playback and we have styled the controls for the media player.

Summary

In this chapter, we appended several media elements in an HTML5 document. We studied the applications of the **audio** element and **video** element and learned how to add custom controls to those elements, as well as how we can style them.

We also learned how to add a text track to provide accessible content for users who can't hear audio and video for various reasons.

We studied several different image formats available in HTML5, including SVG and the **canvas** element and **img** and **picture** elements. These introduced us to how these different options can help make our web pages more responsive and how we can use these tools to program our own icons, art, and animations.

In the next chapter, we will look at how we can animate the content of a web page using CSS. Combining the techniques for adding rich content that we have learned in this chapter with those you will learn in the next chapter will allow you to enhance and enrich web pages further with greater interactivity and visual style.

8

Animations

Overview

By the end of this chapter, you will be able to apply animation to your web page; control animations using keyframes; create a slick menu using transitions; and create a multistep **preloader** for your website. This chapter introduces CSS animations by using transitions and **keyframes** to bring your web page to life with new effects. With the knowledge you will gain from this chapter, you will be able to create CSS animations on your web pages to add another level of interaction to the page for the user to experience.

Introduction

In the previous chapters, we discovered how to structure a web page with HTML and style it with CSS. We also introduced HTML forms, responsive web design, and media elements, including video, audio, and canvas.

In this chapter, we'll look at CSS animations. Introducing these to your web page can be a real strength as they can add valuable feedback and interaction to the page. Humans are naturally drawn to movement, so adding subtle animations can really guide users to the important parts of a web page at any given moment. A good example of this could be an HTML form submission, when the user submits the form if they didn't fill out their email address in the correct format. This means you could use CSS animations to animate the email input box for a second or two (for instance, shake it side to side by a few pixels), alongside showing the error message and highlighting to the user the location of the error that they need to correct before they can move forward with their form submission.

With CSS animations, we'll be exploring how to quickly add animations, starting with one or two lines of code, using transitions. Then, as vwe get more comfortable, we'll dive deeper and master how we can have more control over animations using keyframes. Let's discover more about CSS animations and add more interest to our web page, thereby engaging users in a different way.

CSS Transitions

CSS transitions are used throughout the modern web and, in short, they enable CSS properties to change values, thus creating a simple animation. CSS transitions are the basic fundamentals of CSS animations and are the basis of creating more advanced animations. However, transitions in themselves can create a whole world of effects. A common example of a CSS transition would be that the change in the color of an element on hover but rather than the color changes snapping straight away, they would have a more subtle transition from color **1** to color **2** over **250** or **500** milliseconds, instead of an instant change.

To demonstrate a very simple example with a color change, take a look at the following code snippet, which you can copy and paste into any of your HTML files, or into a new one, to see it in action:

```
<style>
p {
    transition: 250ms;
}
p:hover {
    background-color: darkolivegreen;
    color: white;
}
</style>
<p>This is a very simple example of a transition</p>
```

The preceding code snippet will change the **p** element on its hover state to have a background color of 'darkolivegreen' (from the default of white) and a text color of 'white' (from the default of black). The code says that the change in colors should take **250** milliseconds.

A transition in CSS will describe how a property of a given CSS selector should display the change when given a different value. The **transition-duration** property specifies how many seconds (**s**) or milliseconds (**ms**) a transition effect takes to complete.

A transition will require a minimum of one value, that is, the duration of the animation, which is a numeric value. In our example, we used the most common unit, which is **s** for seconds.

The following screenshot shows the **transition-duration** property with a value of **0.2** seconds:

```
transition-duration: 0.2s;
```

Figure 8.1: Transition duration property

> **Note**
>
> If you're planning to work with JavaScript and keeping the units consistent, you may consider using **ms**, which is short for milliseconds. You can read more about CSS transition duration at https://packt.live/32swmYm.

In addition to **transition-duration**, there are two other important CSS transition properties we can use: **transition-property** and **transition-delay**.

The **transition-property** property specifies what property is going to be involved in the transition. If no **transition-property** is set, then all the CSS properties on the selector will be involved in the transition. To help explain this further, if you just wanted the transition to occur on just the background color, then you would set **transition-property: background-color;**.

The **transition-delay** property specifies when the transition effect will start. The **transition-delay** value is defined in seconds (**s**) or milliseconds (**ms**). Adding a delay to the CSS selector means that the defined time must pass before the transition happens. If you set this value to **2** seconds or more, you would really notice a delay before your transition happens. This would be written as follows: **transition-delay: 2s;**.

Exercise 8.01: Implementing Our First Simple Animation

In this exercise, we want to create a simple tag element and animate its **background-color** and **color** properties. To trigger a transition from an initial value to another value, we'll make use of the **:hover** pseudo-selector.

Follow these steps to master simple CSS animations using transitions:

1. We want to create a new file named **Exercise 8.01.html** within the **Chapter08** folder and create a basic structure for the HTML, like this:

```html
<!DOCTYPE html>
<html lang="en">
  <head>
    <meta charset="UTF-8">
    <title>Simple CSS animations</title>
    <style>
      /* Let's put our style in here */
    </style>
  </head>
  <body>
    <div class="transition-me">
      CSS Animations are fun!
    </div>
  </body>
</html>
```

2. Let's create our CSS selector, `.transition-me`, so that we can target our **div** element by its **class** name:

```
.transition-me {
    transition-duration: 0.2s;
}
```

The following screenshot is the output of the preceding code:

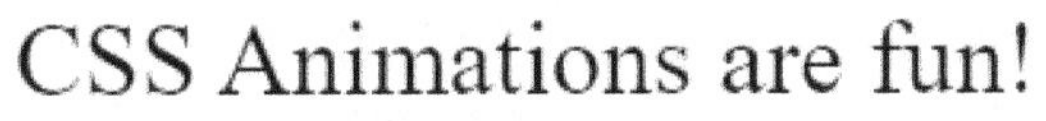

CSS Animations are fun!

Figure 8.2: No animation has been set yet

3. We'll add a bit of styling here so that it's easier for us to view the changes:

```
.transition-me {
    transition-duration: 0.2s;
    padding: 16px;
    cursor: pointer;
    /* those properties will be animated from */
    color: darkorange;
    background: black;
}
```

You will notice that we are using the cursor CSS property here. Having a cursor property with the pointer value will create a pointing hand mouse cursor when the mouse is over the elements in the CSS selector statement (`.transition-me`, in this case). By default, the **cursor** will be the normal mouse cursor on mouseover. Setting the cursor CSS property can define this to be another cursor icon. You can see **cursor: pointer** being demonstrated in the following screenshot. This helps to understand that there's an action on this element (on click or hover):

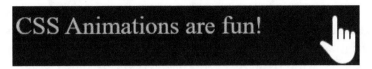

Figure 8.3: Basic styling and properties to animate from

We've added a **padding** with a value of **16px** to separate the content from the box edge so that it's easier to read. We've also added a **color** property with a value of **darkorange**, and a **background** property with a value of **black**. The two colors contrast so that they're easier on our eyes.

4. To create a quick trigger that changes the values of our properties, we'll make use of the `:hover` pseudo selector:

```
.transition-me:hover {
  /* those properties will be animated to */
  background: orangered;
  color: white;
}
```

If you now right-click on the filename in VSCode on the left-hand side of the screen and select **Open In Default Browser**, you will see In the following two screenshots where, you can see the output of an exercise before hovering over the element, and then the output after hovering over the element after the transition has completed:

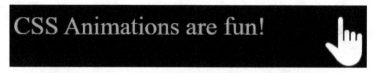

Figure 8.4: The element before the animation starts

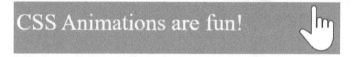

Figure 8.5: The element after the animation ends

All you need to do is move your mouse over the element with the **transition-me** class to see the effect in action.

In this exercise, we noticed that we can animate with CSS quite quickly by using CSS transitions. All we had to do is pass a value in seconds or milliseconds to the **transition** property of a CSS selector and provide some sort of mechanism that would trigger a change of value for the same property to see animations in action. Our two properties are **background** and **color**, and the transition trigger was a `:hover` effect. The background property was animating from **black** to **orangered**, while the color property was animating from **darkorange** to **white**. Our two animations started and ended at the same time and lasted for the same amount of time. What if we want to have a different duration or start time? Let's discover how to do that in our next exercise.

Exercise 8.02: Enhanced Control in CSS Transitions

In this exercise, we want to create a simple tag element and animate its **background-color**, **color**, and **padding-left** properties. To trigger a transition from an initial value to another value, we'll make use of the **:hover** pseudo selector.

Follow these steps to master how to control CSS animations using transitions:

1. We want to create a new file named **Exercise 8.02.html** within the **Chapter08** folder and create a basic structure for the HTML, like this:

```
<!DOCTYPE html>
<html lang="en">
  <head>
    <meta charset="UTF-8">
    <title>CSS transitions enhanced control</title>
    <style>
      /* Let's put our style in here */
      .transition-me {
      }
      .transition-me:hover {
        /* those properties will be animated to */
      }
    </style>
  </head>
  <body>
    <div class="transition-me">
      CSS transitions enhance control
    </div>
  </body>
</html>
```

Our aim is to have one HTML tag, a **div** and add a class to it so that we can target it quickly and painlessly using CSS. We'll also have some random text present inside our element so that we can get away without having to declare its height. In our example, the text's content is **CSS transitions enhanced control**.

2. Inside our style tag, in the `.transition-me` selector, we want to add some spacing between the content and margin using padding with the value of **1rem**. We also want to give a **chocolate color** to the text and a **ghostwhite** color to the background:

```
.transition-me {
  padding: 16px;
  color: chocolate;
  background: ghostwhite;
}
```

The following screenshot shows the `.transition-me` element in its normal state:

CSS transitions enhanced control

Figure 8.6: The element in its normal state

3. Let's continue adding styling to our element for when we move the mouse over it. We want our background to have a **honeydew** color. Our text will receive a color of **brown** and we want to indent it from the left with a value of **32px**:

```
.transition-me:hover {
  /* those properties will be animated to */
  background: honeydew;
  color: purple;
  padding-left: 32px;
}
```

The following screenshot shows the `.transition-me` element in its hover state:

CSS transitions enhanced control

Figure 8.7: The element in its hover state

We want to animate our element with enhanced control. We want to have some transitions starting later. First, we want to start animating our **background** for half a second. Second, as soon as the **background** animation is done, we want to animate the text color for another half a second. Finally, once the text animation is done, we want to animate our text indent for three-tenths of a second. To make our hover effect clearer, we can add a cursor pointer so that the mouse changes when we interact with our element.

4. The code for animating our element with enhanced control is as follows:

```
.transition-me {
  transition-duration: 0.5s, 0.5s, 0.3s;
  transition-delay: 0s, 0.5s, 1s;
  transition-property: background, color, padding-left;
  cursor: pointer;
  /* those properties will be animated from */
  padding: 16px;
  color: chocolate;
  background: ghostwhite;
}
```

If you now right-click on the filename in VSCode on the left-hand side of the screen and select **Open In Default Browser**, you will see the following screenshot that shows the **.transition-me** element in its hover state using the pointer cursor:

CSS transitions enhanced control

Figure 8.8: The element in its hover state with a pointer cursor

It's good to know that CSS offers us great control over our animations with the transition property. One benefit of this is that the code looks clean and easy to write. On the downside, we do have to write a bit more code and we must pay attention to declaring a value or default value for each of the properties we want to animate.

In our exercise, we declared three values, separated by a comma, for **transition-duration**, **transition-delay**, and **transition-property**. It can be incredibly useful to combine a CSS transition for multiple properties at the same time.

In the next exercise, we'll look at the performance of CSS transitions and compare transitions to non-transitions so that you can see the benefits of CSS animations.

Exercise 8.03: CSS Transition Performance

In this exercise, we want to create a simple tag element and animate its **background-color**, **border-color**, and **font-size** properties to observe transitioning of all the properties, and then just transition one property. To trigger a transition from an initial value to another value, we'll make use of the **:hover** pseudo-selector.

Follow these steps to master how to improve CSS animation performance using transitions:

1. Let's start with our HTML document in a file named **Exercise 8.03.html** within the **Chapter08** folder:

```html
<!DOCTYPE html>
<html lang="en">
  <head>
    <meta charset="UTF-8">
    <title>CSS transitions performance</title>
    <style>
      /* Let's put our style in here */
      .transition-me {

      }
      .transition-me:hover {

      }
    </style>
  </head>
  <body>
    <div class="transition-me">
      CSS transitions performance
    </div>
  </body>
</html>
```

2. We want to animate the color of our background over a period of half a second. We can apply a solid border with a width of 10 px and a color of wheat to our elements as well. We'll also give our element a height of 100 px as soon as we experience an unwanted behavior:

```
.transition-me {
  height: 100px;
  background: lightskyblue;
  border: 10px solid wheat;
  transition-duration: 0.5s;
  cursor: pointer;
}
```

The following screenshot shows the `.transition-me` element in its normal state at this point:

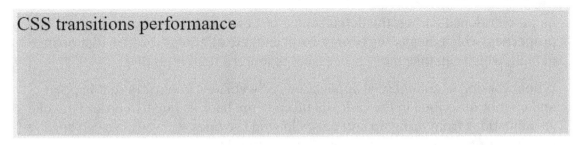

CSS transitions performance

Figure 8.9: The element of the CSS transition performance in its normal state

3. For our animation effect, we'll transition using the `:hover` pseudo-selector. We'll give an orange color to our background and a teal color to our border, while we increase the size of our font to **32px**:

```
.transition-me:hover {
  /* those properties will be animated to */
  background: orange;
  font-size: 32px;
  border-color: teal;
}
```

If you now right-click on the filename in VSCode on the left-hand side of the screen and select **Open In Default Browser**, you will see the following screenshot that shows the .**transition-me** element in its hover state, with the transitions applied:

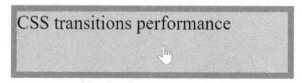

Figure 8.10: The element of CSS transition performance in its hover state

As far as we can see, as soon as we hover over our element, all the properties will be animated. The background color will animate from **lightskyblue** to **orange**. The font size will animate from the default **16px** that the browser has to **32px**, while the border will transition its color from **wheat** to **teal**.

As we mentioned earlier, the default value of **transition-property** is all CSS properties, which means the browser will animate all the properties that change by default, which can take more processing resources than intended.

When working in commercial applications or websites, behaviors that happen and are not described in the task will often come back as bugs. In order to avoid all animations from firing up and only allowing the ones we want; we can specify which transition property or properties we intend to animate.

4. Let's specify a transition property with the value of background and refresh our html page. We can observe that the background animates its color, while the border and font size just jump from one state to another, lacking the animation:

```
.transition-me {
    height: 100px;
    background: lightskyblue;
    transition-property: background;
    transition-duration: 0.5s;
    border: 10px solid wheat;
    cursor: pointer;
}
```

The consequence of this is that the web browser isn't having to work as hard. Previously, when it was animating all the properties, it would be doing around **50** to **70** style recalculations per second, but now, we are only transitioning the background color property, which means it's more in the region of **40** to **60** style recalculations per second. This may seem like a huge saving, but this is a difference of anywhere between **2%** and **4%** extra CPU usage in that moment of processing. If you magnify that over an entire web page full of CSS transitions, you can see how it all adds up very quickly and may cause performance issues for users on a slower device.

In this exercise, we observed a combination of CSS transitions working together, initially with all the properties selected and transitioning together, and then just having the transition on the background color by itself. Aside from the visual difference in the animation's appearance, we've covered the performance difference in animating multiple properties.

In the next exercise, we're going to look at the CSS transition property so that we can write transitions in shorthand when we combine multiple transitions and their properties into a single line of code.

Exercise 8.04: CSS Transition with Multiple Values

In this exercise, we want to create a simple tag element and animate its **background-color**, **color**, and **font-size** properties. We'll be using the shorthand CSS transition property to animate all three properties in one line of code. To trigger a transition from an initial value to another value, we'll make use of the **:hover** pseudo-selector.

Follow these steps to learn how to use the shorthand CSS transition property:

1. Let's create our HTML document in a file named **Exercise 8.04.html** within the **Chapter08** folder:

```
<!DOCTYPE html>
<html lang="en">
  <head>
    <meta charset="UTF-8">
    <title>CSS transition with multiple values</title>
    <style>
      /* Let's put our style in here */
      .transition-me {

      }
      .transition-me:hover {
        /* those properties will be animated to */

      }
    </style>
  </head>
  <body>
    <div class="transition-me">
      CSS transition with multiple values
    </div>
  </body>
</html>
```

2. Let's now add some styles for our normal and hover state, like we did in the previous exercises:

```
    .transition-me {
      height: 100px;
      background: lightskyblue;
      border: 10px solid wheat;
      cursor: pointer;
    }
    .transition-me:hover {
      /* those properties will be animated to */
      background: orange;
      font-size: 32px;
      border-color: teal;
    }
```

3. We want to add a **transition** property. For its value, we'll set a value that holds multiple instructions, separated by a comma:

```
.transition-me {
   height: 100px;
   background: lightskyblue;
   border: 10px solid wheat;
   cursor: pointer;
   transition: background 0.5s, border 0.3s 1s, font-size
      0.5s;
}
```

As can be seen in the preceding snippets, the first value will animate the background for half a second, the second one will animate the border for three-tenths of a second with a delay of one second, and the last one will animate the font size for half a second with a delay of half a second.

In this final step, we can observe how flexible we can be while animating CSS using the shorthand transition property. If you now right-click on the filename in VSCode on the left-hand side of the screen and select **Open In Default Browser**, you will see the final output that is shown in the following image:

Figure 8.11: Output showing the normal state (left) and the hover state (right)

In this exercise, we've learned how to combine the transition properties in the one-line shorthand CSS transition property. So far, we've discovered how we can animate all or some properties, and how to declare the properties and values one by one or in multiples at the same time.

Advanced CSS for Animations

Before diving into the next exercise, we're going to cover a couple of topics to develop our CSS knowledge even further so that we can build more advanced animations.

CSS Positioning

First, we're going to review the CSS **position** property. This is an important property that we use when animating as it allows us to move the elements using the **top**, **left**, **bottom**, and **right** CSS properties to create a movement animation for our element(s). The position property also defines the start point for the **top**, **left**, **bottom**, and **right** coordinates. The position property can have several values. For this chapter, we're going to review two of those possible values: **position: relative** (this denotes any child elements with absolute positioning starting from this element's coordinates) and **position: absolute** (this denotes that any child elements positioned under the absolute position element will also use this element's coordinates for the start point of its position change, but the element itself will be relative to the nearest parent element with relative positioning).

The following diagram illustrates this further:

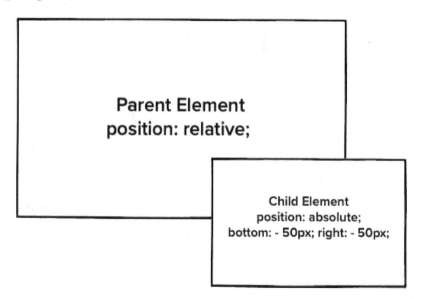

Figure 8.12: CSS positioning simple example

As you can see in the preceding diagram, the parent element is relatively positioned and that the child element is absolutely positioned with the parent element by the **bottom: -50px;** and **right: -50px;** coordinates. This causes the child element to go minus **50** pixels outside the right and bottom boundaries of the relatively positioned parent element.

Using the CSS position property can be really useful when working with CSS animation as it allows you to move items' coordinates. For example, you can move an element from the left-hand side of the screen to the right-hand side of the screen.

Overflow

Overflow is a useful property for adding to a parent element as it makes all the child elements only visible within the parent element. It doesn't allow child elements to "**overflow**" outside of the parent element's boundaries. The syntax for overflow is as follows if we wish to hide overflowing content:

```
overflow: hidden;
```

To restore overflowing content (the default value), we can use the following syntax:

```
overflow: visible;
```

To see the hidden overflow in action, take a look at the following diagram:

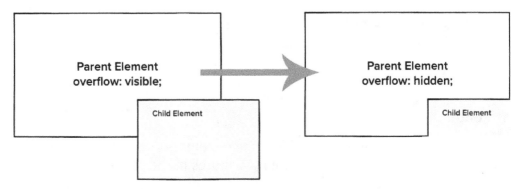

Figure 8.13: Overflow visible and hidden in action

Opacity

Another CSS property that's important to consider using with CSS animations is opacity. This ranges on a scale of possible values between **0** and **1**, and all the decimals within. The syntax for opacity is as follows:

```
opacity: 0.8;
```

The preceding code indicates that the opacity is at **0.8** (or **80%**). Once opacity reaches **0**, the element won't be visible at all. This is illustrated in the following diagram:

Figure 8.14: Opacity from 1 to 0

Blur

Blur is a useful CSS property to animate when we wish to change an element or lose focus on an element. It causes the contents of the element it's been applied to either **blur** or unblur its contents, depending on the value that was assigned. The greater the value that's assigned in pixels, the more **blur** occurs, and to unblur, you would simply give it a value of **0** to reset it. The syntax for **blur** is as follows:

```
filter: blur(5px);
```

The preceding code would blur the element by **5** pixels. The following diagram illustrates **blur** in action:

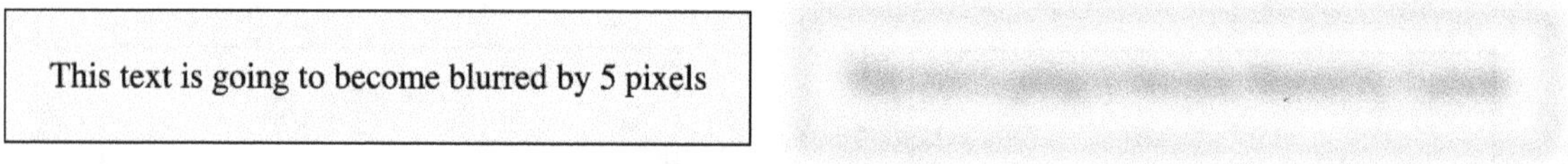

This text is going to become blurred by 5 pixels

Figure 8.15: Blur in action

Inserting Content with attr()

If you want to change content in CSS using pseudo elements such as **:before** and **:after**, then we can get the data from other attributes in the element HTML, such as the title. This is where the **attr()** function in CSS comes into play with the content of pseudo elements. Take a look at the following syntax example of the **:before** part of the selector, which is using the element's title for its content:

HTML:

```
<a class="selector" title="Content Here Will Appear In :before">
Selector Text</a>
```

CSS:

```
.selector:before {
    content: attr(title);
}
```

We've now covered how to briefly use the position, overflow, opacity, blur, and inserting content CSS properties in the CSS from attributes using **attr()**. This knowledge will help us in the exercises to come. In the next exercise, we'll look at animating a website menu and changing its background color and position when hovering over the menu items.

Exercise 8.05: Animating a Website Menu

In this exercise, we want to create a simple menu with three links in it and animate the **background-color** and **position**. To trigger a transition from an initial value to another value, we'll make use of the **:hover** pseudo-selector. The following diagram is a wireframe of the menu we are going to build:

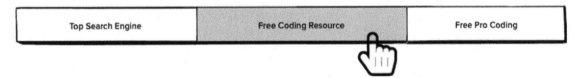

Figure 8.16: Wireframe of the website menu

Follow these steps to master how to build CSS animations using transitions for a menu navigation:

1. Let's create our HTML document in a file named **Exercise 8.05.html** within the **Chapter08** folder:

```
<!DOCTYPE html>
<html lang="en">
  <head>
    <meta charset="UTF-8">
    <title>CSS transition with multiple values</title>
    <style>
      /* Let's put our style in here */
    </style>
  </head>
  <body>
  </body>
</html>
```

2. Inside our body, let's add some navigation and links. We'll represent the navigation with a **nav** tag and inside it, we'll add three anchor tags, represented by the **a** tag. For each anchor tag, we want to have a class, href, and title attribute:

```
<body>
  <nav class="top-navigation-menu">
    <a class="top-navigation-link" href="https://www.google.com"
       title="Top Search Engine"></a>
    <a class="top-navigation-link" href="https://www.packtpub.com/
       all-products" title="Free Coding Resource"></a>
    <a class="top-navigation-link" href="https://www.packtpub.com/
       free-learning" title="Free PRO Coding"></a>
  </nav>
</body>
```

3. In our style tag, let's add a color for the background and provide some dimension values for our links:

```
body {
    background-color: silver;
}
.top-navigation-link {
    height: 28px;
    position: relative;
    width: 225px;
    display: inline-block;
}
```

In the preceding code, we are using the height and width CSS properties to create a box since our anchor tags lack any content at the moment. We'll use a position of relative as a stopper for other elements inside the anchor tag that will have a position of absolute later on. When we use **position: relative**, we'll take the element out of its original flow and position it relative to its original position.

The following screenshot shows **.top-navigation-link** highlighted in Chrome's Developer Tools. It is currently empty because we haven't added content to it yet:

Figure 8.17: Website menu example; empty link

Looking at our code in the browser, we can see that the box has been properly created with a width of **225px** and a height of **28px**. We can confirm this for the other links by using the inspector and hovering over the other anchor tags inside the **Elements panel** of the inspector.

4. To fill our links with content, we'll call **attr()** on our title attribute to grab its value and assign it to the content property of our anchor tag:

```
.top-navigation-link:before,
.top-navigation-link:after {
   text-align: center;
   line-height: 28px;
   content: attr(title);
   position: absolute;
   left: 0;
   right: 0;
   transition: 250ms;
}
```

:before and **:after** act as additional boxes inside the original element, which can receive separate styles and content without increasing the DOM node's number. When building a complex HTML page, you want to keep the number of tags you use to render the application or website as low as possible.

This is what our menu looks like so far. The before and after have a position of **absolute**, which means it'll look up the document tree until it finds a parent with a position of relative, fixed, or absolute, and it'll have its left and right related to the first parent element found to satisfy the condition.

We've also positioned our text to be centered and we've added a transition of **250ms**, which we'll see in action as soon as we implement a hover effect. The following screenshot shows the website menu with the content that was imported from the element title attribute in CSS using **content: attr(title)**:

Top Search Engine	Free Coding Resource	Free PRO Coding

Figure 8.18: Website menu example; links with content

Note that the before and after are sitting one on top of the other as they have the same properties and to our eyes, it looks like we can only see one. If we were to play with their properties for a moment, just to confirm what really happens under the hood, we can clearly see that there are two of them – two boxes with content inside our anchor tags. We want to get rid of this code as soon as we are done testing. There is no reason to leave the **red** and **greenyellow** text styled like this:

```
.top-navigation-link:before {
  top: -10px;
  left: -10px;
  color: greenyellow;
}
.top-navigation-link:after {
  color: red;
}
```

The following screenshot shows the website menu with the links using before and after for the content to appear before we hide one of them. This is done so that you can only visually see one at a time, that is, when the animation is coded:

Figure 8.19: Website menu example; links with before and after

5. On hover, we want to have a transition where the original text flies up, blurs, and disappears, and another one flies in. We have **:before** visible, not blurred, sitting at the top. **:after** is not visible and is positioned **28px** away from the top with some blur filter added as well. The filter property defines visual effects (such as **blur** and saturation) to an element (often **<img>**):

```
.top-navigation-link:before {
  top: 0;
  filter: blur(0);
  opacity: 1;
  background-color: cadetblue;
}
.top-navigation-link:after {
  top: 28px;
  filter: blur(5px);
  opacity: 0;
  background-color: firebrick;
}
```

The following screenshot shows the website menu with the initial styling before the transitions occur:

| Top Search Engine | Free Coding Resource | Free PRO Coding |

Figure 8.20: Website menu example; initial styling before transition

6. On mouse over, we want to transition **:before** from **top 0** to negative **28px**. This will make it fly up:

```
.top-navigation-link:hover:before {
  top: -28px;
  filter: blur(5px);
  opacity: 0;
}
.top-navigation-link:hover:after {
  top: 0;
  filter: blur(0);
  opacity: 1;
}
```

The animation effect has been improved. We also applied a blur filter to it and reduced its visibility to zero. We did the opposite for **:after**. We flew it in from the bottom to **top 0**, removed the **blur**, and made it visible. A blur effect was applied to the image. A larger pixel value would create more blur, whereas a value of **0** would have no blur.

The following screenshot shows the website menu with the initial styling after the transition occurs on hover:

Figure 8.21: Website menu example; initial style with transition

7. To add a finishing touch, we want to set the color of the content so that it contrasts with the background with the value of **whitesmoke**. We also want to convert all the characters into uppercase, while making the text more readable using an **Arial** font. We want the elements inside to stick to one another horizontally and we want to add a display with the value of **flex**:

```
.top-navigation-menu {
  display: flex;
  text-transform: uppercase;
  font-family: Arial;
}
.top-navigation-link:before,
.top-navigation-link:after {
  text-align: center;
  line-height: 28px;
  content: attr(title);
  position: absolute;
  left: 0;
  right: 0;
  transition: 250ms;
  color: whitesmoke;
}
```

The following screenshot shows the final website menu with the completed styling and selected link, "**Free Coding Resource**", after the transition has occurred on hover:

Figure 8.22: Website menu example; final

Let's talk about fixing bugs. Sometimes, we position our mouse on top of the content with **opacity 0**, that is, the one that is positioned **28px** from the **top**. When this happens, the hover effect takes place, and the box with the content I've just mentioned starts transitioning up. At some point, it'll slide out and no longer be under the mouse, which will stop the mouse over effect. This means it will go back to its original state, which puts it back under the mouse. Then, the hover takes over, and everything starts again and again. This issue is a common one and I've personally encountered it many times.

8. There are several fixes for this, and one of them is to instruct the browser to look for hover effects on the anchor tag and not on the **:before** or **:after** boxes. This can be done by adding the **pointer-events** property with a value of **none**:

```
.top-navigation-link:before,
.top-navigation-link:after {
    text-align: center;
    line-height: 28px;
    content: attr(title);
    position: absolute;
    left: 0;
    right: 0;
    transition: 250ms;
    color: whitesmoke;
    pointer-events: none;
}
```

If you now right-click on the filename in VSCode on the left-hand side of the screen and select **Open In Default Browser**, you will see the following screenshot that shows the final website menu during the transition:

Figure 8.23: Website menu example; final – fixing bugs

In this exercise, we have learned how to apply more advanced CSS transitions using a background color and position change alongside effects such as blur to bring the website menu to life. We've experienced what CSS transitions can do. We'll be developing our knowledge so that we can look at CSS animations with keyframes in the next part of this chapter.

Transition Duration Sweet Spot

So far, while playing with our code, we've used different durations for our transitions. When animating menus, panels, boards, or other elements on the screen, a sweet spot to set our duration for is between **250ms** and **300ms**. Smaller values would make the animation happen too quickly, while bigger ones would slow it down too much. This is the case for most scenarios; however, if your animation looks better at **750ms**, just go for it. Most of the time, you want to keep it under one second.

Slowing Animations Down

When you develop the animations for your website, sometimes, you may want to slow them down so that you can inspect them visually and adjust them until you're happy with the outcome. Instead of playing with the values while going back and forth, you can use the developer tools in browsers that support this feature, such as Chrome and Firefox.

In Chrome, follow these steps to get to the interface that allows you to slow down the animations:

1. Click the `triple vertical dots` that you can find at the far top right of the Chrome browser.

2. Select `More tools`.

3. Select `Developer tools`.

   A better way to do this is to remember the shortcut and press CTRL + SHIFT + i on Windows, or ALT + CMD + i on a Mac.

   The following screenshot illustrates how to open the `Developer` Tools in the Chrome web browser:

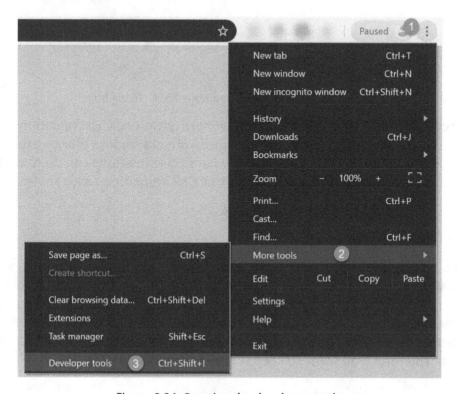

Figure 8.24: Opening the developer tools

4. Find the **three vertical dots** in the developer tools and click on it.

5. Select **More tools**.

6. Select **Animations**.

The following screenshot shows how to open the **Animations** tab with the **Developers** Tools in the Chrome web browser:

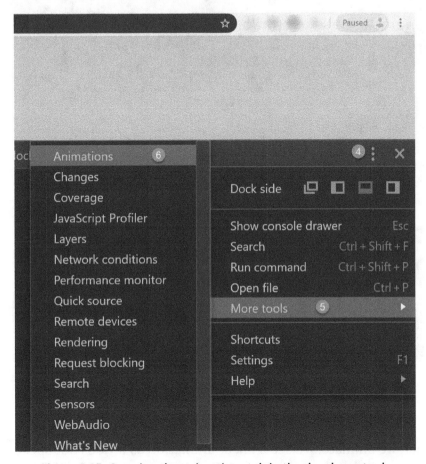

Figure 8.25: Opening the animations tab in the developer tools

We can slow down the animations to **25%** or **10%**. We can pause all and resume play as well.

The following screenshot shows the **Animation** tab in Chrome's developer tools while running the animation at **25%** of its normal speed:

Figure 8.26: Slowing down the Animations tab in the developer tools

Animation Acceleration and Deceleration

By default, when we animate, we see a linear animation as it does not speed up or slow down. It starts and ends at the same speed. We can customize this by using **transition-timing-function** in CSS. If we want our animation to start slowly, then fast, and then end slowly, we give it a value of **ease**:

```
.top-navigation-link:before,
.top-navigation-link:after {
  transition-timing-function: ease;
}
```

If we want it to start slow, we give it a value of ease-in. If we want it to end slow, the value would be ease-out.

For even more control over the acceleration and deceleration of our animations, we can use **cubic-bezier**. A good place to start playing with values for **cubic-bezier** is the website with the same name, https://packt.live/34AI9Wb, where you can play with draggable handles and preview your animation until you find one that meets your expectations:

```
transition-timing-function: cubic-bezier(.93,.45,.73,1.3);
```

The following screenshot shows the **cubic-beizer** animation playground:

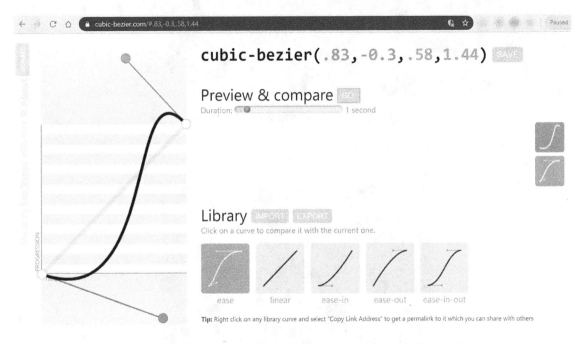

Figure 8.27: https://cubic-bezier.com animation playground

You can play with the **cubic-bezier** values directly in the browser as well. The following screenshot shows an example of using Chrome to preview animations using **cubic-bezier**:

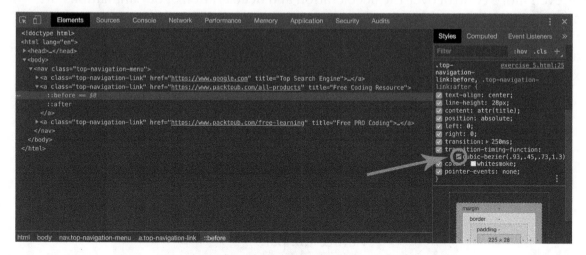

Figure 8.28: cubic-bezier animation playground in Chrome icon

In the following screenshot, you can see how to update the **cubic-bezier** animation in Chrome's developer tools:

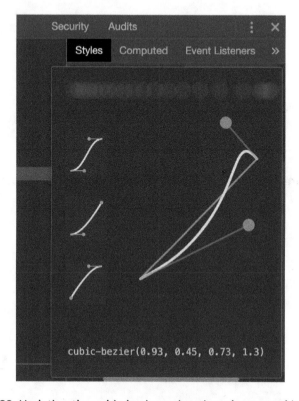

Figure 8.29: Updating the cubic-bezier animation playground in Chrome

Keyframe Animations in CSS

So far, we've discovered how to animate from one starting point to an ending point using transitions in CSS. There are scenarios where we may want to animate using more than two points of control, that is, more than a starting and ending point. To do this, we can use keyframes.

The syntax looks like this:

```
@keyframes animationName {keyframes-selector {css-styles}}
```

The **@keyframes** word indicates to CSS that you're about to write code for a keyframe-based CSS animation. The next one, **animationName**, can be anything that describes the animation you intend to create. An example of keyframes to animate the opacity property is as follows:

```
@keyframes showHide {
   0%   { opacity: 0; }
   100% { opacity: 1; }
}
```

The preceding code is a simple keyframes code snippet to show or hide an element (depending on whether animation is running forward or backward).

Inside the first set of curly braces, we want to add one or more steps. An animation step starts with a keyframe selector that can be **from**, **to**, or a percentage value such as **0%**, **45%**, and **78%**. The **from** keyword has the same effect as **0%**, while the **to** keyword has the same effect as **100%**. Inside the second pair of curly braces, we want to declare one or more keys: value sets, just like in our normal CSS selectors.

Using the CSS Animation Property

Before we start the next exercise, we're going to look at the CSS animation property. This is a very powerful property and it can achieve a lot. Let's start by looking at its syntax, as follows:

```
animation: name duration timing-function delay iteration-
   count direction fill-mode;
```

You can use all or some of the values in the preceding syntax example and all you have to do is replace the placeholders with the appropriate values. If you want to write an animation that only has a name, duration, and iteration count, you can do something like this:

```
animation: rotateBall 3s infinite;
```

The timing function property can have various values, for example, you could use **ease-in-out**, which means the animation will speed up in the middle, and **push-ease-off** (slow down) at the start and end of the animation. You could also use **steps()** timing-function, which means the animation will no longer transition from one initial value to the ending value – instead, it will jump directly. **steps()** takes one argument, that is, how many jumps from one point to another the animation should have. Since we've passed 1 as the argument for **steps()**, like **steps(1)**, it will jump directly from the initial value to the end value of that step. For example, if we had **top: 0** and **top: -50 px**, using **steps(1)**, it will directly jump from the first value to the second.

By default, once the animation finishes, it jumps to the first frame. To allow our animation to retain the values of the last frame, we would want to give a value to the **animation-fill-mode** property. That value is **forwards**.

Now that we've covered some of the basics of CSS animation properties, we'll begin our next exercise and build a preloading animation.

Exercise 8.06: CSS Preloader Using Keyframes

In this exercise, we want to create a simple preloader that animates its width from zero to a quarter, and then from a quarter to its full width. The following diagram shows an example of the preloader that you are going to create using animation keyframes:

Figure 8.30: Preloader using keyframes

Follow these steps to master how to build a CSS animated preloader using keyframes:

1. Let's create our HTML document in a file named **Exercise 8.06.html** within the **Chapter08** folder and type in the following code:

```
<!DOCTYPE html>
<html lang="en">
  <head>
    <meta charset="UTF-8">
    <title>CSS animations preloader</title>
    <style>
      /* Let's put our style in here */
    </style>
  </head>
  <body>
    <div class="preloader-wrapper">
      <div class="preloader-bar"></div>
    </div>
  </body>
</html>
```

2. We want to give the **preloader** wrapper a height of **32px** and a border radius with the value of half of the height value to produce what looks like a *pill*. We also want to hide the contents of the wrapper that may overflow over the corners. To hide everything inside the wrapper's visible area, we'll use **overflow: hidden**. We also want to give it a background color of **dodgerblue**:

```
.preloader-wrapper {
  border-radius: 16px;
  height: 32px;
  overflow: hidden;
  background-color: dodgerblue;
}
```

In the following diagram, we are showing **.preloader-wrapper** with its rounded corners:

Figure 8.31: Preloader wrapper with rounded corners

Note that you want to pay attention to the overflow hidden key: value pair. This is something you'll want to remember and understand as it's a common scenario where you want to display only the visible contents of a container, that is, of a box. So, whatever flows outside the box area will not be visible. You want to do this when you use border radius, or when you have a defined width or height and the contents inside your box exceed the length of the width or height of the box.

3. Let's style the **preloader** bar and animate it as well. Since the parent of the **preloader** bar does have a height declared, we can use a percentage in the direct child, or first level child, for the value of the height property, for example, **height: 100%;**. Let's add a dark gold background color to it, assign the **preloader** animation so that it runs for five seconds, but just once, and the animation will start slow and end slow:

```
.preloader-bar {
  height: 100%;
  background-color: darkgoldenrod;
  animation: preloader 5s ease-in-out 1s;
}
@keyframes preloader {
  0% { width: 0 }
  50% { width: 25% }
  100% { width: 100% }
}
```

The **preloader** animation has three points declared: the start at **0%** with no width for the **preloader** bar, the second point at half of the animation, which gives the **preloader** bar a width of **25%**, and the last point, which animates the width from **25%** to **100%** in the other half of the animation.

4. It would be nice if we could add some indication of the percentage the animation is at. Let's do that by adding the following structure to our **Exercise 8.06.html** file, within the **.preloader-wrapper** parent element, just after the **.preloader-bar** child element:

```
<div class="preloader-percentage">
  <div class="preloader-value">0%</div>
  <div class="preloader-value">25%</div>
  <div class="preloader-value">100%</div>
</div>
```

We've added the div with the `.preloader-percentage` class, just below the div with the `.preloader-bar` class. Preloader percentage and preloader bar are siblings, at the same level, inside the div with the `.preloader-wrapper` class. The idea is to show only one `preloader` value `div` at one time. We can do that by stacking them either horizontally or vertically. I chose to stack them vertically so that when we want to show a different value, all we have to do is move the first parent that holds the `preloader` value's boxes to a new one. We can find the new value by subtracting the height of the `preloader` from the current position on the **y** axis.

5. Let's style the `preloader` percentage and value. As we mentioned earlier, we'll start our animation from position 0 on the y axis. The next step would be **"0 – 32" px = -32 px**. The last step would be **"-32 px – 32 px" = -64 px**. We can also name the first step instead of using **0%** by using the word **from**. For the last step, that is, **100%**, we can name it using the word **to**:

```css
.preloader-wrapper {
  border-radius: 16px;
  height: 32px;
  overflow: hidden;
  background-color: dodgerblue;
  position: relative;
}
.preloader-percentage {
  position: absolute;
  top: 0;
  left: 0;
  right: 0;
  height: 100%;
  animation: preloader-value 5s steps(1) forwards;
}
.preloader-value {
  height: 100%;
  line-height: 32px;
  text-align: center;
  color: whitesmoke;
}
@keyframes preloader-value {
  from {top: 0}
  50% {top: -32px}
  to {top: -64px}
}
```

Giving a **position** of **absolute** for our **preloader** percentage will take the element out of the normal document flow and allow us to position the element using the **top** property on the **y** axis, in relation to the first parent that has a **position** of **fixed**, **absolute**, or **relative**. That's why we added **position: relative;** to our **preloader** wrapper.

If you now right-click on the filename in VSCode on the left-hand side of the screen and select **Open In Default Browser**, you will see the following diagram, you can see the output of this exercise in the web browser, showing its three key stages at the start of the animation, at **25%** of the animation, and at **100%** of the animation completed:

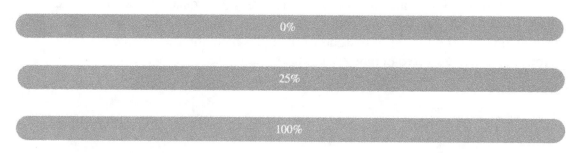

Figure 8.32: Final output of the web browser, shown in its three key stages

More CSS Tips and Tricks

When you want to position a container in the center horizontally without having to use the **text-align** property with a value of **center**, you can give it a **margin** of **left** and **right** with a value of **auto**:

```
margin: 0 auto;
```

When using transformation, you may be tempted to apply multiple transformation key-value pairs inside the same selector, assuming that all of them will take effect. Unfortunately, CSS doesn't work that way. Only one will be applied at most. If you have more than one, the one that's the last in your selector will be applied.

You can make an element smaller or bigger and shrink it or expand it using the **scale** *transform function*. So, if you want to make an element taller, you'd want to use **scaleY** with a value greater than **1**, such as **1.25**, which means it'll be **25%** bigger than the original. To make an element wider, you'd want to use **scaleX**, with a value bigger than **1**. Using a value smaller than **1**, which you can think of as **100%**, will transform the element into a smaller size than the original one.

Using the **scale** transform function will apply the value that was passed to it to both the **X** and **Y** axis. Using **scaleZ** will apply a transformation on the **Z** axis. When you want to apply transformations on all the axes, we can use **scale3d** and pass to it three arguments or values: the value for **scaleX**, the value for **scaleY**, and the value for **scaleZ**. Here is an example of how we can make our element bigger by **10%** on all the axes:

```
transform: scale3d(1.1, 1.1, 1.1);
```

When we want to control the acceleration of entering and exiting the animation, just like in our previous example of using the **cubic bezier** in transitions, **transition-timing-function: cubic-bezier(.93,.45,.73,1.3);**, we can use **animation-timing-function**:

```
animation-timing-function: cubic-bezier(0.2, 0.61, 0.35, 1);
```

The following diagram shows the **X**, **Y**, and **Z** axes of an element, which is useful to help us understand how to scale items with 3D transformations:

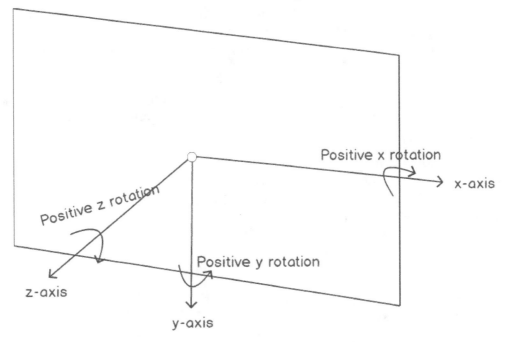

Figure 8.33: The X, Y, and Z axes of an element

The top left-hand corner of an element is the **0, 0, 0** origin of the **X**, **Y**, and **Z** axes.

Now that we are comfortable with making elements bigger or smaller using **scale**, **scaleX**, **scaleY**, **scaleZ**, or **scale3d**, we can easily learn how to move an element away from the **0, 0, 0** origin while keeping its size the same. We can use translate, **translateX**, **translateY**, **translateZ**, and **translate3d** to do this. Passing a positive value to **translateX** will move the element away from its original **X** axis position to the right, while using a negative value will move it to the left, away from its original position.

When we want an animation to be stopped, be it before it starts or while it's in progress, we can assign the **paused** value to the **animation-play-state** property, as follows:

```
animation-play-state: paused;
```

When we want to instruct an animation to resume from its paused play state, we want to assign the **running** value to its **animation-play-state** property, like so:

```
animation-play-state: running;
```

So far, we've completed six exercises and learned about CSS transitions and CSS animations so that we can create various different effects, including a simple color change on hover, to running multiple transitions together, and then observing the difference between having and not having a property included in the transition in terms of performance and appearance.

After this, we worked with using the shorthand CSS transition property and then made an animated menu using CSS transitions to create some new effects, including blur on transition.

Our final exercise looked at using CSS animation properties, in shorthand, to achieve keyframe animations to allow you to code more detailed animations. We are now ready to start our next activity, where we will bring our video store home page to life by adding some CSS transitions and animations to the web page.

Activity 8.01: Animating Our Video Store Home Page

In this activity, we're going to take our video store home page that we used in *Chapter 6, Responsive Web Design and Media Queries*, and then add some transitions and animations to bring the web page to life. Take a look at the following screenshot to remind yourself of the web page before we make any changes to it:

Figure 8.34: Video store home page without animation

1. First, take a copy of **Activity 6.01.html** from your **Chapter06** folder and save this in your **Chapter08** folder as **Activity 8.01.html**.

2. Now, we're going to edit the navigation links so that we have a CSS transition on the active hovered navigation link item by moving it slightly upward and then smoothing the transition of its hover color to red. Both transitions will be completed over **500** milliseconds. This will be triggered upon hovering the mouse over the link. Note that we only want to apply this to our desktop screen sizes since this could impair the usability of the mobile layout for users otherwise.

3. Next, we're going to use keyframes to create a hover effect on the product cards as the mouse hovers over each of them. On the animation, we're going to change the background color of the card, add a thicker border to the card (while reducing the padding simultaneously so as not to increase the overall width of the box), and then change the color of the text so that it works with the new background color.

   We will apply this style to mobile and desktop screen sizes, so it doesn't need to be desktop-specific this time.

4. When observing your **Activity 8.01.html** file in your web browser, you should see that it has a vertically rising link and color change on hover when viewing on a desktop, and, on mobile and desktop, the product cards should animate to what's shown in the preceding screenshot when hovered over or tapped on a touch device. Take a look at the following screenshot for the expected end result, with the two animated hover states shown:

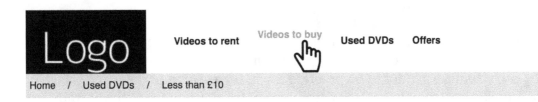

Figure 8.35: The complete output with both hover states shown

> **Note**
>
> The solution to this can be found on page 618.

Summary

Well done for following along and mastering how to quickly write transitions and create more complex animations to spice up the journey a user has while making use of websites you've built. The more you practice them, the easier and quicker it'll be to write code for slick animations that will make your users smile and feel comfortable as they feel that you care for them.

I encourage you to quickly write three to five basic ideas or update a piece of code you've already written where you animate some parts of your interface through transitions of keyframes and share it with your friends, workmates, or clients. Feedback is always welcome, as well as recognition for being an awesome frontend developer.

The next chapter is all about accessibility. Here, we will look at how to make our web page more accessible to technologies such as screen readers, thereby allowing users with visual impairments to still use and navigate our web page.

9

Accessibility

Overview

By the end of this chapter, you will be able to identify various criteria required to build accessible web pages; create accessible images on a web page; create accessible forms on a web page; and use Axe to identify accessibility issues in web pages. In this chapter, we will look at the importance of accessibility in developing web pages. We will look at the techniques and tools available to a web developer to make the content reach as wide an audience as we possibly can, and this means doing what we can to make text, forms, images, and interactions accessible.

Introduction

In the previous chapter, we looked at how we can add animation to our web pages using HTML and CSS. We learned how we can use motion and transitions to great effect to enrich our web pages and add visual flair.

In this chapter, we are going to look at a very important but sometimes overlooked aspect of web development, that is, accessibility. This chapter is about making a web page available to everyone. It is about removing barriers by making our code as easy to use as possible for as wide a range of users as possible. The web is an open platform, and we should strive to make its content accessible to as many people as we can. The web should work for everyone.

As a web developer, this means we have a great responsibility to use the tools available to us to make accessible websites. HTML has a lot of features that are designed to make web pages accessible and, in this chapter, we will look at several of these – `alt` attributes, form labels, and fieldsets, to name a few.

We will start by identifying some common issues that can cause web pages to become inaccessible. We will look at how a combination of thinking with an accessibility mindset from the outset of our project and using semantic HTML as intended will solve a lot of these issues.

We will also look at some of the tools that are available to help us improve the accessibility of our web pages. We will be looking at a tool called Axe that can automatically flag accessibility issues.

Finally, as an activity, we will look at a page with some accessibility issues and use what we have learned in this chapter to identify and fix those issues.

> **Note**
>
> Accessibility is a long word often shortened to the **numeronym** (a word abbreviated with numbers) **a11y**. The term a11y is a useful one to recognize as it is widely used when talking about accessibility and, if you are researching the subject, a web search for a11y will return useful results.

What Is Accessibility?

Accessibility is a very important subject for the web and web developers, but it is one that isn't always that well understood. By learning about the accessibility improvements, we will look at in this chapter, we can help a lot of users. We can remove barriers for the following:

- Those with visual impairments who cannot get information from images that do not have a text alternative

- People who have hearing impairments and cannot get information from media (audio or video)

- Those with physical impairments that prevent them from using a mouse

By making our web pages accessible to those who are differently abled, we also make the pages more usable by those who may face technical limitations. Accessibility issues are not only limited to the aforementioned scenarios but also include users on mobile devices, users who are on a website in a location with a lot of background noise or during a presentation or on a monitor that is not well calibrated, and where color contrast may be an issue. Simply put, accessibility improves usability.

The web is a great platform for distributing content, and the main purpose of that content is to reach as many people as possible. Whether our reasons are due to ethics, profit margins, fear of litigation, or pure empathy, there are no good reasons to ignore the accessibility needs of our users.

You'll be happy to hear that we've already learned about a lot of the techniques we need to make our web pages as accessible to as many people as we can. Before we progress, let's review some of the accessibility concepts we have covered in the preceding chapters and how we will be building upon them in this chapter.

In *Chapter 1, Introduction to HTML and CSS*, and *Chapter 7, Media – Audio, Video, and Canvas*, we talked briefly about the **alt** attribute on the **img** tag. In the next section of this chapter, we will look at how we can make images accessible to visually impaired users (and those with image loading disabled) using the **alt** attribute in the right way in more detail. The context in which an image is used and the information it conveys means we should use different approaches for **alt** text.

In *Chapter 2, Structure and Layout*, we learned about some of the semantic blocks for structuring a web page. In the *Semantic HTML* section in this chapter, we will further explore how we can use HTML tags appropriately to provide a good accessible experience for our users.

In *Chapter 4, Forms*, we looked at creating HTML forms. We will reinforce some of that learning by focusing on the fundamentals for making these forms accessible, including connecting labels to inputs correctly and keyboard accessibility.

In *Chapter 7, Media – Audio, Video, and Canvas*, we also learned about techniques for adding captions to our audio and video content and making that content available to users who can't hear it (because they have hearing difficulties, their speakers don't work, or they are in, for example, a library and can't use their speakers).

An area we have not covered previously and one that is specific to enhancing the accessibility of web pages is the **Web Accessibility Initiative's set of web standards for Accessible Rich Internet Applications (WAI-ARIA)**.

The WAI-ARIA standards are extensive and are often used to enhance web pages with a lot of dynamism. Put simply, where a web application uses JavaScript to update the page often, WAI-ARIA can help maintain the meaning of the page, keep a user's focus in the right place, and keep behavior accessible.

A good example of WAI-ARIA in action is if we have a JavaScript triggered modal with an alert message in it. If the modal trigger is not properly implemented, a screen reader navigating a page may miss the update to the page when the modal appears and the screen reader may, unexpectedly, continue to browse the page beneath the modal. With WAI-ARIA, we can tell the browser that an element should be treated as a modal with the `aria-modal` attribute. This will add focus to the modal content and prevent interaction with content outside the modal when that modal is triggered.

Without going into great detail, we will mention a few of the attributes that WAI-ARIA provides, where they are useful, and how to supplement them or correspond them to the accessible HTML implementation we are learning about.

First, we will look at how we can make our HTML images accessible.

Accessible Images

As the saying goes, a picture is worth a thousand words, and images can add a lot to a web page. Some pictures are decorative, whereas others are an important piece of content that gets your web page's message across with great impact.

Every time we add an image to a web page, there are accessibility considerations we should take into consideration in order to make an accessible image. Not all users may be able to see an image. For example, if a user requires a screen reader or other form of non-visual browser to be able to navigate through a web page, they will require a textual description of an image to be able to garner any meaning from it. It can also be the case that a user does not download images because of limitations on network bandwidth or for security reasons. All of these users will benefit from an alternative text description of an image.

The way we can provide this information is through the `alt` attribute, which we learned about in *Chapter 1*, *Introduction to HTML and CSS*, and again in *Chapter 7*, *Media – Audio, Video, and Canvas*. An `alt` attribute should provide a meaningful text alternative to an image.

For example, if we were to add an image to our web page that showed an infographic with some important business data, such as the budget for this and last year, the information – any text, labels, or the numbers the bars represent – would not be accessible to a non-visual browser if they were included in the image. We would need to add an **alt** tag that expressed that information.

The following **img** tag could be used to provide an image and an appropriate **alt** text:

```
<img src="bar-chart.png" alt="Bar graph of profits for 2019 (£40,000),
    which are up £20,000 on profits for 2018 (£20,000)" />
```

The following is the image of our infographic:

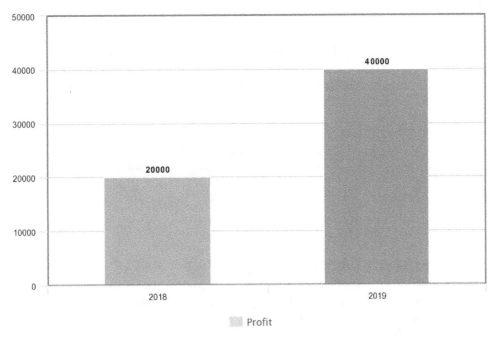

Figure 9.1: Image of a bar graph

The following screenshot shows the **alt** tag (with the image not loaded). The **alt** tag can also be read by a screen reader:

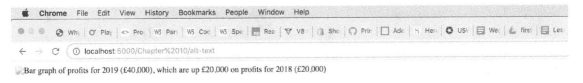

Figure 9.2: The alt text for the bar graph image

There are a few decisions we should consider when we add an **alt** attribute, and these depend on the nature of the image (or images) on our web page.

> **Note**
>
> The W3C provides a decision tree that can be very useful for making a decision about what content you need to provide in the alt attribute of an image on your web page. You can find it at https://packt.live/33CzxxW.

As we have shown in the previous example, informative images need a text alternative. Where an image provides the user with information in the form of a diagram, graph, photo, or illustration that represents data or a concept, there should also be an **alt** text attribute. We should, at least, provide a short text description of the essential information the image conveys. Often, and this especially applies where the information in the image is complex, we need to back this up with more text content or with a table of data.

If an image is purely decorative, we should still provide an **alt** attribute, but we don't need to add any text. We just add an **alt** attribute with an empty string (**alt=""**). Decorative images include those that are used for visual styling and effects.

Images that are described by surrounding text on the page can also be considered decorative. If the information that's conveyed by an image is also described as text on a web page, you can again add an empty **alt** tag.

If an image has a functional role, for example, it is an icon that's used as a button or a link, we need to provide alternative text that describes the function rather than the image. For example, where an image of a floppy disk is used as a button for saving a file, we would provide the **alt** text "Save" or "Save file" rather than something such as "Floppy disk icon", which would describe the image but would not be useful for a user who wanted to use that functionality.

Where multiple images are used to convey a single piece of information, we can add alt text to the first image to provide a description. The rest of the images can then have an empty **alt** attribute. For example, in the forthcoming exercise, we will create an element for showing a rating using five separate star images. By adding an **alt** text description to the first image and empty alt attributes to the other images, we inform the user of the rating, even when the images are not visible.

Here is the relevant code:

```
<div class="rating">
    <img src="images/full-star.png" alt="Rated 3 and a half out of 5
      stars" />
    <img src="images/full-star.png" alt="Rated 2 and a half out of 5
      stars" />
    <img src="images/full-star.png" alt="Rated 1 and a half out of 5
      stars" />
    <img src="images/half-star.png" alt="Rated half out of 5 stars" />
    <img src="images/empty-star.png" alt="Rated empty out of 5 stars" />
</div>
```

Having seen the basics of the **alt** attribute, we will implement this in the following exercise, where we will make the rating section of a typical product page accessible. We will make the group of images accessible with the appropriate use of **alt** text.

Exercise 9.01: Accessible Ratings

In this exercise, we are going to create a product page. We are going to include an element for showing a product rating. This is a common UI pattern seen on eCommerce sites or anywhere users can rate cultural artifacts such as books and films. The rating will be represented by 5 stars, and a rating can be any value from 0 to 5 stars, rising in increments of a half.

To create the rating element, we will use five separate images. We will make the group of images accessible by providing **alt** attributes to express the information of our set of images. In other words, we will describe the rating given.

The following screenshot is what the product page will look like when it's finished. For this exercise, we will give the product a rating of 3 and a half stars:

Product Description

Lorem ipsum dolor sit amet, consectetur adipiscing elit. Phasellus scelerisque, sapien at tincidunt finibus, mauris purus tincidunt orci, non cursus lorem lectus ac urna. Ut non porttitor nisl. Morbi id nisi eros.

Figure 9.3: Product page with ratings UI

The steps are as follows:

1. First, create and save a file named **Exercise 9.01.html** within the **Chapter09** folder. In **Exercise 9.01.html**, copy the following code to set up a basic HTML page:

```
<!DOCTYPE html>
<html lang="en">
    <head>
        <meta charset="utf-8">
        <title>Exercise 9.01:Accessible rating</title>
        <style>
        html {
            box-sizing: border-box;
        }
```

```
        *, *:before, *:after {
            box-sizing: inherit;
        }
        body {
            padding: 0;
            margin: 0;
            font-family: Arial, Helvetica, sans-serif;
            font-size: 16px;
            line-height: 1.7778;
        }
    </style>
  </head>
  <body>
  </body>
</html>
```

2. In the **body** element, add a **div** with a class attribute with the **"container"** value. This is where our product markup will be hosted:

```
<div class="container"></div>
```

3. We want the product to be centered on the page, so we will use flexbox to center the contents of the container in CSS. By setting the flexbox flow to a column with no wrapping, the elements within that container will flow vertically, one after another. We add the following declaration block to the **style** element:

```
.container {
    display: flex;
    align-items: center;
    justify-content: center;
    flex-flow: column nowrap;
}
```

4. We are going to create a product item page that shows an item with a title, a description, and a rating. We can add the initial markup for the product – a section element with the **"product"** class attribute and child elements for a product image, as well as a heading and a paragraph for the description. We add this to the **body** element:

```
<section class="product">
    <img class="product-image" src="images/product.png"
      alt="Product" />
    <h2 class="product-heading">Product Description</h2>
    <p class="product-description">
```

```
        Lorem ipsum dolor sit amet, consectetur adipiscing
        elit. Phasellus scelerisque, sapien at tincidunt
        finibus, mauris purus tincidunt orci, non cursus
        lorem lectus ac urna. Ut non porttitor nisl.
        Morbi id nisi eros.
    </p>
    <hr class="divider" />
</section>
```

5. To style the product item, we will add the following CSS declarations. This defines a box for the product with a shadow around it and the content centered:

```css
.product {
    width: 50vw;
    min-width: 640px;
    margin: 2rem;
    padding: 1rem 2rem;
    display: flex;
    justify-content: center;
    flex-flow: column nowrap;
    border-radius: 3px;
    box-shadow:
        rgba(0, 23, 74, 0.05) 0px 6px 12px 12px,
        rgba(0, 23, 24, 0.1) 0 6px 6px 6px,
        rgba(0, 23, 24, 0.3) 0 1px 0px 0px;
}
.product-heading {
    margin: 0;
}
.product-image {
    padding: 0;
    margin: 2rem auto;
    width: 60%;
    height: 100%;
}
.product-description {
    width: 100%;
}
.divider {
    width: 100%;
}
```

6. To create the UI element for showing the ratings, we will add a **div** element with a **rating** class attribute and five image elements as children. Each of these images will either show a star, a half-filled star, or an empty star. We will add this below the description at the end of the product section:

```
<div class="rating">
    <img src="images/full-star.png" />
    <img src="images/full-star.png" />
    <img src="images/full-star.png" />
    <img src="images/half-star.png" />
    <img src="images/empty-star.png" />
</div>
```

7. We need to add some more CSS declarations to the style element to center the ratings and to keep the size of the rating responsive while the container is resized due to window width. We have added **flex-shrink** to each rating image as this will allow the image to shrink in order to fit the available space:

```
.rating {
    margin: 1rem auto;
    width: 60%;
    display: flex;
    flex-wrap: row nowrap;
}
.rating img {
    width: 20%;
    flex-shrink: 1;
}
```

8. Next, we will add alt attributes to each image element with an empty string. This means that the images will be treated as decorative:

```
<div class="rating">
    <img src="images/full-star.png" alt="" />
    <img src="images/full-star.png" alt="" />
    <img src="images/full-star.png" alt="" />
    <img src="images/half-star.png" alt="" />
    <img src="images/empty-star.png" alt="" />
</div>
```

The five image elements are required to describe some information that our users may find useful. At the moment, our rating element does not describe that information in a text-based format. A screen reader user would not get information about the product rating and that could hamper their choice.

9. By adding the rating information as the value of an **alt** tag for the first image, we provide the necessary information for non-visual users:

```
<div class="rating">
        <img src="images/full-star.png" alt="Rated 3
          and a half out of 5 stars" />
        <img src="images/full-star.png" alt="" />
        <img src="images/full-star.png" alt="" />
        <img src="images/half-star.png" alt="" />
        <img src="images/empty-star.png" alt="" />
</div>
```

We've added alt attributes to each of the images, but only the first one describes the information portrayed by the set of images. This provides information for a user who cannot see the visual representation of the images and means that the ratings are accessible and useful.

If you now right-click on the filename in VSCode on the left-hand side of the screen and select **Open In Default Browser**, you will see the final result that will look like the screenshot that was provided at the beginning of this exercise.

The following screenshot shows the product page without images. The alt text shows that we can still get the information provided by the images when the images are not available. The rating is shown as "**Rated 3 and a half out of 5 stars**" so that a user who has a screen reader could make a decision about this product with the same information as a user who can see the images:

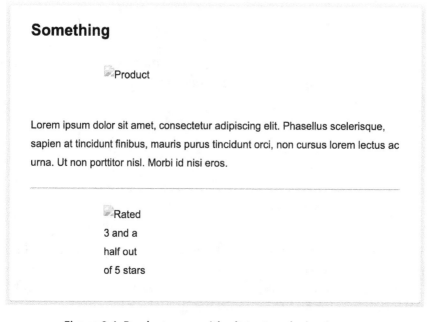

Figure 9.4: Product page with alt text replacing images

In the next section, we are going to look at what we can do to make HTML forms accessible to users.

Semantic HTML

As web developers, one of the biggest opportunities we have for creating accessible web pages is the use of semantically correct HTML. What this means is that we should make use of the correct HTML tags to define the meaning of the content of our page.

Instead of using **div** elements to create the sections of a page, we should use the elements provided by HTML5 – **section**, **header**, **footer**, **main**, **aside**, and **nav**. Also, we should use h1-h6 and use the right level of heading, depending on our document structure and nesting.

You will sometimes hear the structural elements of the web page being called **landmarks**, and that is because they provide structural points of reference that a screen reader can move between on a web page. This helps speed up browsing and navigation for a non-visual user. Keeping this structure well defined and simple will be of real benefit to screen reader users.

We've covered this throughout the previous chapters. In particular, in *Chapter 2, Structure and Layout*, we learned about structuring a web page with the appropriate elements. By structuring the document properly, we can make sure the users browsing our web pages can understand the structure and navigate through it as easily as possible.

Accessible Forms

We learned about HTML forms in *Chapter 4, Forms*. Making forms accessible is very important because forms are one of the key areas where users will interact with your site. This is where users will sign up, send feedback, or pay for goods.

Making forms accessible takes some thinking, and there are fundamental practices we should follow when we are creating forms for a web page. By following these practices, we will have gone a long way toward making accessible forms and web pages.

The techniques we will look at here are as follows:

- Labels and input fields
- Fieldset

A common mistake that's found in forms across the web can be seen here:

```
<p>First name:</p>
<br />
<input type="text" id="first-name" />
```

The following screenshot shows the result of this markup:

First name:

Figure 9.5: Form markup

Visually, this markup may look fine. This may be how we've designed the **input** and **label**. However, the problem for a user of a screen reader is that the text First name: is not associated with the input. A screen reader would not read out the text when navigating to the form field.

We need to provide labels for our form inputs, and we need to make sure that the labels are correctly associated with those input fields.

To make the preceding example accessible, we can do one of two things:

- We can associate a text label with a form **input** field using a **label** element with a **for** attribute. The value of the **for** attribute must match an **id** attribute value on the appropriate form field. For example, the following code provides the label First name and it is associated with the input with the ID "first-name".

```
<label for="first-name">First name</label>
<input type="text" id="first-name" />
```

- We can wrap the input with the label, again creating an association between the two. This would look like this:

```
<label>First name: <input type="text" /></label>
```

In both cases, we now have a form field associated with a label. When a screen reader user navigates to the input element with the ID "`first-name`", the screen reader will read out the label First name. This explains what information the `input` field is expecting without the user relying on a visual association.

When forms get more complex with a large number of `input` fields, they can be difficult to navigate through and understand. We often use white space and visual groupings to break up the form, but that doesn't work if a user is browsing with a screen reader. To meaningfully break the information up into understandable groupings, we can use the `fieldset` and `legend` elements.

A `fieldset` element wraps a set of form `input` fields. You can add a `legend` element nested in the `fieldset` element to provide a textual caption for the `fieldset`. As a screen reader user, this grouping and text caption helps the user understand what they are being asked to input and provides context.

As an example, the following code creates a form with two fieldsets. The first set of fields asks for the user's address and the second set asks the user to choose their favorite color. The use of a `fieldset` helps all users understand that the two fieldsets are grouped separately:

```html
<form>
    <fieldset>
        <legend>Provide your address:</legend>
        <label for="house">House</label>
        <input type="text" id="house" />
        <br />
        <label for="street">Street</label>
        <input type="text" id="street" />
        <br />
        <label for="zipcode">ZIP code</label>
        <input type="text" id="zipcode">
    </fieldset>
    <fieldset>
        <legend>Choose a favorite color:</legend>
        <input type="radio" value="red" id="red" name="color">
        <label for="red">Red</label>
        <input type="radio" value="green" id="green" name="color">
        <label for="green">Green</label>
        <input type="radio" value="blue" id="blue" name="color">
        <label for="blue">Blue</label>
    </fieldset>
</form>
```

The result of the preceding code (with default user agent styling) is shown in the following screenshot:

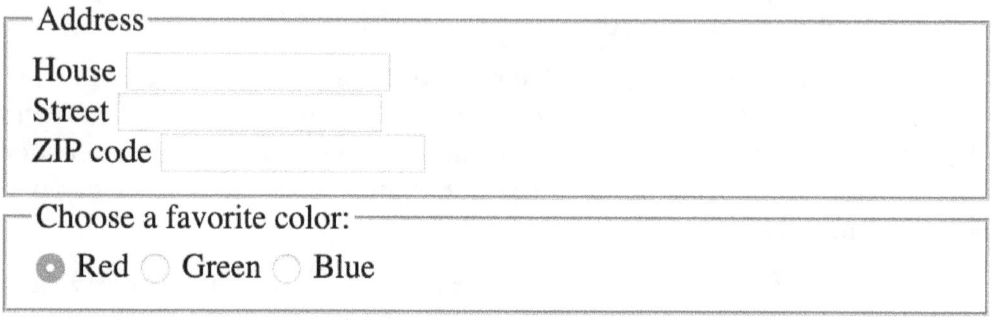

Figure 9.6: Form with fieldsets and legend

Visually, the two different fieldsets are obvious and thematically relate the input fields. For a screen reader, the second **fieldset** might be read as follows:

Choose a favorite color radio button Red, radio button Green, radio button Blue.

Having looked at some of the fundamental techniques we can use to make forms accessible, we will put these techniques into practice in the next exercise by creating an accessible signup form.

Exercise 9.02: Accessible Signup Form

We are going to create an accessible HTML form in this exercise. The form will be a simple example of a signup form and we will focus on making it accessible by providing the appropriate fields with labels and grouping them as fieldsets. By doing this, we will learn how to make an accessible form.

We can see the wireframe of the form in the following diagram. This is a signup form; it is simple and functional, and we will particularly focus on making sure it is accessible by making use of the **label** and **fieldset** elements:

SIGN-UP FORM

ADD USER'S DETAILS

FIRST NAME

LAST NAME

E-MAIL

SET A PASSWORD

PASSWORD

CONFIRM PASSWORD

SIGN-UP

Figure 9.7: Wireframe of the signup form

As we can see from the wireframe, the form will have a heading (**Sign-up Form**), two fieldsets each with a legend (**Add user's details** and **Set a password**), a submit button with the **Sign-up** label, and the five input fields with corresponding labels for **First name**, **Last name**, **E-mail**, **Password**, and **Confirm Password**.

The steps are as follows:

1. First, create and save a file named **Exercise 9.02.html** within the **Chapter09** folder. In **Exercise 9.02.html**, copy the following code to set up a basic HTML page:

```
<!DOCTYPE html>
<html lang="en">
    <head>
        <meta charset="utf-8">
        <title>Exercise 9.02: Accessible signup form</title>
```

```
                    <style>
                    html {
                        box-sizing: border-box;
                    }
                    *, *:before, *:after {
                        box-sizing: inherit;
                    }
                     body {
                        padding: 0;
                        margin: 0;
                        font-family: Arial, Helvetica, sans-serif;
                        font-size: 16px;
                        line-height: 1.7778;

                     }
                    </style>
                </head>
                <body>

                </body>
            </html>
```

2. We want to create a form, so we'll add a class attribute with the "**signup**" value, which we will use for styling. We will add a **h2** heading element to the form with the text content **Sign up form** so that we know what the form is for:

```
                    <form class="signup">
                        <h2>Sign up form</h2>
                    </form>
```

3. Next, we'll add our **input** fields to the form. The information we want to capture is the user's first and last name, an email address, and a password. We also need a button to **submit** the form. We will use the **label** element to provide the appropriate labeling to each field. To make sure each **label/input** pair is on a separate line, we are going to use the **br** element. We have connected each label to an input using the **for** attribute and a corresponding **id** on the **input**:

```
                    <form class="signup">
                        <h2>Sign up form</h2>
                        <label for="first-name">First name:</label>
                        <input type="text" id="first-name" />
                        <br />
                        <label for="last-name">Last name:</label>
                        <input type="text" id="last-name" />
                        <br />
```

```
<label for="email">E-mail:</label>
<input type="email" id="email" />
<br />
<label for="password">Password:</label>
<input type="password" id="password" />
<br />
<label for="confirm-password">Confirm password:
  </label>
<input type="password" id="confirm-password" />
<br />
<button type="submit" class="signup-button">Sign-up
</button>
</form>
```

The following screenshot shows our form at this stage:

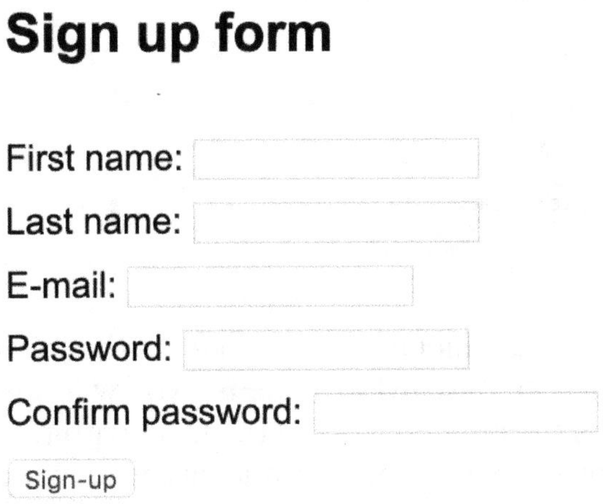

Sign up form

First name:

Last name:

E-mail:

Password:

Confirm password:

Sign-up

Figure 9.8: Unstyled form with inputs and labels

4. We can improve the layout of the form to help make the experience better for visual users. Add the following CSS to the style element in order to add some whitespace around the form, style the Sign-up button so that it's more visually interesting and so that it stands out more, and increase the size of the **input** fields so that they take up a whole line:

```
.signup {
    margin: 1rem;
}
.signup-button {
    -webkit-appearance: none;
    appearance: none;
    padding: .5rem 1rem;
```

```css
        background: #7e00f4;
        color: white;
        font-size: 1rem;
        border-radius: 3px;
        margin: 1rem 0;
        outline: none;
        float: right;
        cursor: pointer;
    }
.signup-button:hover,
.signup-button:focus {
        background: #500b91;
    }
.signup label {
        display: inline-block;
        width: 150px;
    }
.signup input {
        width: 100%;
        height: 2rem;
        margin-bottom: 1rem;
    }
```

If you now right-click on the filename in VSCode on the left-hand side of the screen and select **Open In Default Browser**, you will see the following screenshot that shows the effect of styling the form with the preceding CSS code. It makes the form more visually appealing and easier to work with:

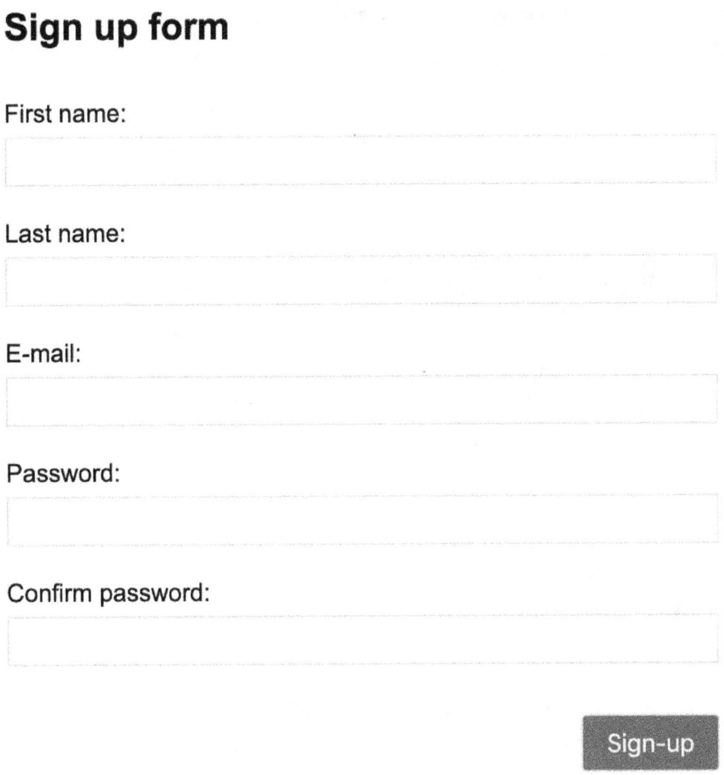

Sign up form

First name:

Last name:

E-mail:

Password:

Confirm password:

Sign-up

Figure 9.9: Styled signup form

5. All our users will benefit from partitioning the form thematically. It will make it easier to understand for both, visual and non-visual browsers alike. We will do this by separating the password and user data portions with the **fieldset** element. We will give each part a **legend** element to describe the information being requested.

    The first **fieldset** is for user details:

```
<fieldset>
    <legend>Add user's details:</legend>
    <label for="first-name">First name:</label>
    <input type="text" id="first-name" />
    <br />
    <label for="last-name">Last name:</label>
    <input type="text" id="last-name" />
    <br />
    <label for="email">E-mail:</label>
    <input type="email" id="email" />
</fieldset>
```

The second **fieldset** is for setting a password:

```
<fieldset>
    <legend>Set a password:</legend>
    <label for="password">Password:</label>
    <input type="password" id="password" />
    <br />
    <label for="confirm-password">
      Confirm password:</label>
    <input type="password" id="confirm-password" />
</fieldset>
```

6. To finish, we add some styling to the **fieldset** element by adding the following declaration to the style element:

```
.signup fieldset {
    margin: 1rem 0;
    background: #f5ffff;
    border: 2px solid #52a752;
}
```

The end result will be similar to what is shown in the following figure. The screenshot shows the final output – an accessible signup form:

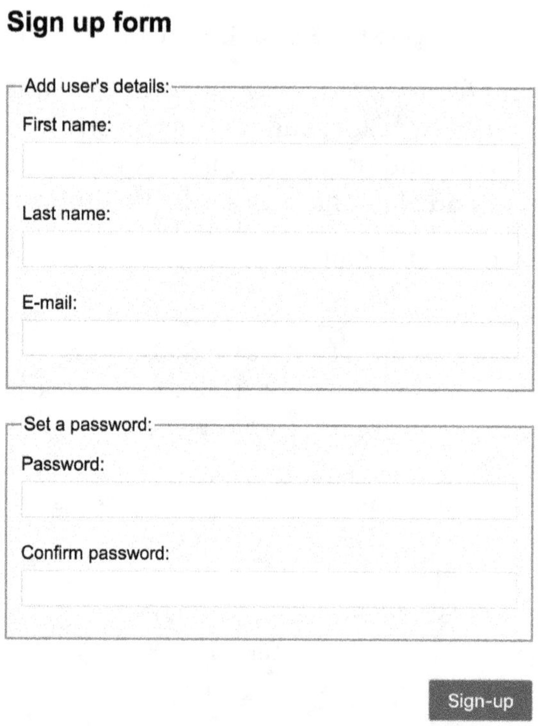

Figure 9.10: The accessible signup form we have created

This form has been labeled and the labels are paired appropriately, to form fields. The form fields are grouped to help a user make sense of them. All users will benefit from these structural grouping and labeling techniques and they will allow screen reader users to sign up successfully.

In this exercise, we have created an accessible form and learned about grouping parts of a form and associating labels with inputs to make sure users can access the form using a screen reader.

In the next section, we will look at keyboard accessibility. The keyboard provides input for many users who cannot use a mouse. As developers, we cannot take mouse usage for granted. Many users find the mouse difficult or impossible to use and many use the keyboard out of necessity or as a preference.

Keyboard Accessibility

For some of our web page's users, a mouse may not be of much use. It requires the ability to follow a visual pointer on screen, a certain amount of sensitivity of touch, and fine motor skills.

Whenever we test our web pages, we should check that we can use the keyboard to get to all the content and that we can interact successfully without using a mouse.

We can navigate through a web page using the following keys on the keyboard:

- *Tab*
- *Shift + Tab*
- *Enter*

To navigate a web page, we use the *Tab* key to cycle through elements on the web page. We can see this if we open the **Exercise 9.02.html** file and use the *Tab* key. *Shift + Tab* will cycle in the reverse order. We can use the two in combination to move back and forth through the elements of the web page.

In the following screenshot, you can see the default focus ring on the `First name` input field:

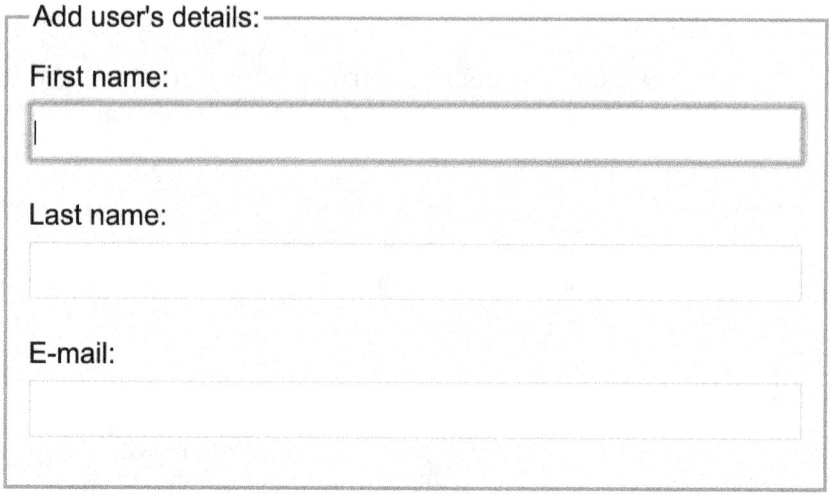

Figure 9.11: Form with focus on the First name field

In the following screenshot, we see that the focus ring has moved to the Last name input. This is because we pressed the *Tab* key:

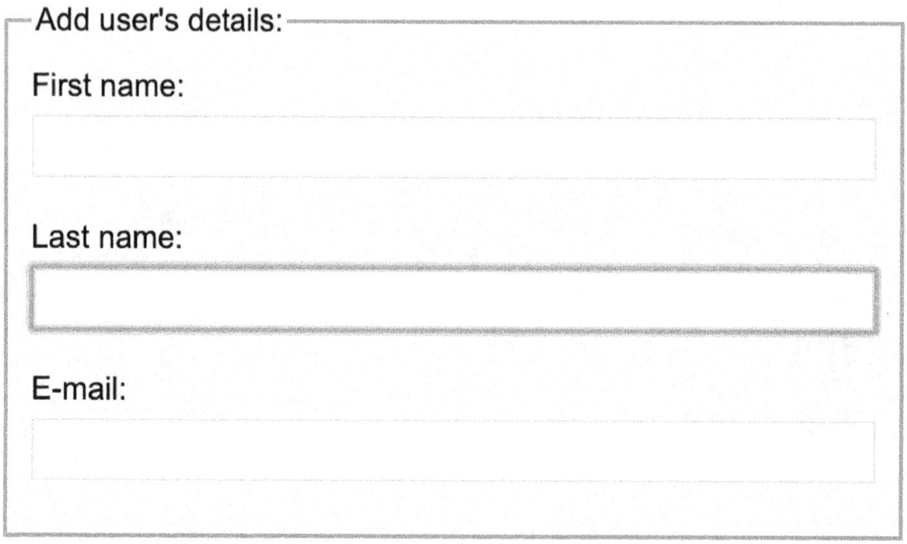

Figure 9.12: Form with focus on the Last name field

To submit a form or interact with a link or button, we can use the *Enter* key. In the case of buttons, we can also use the *Spacebar* to use the button.

To toggle the `radio` and `checkbox` input fields, we can use the *spacebar*.

Checking that styling or JavaScript has not made the tab order confusing is very important for keyboard accessibility.

We should also make sure that our interactive elements, such as buttons and input fields, have some differentiation for when they have focus. We can do this with the `:focus` pseudo-class in CSS.

Making sure you can consistently interact with the elements on your web page using a keyboard helps you know with confidence that a user will not find barriers when using your web page.

Accessible Motion

CSS animations and transitions have become a massive part of creating rich and visually interesting web pages. We learned about animating our web pages in *Chapter 8, Animations*. Motion can add a lot to a web page and can help us highlight content or make content changes more obvious.

With power comes responsibility, and in the case of motion, we should consider whether repeated transitions and animations may cause a user distraction and, in some cases, irritation or make for a difficult user experience.

In particular, repetitive motion, parallax effects, and flicker effects can be really bad for people with vestibular disorders, where such motion can cause severe discomfort and even nausea.

As browsers get better integration with operating systems, they can make use of more fine-grained accessibility configurations. In the case of motion, a media query has been added to most modern browsers, which we can use to check whether the user prefers to have reduced levels of motion. We can combine this, responsively, with our CSS transitions and animations to give a user the appropriate experience.

The media query is `prefers-reduced-motion` and it has two possible values: `reduce` or `no-preference`.

For example, the following CSS would apply an animation only if the user has no preference regarding the **prefers-reduced-motion** media query:

```
<div class="animation">animated box</div>
<style>
    .animation {
        position: absolute;
        top: 150px;
        left: 150px;
    }
    @media (prefers-reduced-motion: no-preference) {
        .animation {
            animation: moveAround 1s 0.3s linear infinite both;
        }
    }
    @keyframes moveAround {
        from {
            transform: translate(-50px, -50px);
        }
        to {
            transform: translate(50px, 100px);
        }
    }
</style>
```

We can test this media query by configuring the accessibility setting in our operating system. To do so, do the following:

- On Mac, go to **System Preferences** > **Accessibility** > **Display** and toggle the **Reduce motion** checkbox.

- On Windows, this preference can be controlled via **Settings** > **Ease of Access** > **Display** > **Show Animations**.

The following screenshot shows the Accessibility Display preferences on Mac with the setting for **Reduce motion** selected:

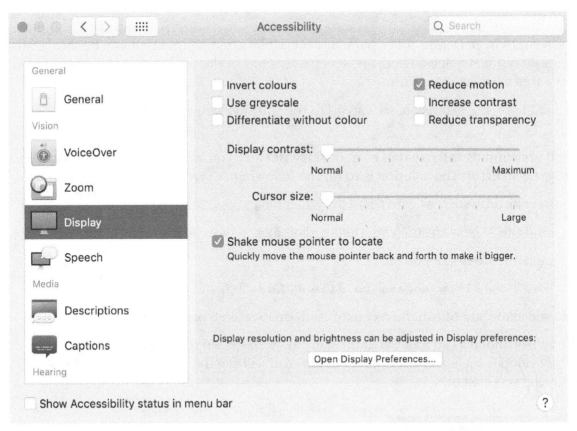

Figure 9.13: Mac display preferences with the Reduce motion toggle

In the next section, we will look at how we can use tools to help understand and audit a web page for accessibility problems.

Accessibility Tools

There are a lot of in-built and third-party tools available that can help with different aspects of accessibility.

A great number of tools have been created that can help us with the accessibility of our web pages. Some are good for diagnosing issues and auditing pages for structural issues, some help us check tab order and keyboard accessibility, and some help us with design decisions around color contrast and text legibility.

In this section, we are going to look at the Axe accessibility checker, which can be used to audit a web page or site to highlight issues in our semantics and HTML structure that may cause accessibility issues.

Axe Tool

Axe is a tool from Deque. It is a popular accessibility testing tool that can be used to flag issues on our web pages. There are several versions of the tool, but we are going to look at the free Chrome extension.

To install the Axe extension, we can go to the Chrome web store at https://packt. live/2WSY7lz.

If you are running the Axe tool on a local file (file://), you may have problems with Axe throwing an error. The solution is to do the following:

1. In your browser, navigate to https://packt.live/35yZYXl.

2. Find the Axe extension (with the heading **axe - Web Accessibility Testing**).

3. Click the **details** button.

4. Locate the **Allow access to file URLs** switch and enable it.

This will allow you to run the Axe extension on local web pages.

The tool is added as an **axe** tab in the Chrome web developer tools. Open the web developer tools and select the Axe tab. You will see the panel shown in the following screenshot:

Figure 9.14: Axe accessibility checker Chrome extension

To run the accessibility checker, we click the ANALYZE button in the left-hand panel. This will check the site we are currently on and report any issues it has detected from analyzing the markup.

The following screenshot shows the reported results from running Axe on a web page. In the left-hand panel, we can see a summary of all the issues that were found. These can be anything from semantic HTML issues, through incorrect levels of nested headings, to color contrast issues. The right-hand panel provides a more detailed description of an issue, including the location of the issue and what can be done to fix the issue:

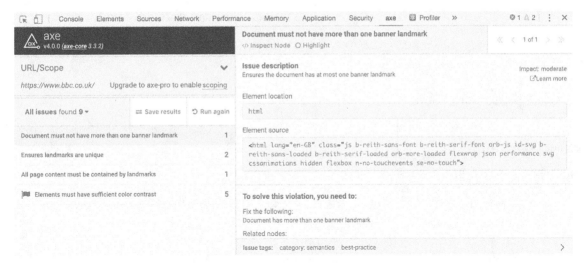

Figure 9.15: Axe accessibility checker report

Taking the preceding screenshot as an example, we can see that four different issues have been reported on in the analyzed page. The issues are as follows:

- Document must not have more than one banner landmark

- Ensures landmarks are unique

- All page content must be contained by landmarks

- Elements must have sufficient color contrast

Several of these refer to landmarks, which we learned about in the *Semantic* HTML section. The results are flagged for the following reasons:

- An HTML page should only have one banner landmark (meaning, the HTML header element or the WAI-ARIA role=**"banner"** attribute) this is because it makes the web page easier for non-visual users to navigate and find their way around.

- A landmark should be unique, which means it should have a unique combination of role and label. This helps to distinguish between landmarks. Again, this improves navigation around the web page.

- All the content of a page should be contained within a landmark (such as a header, footer, nav, or role=**"banner"**), and again, this is important for web page navigation.

- Strong color contrast helps users to distinguish text from its background and helps all visual users, especially those with visual impairments. Axe can detect low contrast levels between a foreground (text color) and background color and will flag such issues.

These are just a few of the many topics that the Axe tool can and will flag for you.

In this section, we've looked at the Axe tool and some of the results it flags. Next, we will try it out on a web page of our own.

Exercise 9.03: Using Axe

In this exercise, we will run the Axe tool on a web page and analyze the findings in the report it creates.

The steps are as follows:

1. First, create and save a file named **Exercise 9.03.html** within the **Chapter09** folder. In **Exercise 9.03.html**, copy and paste in the following code to set up an HTML page. The page is the bare bones of quite a simple layout and contains some landmarks (such as **nav**, **header**, and **footer**). We will open this page in the browser where we have the Axe extension. The following screenshot shows this page:

Site Heading

Content Heading

Lorem ipsum...

Site Links:

Nav 1   Nav 2   Nav 3

Figure 9.16: Site structure

2. Next, open the developer tools and select the Axe tab. Then, run the Axe analysis tool by clicking the **Analyze** button.

3. The results of running the tool on this page will be a set of five flagged issues on the page. The resulting issues are as follows:

    Elements must have sufficient color contrast

    Document must not have more than one banner landmark

    Document must have one main landmark

    Ensures landmarks are unique

    All page content must be contained by landmarks

4. We will look at each issue one by one. The details of the first result are selected by default.

    As shown in the following screenshot, we have several elements with insufficient color contrast. From the issue description, we can see that this is because the top nav has a yellow background and white text. The text is barely visible:

Figure 9.17: Details of the "Elements must have sufficient color contrast" issue

5. Select the next result (Document must not have more than one banner landmark). This issue is caused by having both a header element and a div element with **role="banner"**. Both of these landmarks have the same purpose and role in an HTML page. We can see the suggested solutions in the following screenshot:

Figure 9.18: Details of the "Document must not have more than one banner landmark" issue

6. Next, we select the third result (Document must have one main landmark). This issue is caused by having no main element or element with a **role="main"** included on the page. We have a div with **class="content"**, but we could benefit our users by making this a main element. Again, we see the suggested solution in the following screenshot:

Figure 9.19: Details of the "Document must have one main landmark" issue

7. Next, we select the fourth result (Ensure landmarks are unique). This issue is similar to the second one and is caused by the same problem (that we have a header and an element with **role="banner"**). This means that the two landmarks are not unique. We can see the solution in the following screenshot:

Figure 9.20: Details of the "Ensure landmarks are unique" issue

8. Next, we select the final result (All page content must be contained by landmarks). The cause of this issue is the content being in a div that is not a landmark (the div with class content). Again, we can fix this by changing this div to a main element. We can see the solution in the following screenshot:

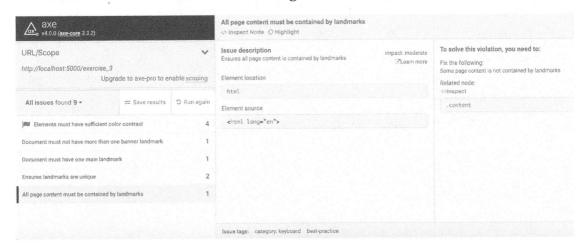

Figure 9.21: Details of the "All page content must be contained by landmarks" issue

By running the Axe tool, we have found some issues early in the development of a web page, when they can be easily solved. The Axe tool has helped us discover these issues quickly and has suggested solutions.

In this exercise, we have seen how the Axe web accessibility checker can help us identify structural, semantic, and visual accessibility issues in our web pages.

To put into practice all the skills we have learned in this chapter, we will finish with an activity in which we will use the Axe tool to diagnose some issues with a web page. We will then fix the issues that have been flagged for us.

Activity 9.01: Making a Page Accessible

You have been asked by a client to look at the accessibility of their product feedback page. The client has received several complaints from users trying to send feedback. The client has provided the source code for the page and wants you to make changes to improve the accessibility of the page.

The steps are as follows:

1. First, make a copy of the **activity_1_inaccessible.html** file within the **Chapter09** folder, and rename it **Activity 9.01.html**. The code for **activity_1_inaccessible.html** is at https://packt.live/2NqdLYY.

2. Run the Axe accessibility checker tool on this page.

3. In the Axe tool's summary, you should see five types of issues flagged by the Axe tool.

4. Fix each of the issues with the techniques you have learned about in this chapter and the hints that are given by the Axe tool.

5. Then, run the Axe tool again to check that the issues have been fixed.

> **Note**
>
> The solution to this activity can be found in page 622.

Summary

In this chapter, we learned about accessibility. We looked at some of the simple techniques we can employ to make our web pages accessible. We also learned about the right ways to use alt text and text descriptions to make the images we embed in a web page accessible to users who can't see the image. We then learned about making forms accessible and about controlling a web page via the keyboard.

Most importantly, we raised awareness about a topic that affects a lot of people and an area that all web developers should champion.

In the next chapter, we will look at how build tools can help us with developing complex modern websites and how we can use preprocessors to expand the capabilities of what we can do in web development.

10

Preprocessors and Tooling

Overview

By the end of this chapter, you will be able to explain CSS preprocessing; use CSS preprocessing techniques; write maintainable code; write SCSS code using nesting, import, control directives, and mixins; and compile SCSS code in CSS in different output styles. This chapter introduces the preprocessing of CSS and the tooling used to compile it, enabling you to understand and write **Sassy Cascading Style Sheets (SCSS)** preprocessed code and compile it in CSS3 successfully, with the aim of saving you time by writing less and achieving more.

Introduction to CSS Preprocessors

In the previous chapters, we learned about many aspects of HTML5 and CSS3 web development, including layouts, themes, responsive web design, media, animation, accessibility, and so on.

Inherently, CSS has its own issues with maintenance, particularly with large or complex projects, not to mention that it's limited in terms of its ability to process functions, logic, and variables. CSS preprocessors came about to address such issues and more, extending the capabilities of CSS to help the developer achieve more with less while keeping the code maintainable.

Now, you may have heard of CSS preprocessors such as **Syntactically Awesome Style Sheets (SASS)**, **Leaner Style Sheets (LESS)**, Stylus, and others. These are all scripting languages that allow you to achieve more in CSS by writing less. All these CSS preprocessors have some logic in common, such as the ability to have variables, nesting styles, math calculations, reusable **mixins**, and so on. The idea of CSS preprocessing is based upon the **Don't Repeat Yourself (DRY)** software development principle, which aims to minimize repetition and create maintainable code.

Each CSS preprocessor has its own syntax, which is then compiled by tooling into CSS that the web browser can understand. In this chapter, we will explore the CSS preprocessor scripting language called SCSS, a version of SASS. The syntax is closer to CSS3 than the likes of most other CSS preprocessors, such as **Stylus** or regular SASS, for example, making a smoother upgrade to your existing CSS3 knowledge. A simple example of this is the curly braces that wrap styles. These are required in SCSS (just like in CSS3) but are not required in preprocessor styling languages such as Stylus and regular SASS.

In this chapter, we're going to look at CSS preprocessors to understand what they are and learn how to write a CSS preprocessed scripting language called SCSS.

Getting Started with Node.js, npm, and SASS

In order to compile SCSS files in CSS, we'll need to install the Node.js software, **Node Package Manager (npm)** (a software package manager for Node.js) and the npm SASS package to run the compilation script itself.

First, we'll install Node.js. If you navigate to https://packt.live/2CmLXhz in your web browser, you will find the download links for the Node.js installation package for your operating system, as shown in the following screenshot:

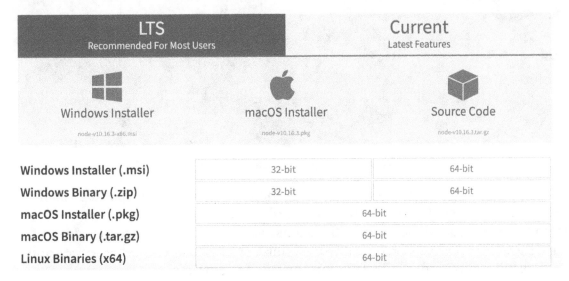

	LTS Recommended For Most Users		Current Latest Features	
Windows Installer (.msi)	32-bit		64-bit	
Windows Binary (.zip)	32-bit		64-bit	
macOS Installer (.pkg)	64-bit			
macOS Binary (.tar.gz)	64-bit			
Linux Binaries (x64)	64-bit			

Figure 10.1: Node.js Downloads page

We recommend selecting the **Long-Term Support (LTS)** download package for your operating system, as this is usually the better supported and more stable version of Node, so you are less likely to encounter installation or runtime issues with this version. Once the installation package has downloaded, you can run it, and you'll see a window similar to the following screenshot:

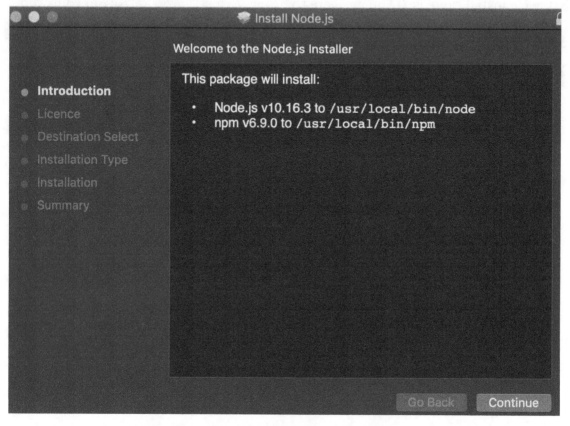

Figure 10.2: Node.js installation package

You will notice that the Node.js LTS version is bundled with the npm package manager too, so this is one less thing to install afterward. Continue through the installation by clicking `Continue`, and then accept the license agreement on the next screen (it's free to use the software). You don't need to customize the installation path. Proceed and select install and wait for the installation package to complete.

Once the installation of Node.js and npm is complete, you will need to open your Terminal window (Command Prompt in Windows). At the prompt, you can run the **node -v** and **npm -v** commands. Both of these should return a version number, as shown in the following figure:

```
Matthews-MacBook-Pro:~ matthewpark$ node -v
v10.16.3
Matthews-MacBook-Pro:~ matthewpark$ npm -v
6.9.0
Matthews-MacBook-Pro:~ matthewpark$
```

Figure 10.3: Checking that Node.js and npm are installed

This means that Node.js and npm are installed on your machine and ready to use.

Next, we will prepare our project folder with the SASS module required to build the exercises and activities in this chapter by creating a new folder on our machine's desktop called **Chapter10**, and then we'll navigate to it in our Terminal window. You can do this with the **cd path/to/directory** command; that is, **cd ~/Desktop/ Chapter10** (the tilde, ~, refers to your user directory, so this command means, access the **Chapter10** subfolder within my desktop folder within my **user** directory). Once we are within the **Chapter10** directory, we can create our **package.json** file. This is a file that contains information about the software dependencies your project requires, so it's manageable by other developers too. To create our **package.json** from within the **Chapter10** directory, we'll need to run the **npm init** command. This will prompt you for the following (note that you can just press the *Enter* key to accept the default value):

- **Package name**: You can leave this set to the default (**Chapter10** or your folder name). You cannot have spaces in the package name, for instance, use **Chapter10**, not **Chapter 10**.

- **Version**: You can leave this set to the default, **1.0.0**.

- **Description**: You can give your project for this chapter a description, for example, **HTML5 & CSS3 Workshop Chapter10 Exercises**.

- **Entry point**: Leave this as the default, **index.js**.

- **Test command, GIT repository, and keywords**: These can be left blank at this point.

- **Author**: You can write your name here.

- **License**: You can leave this blank.

You can then confirm everything is okay with your **`npm init package.json`** configuration values. See this demonstrated in the following screenshot:

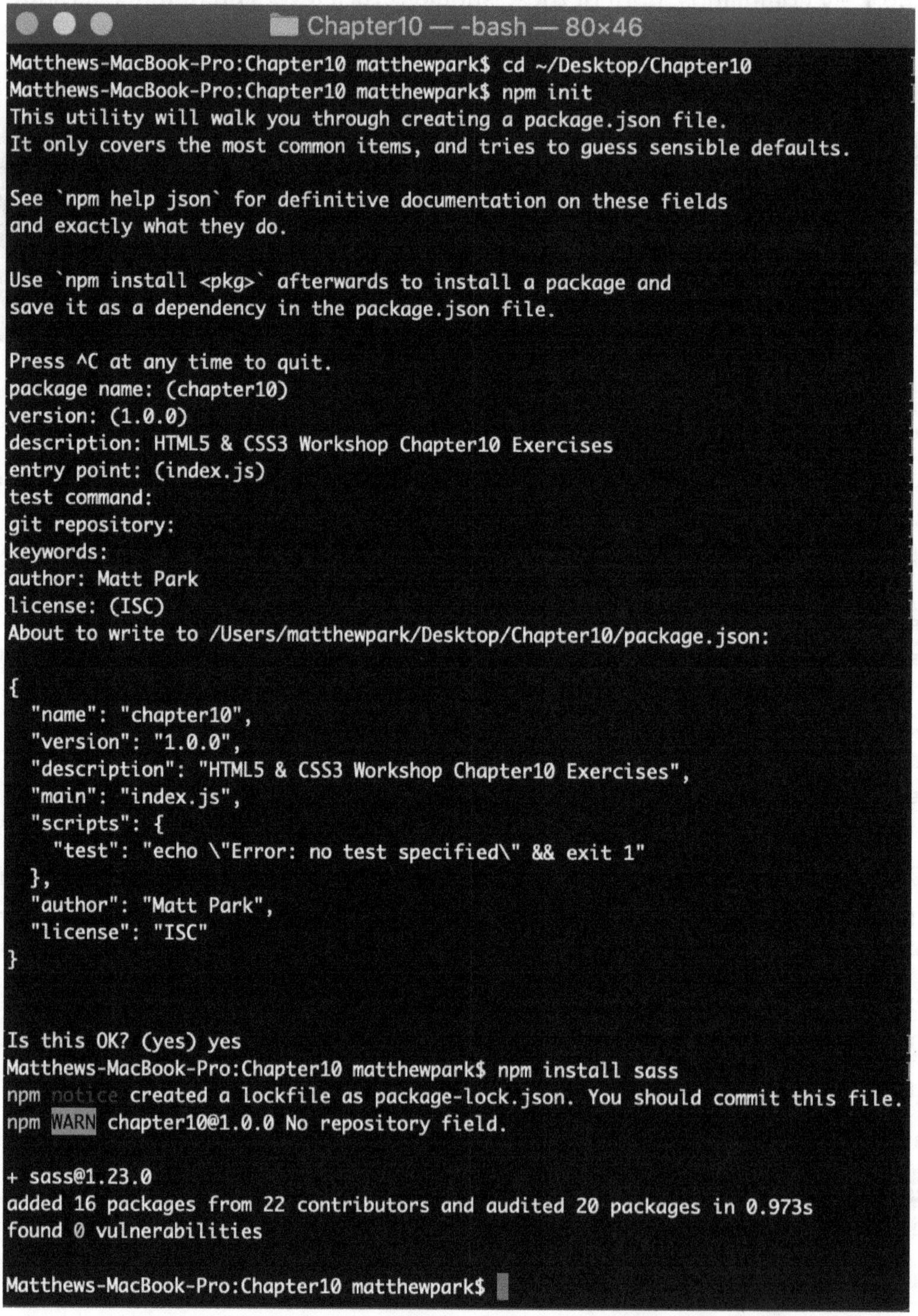

```
Matthews-MacBook-Pro:Chapter10 matthewpark$ cd ~/Desktop/Chapter10
Matthews-MacBook-Pro:Chapter10 matthewpark$ npm init
This utility will walk you through creating a package.json file.
It only covers the most common items, and tries to guess sensible defaults.

See `npm help json` for definitive documentation on these fields
and exactly what they do.

Use `npm install <pkg>` afterwards to install a package and
save it as a dependency in the package.json file.

Press ^C at any time to quit.
package name: (chapter10)
version: (1.0.0)
description: HTML5 & CSS3 Workshop Chapter10 Exercises
entry point: (index.js)
test command:
git repository:
keywords:
author: Matt Park
license: (ISC)
About to write to /Users/matthewpark/Desktop/Chapter10/package.json:

{
  "name": "chapter10",
  "version": "1.0.0",
  "description": "HTML5 & CSS3 Workshop Chapter10 Exercises",
  "main": "index.js",
  "scripts": {
    "test": "echo \"Error: no test specified\" && exit 1"
  },
  "author": "Matt Park",
  "license": "ISC"
}


Is this OK? (yes) yes
Matthews-MacBook-Pro:Chapter10 matthewpark$ npm install sass
npm notice created a lockfile as package-lock.json. You should commit this file.
npm WARN chapter10@1.0.0 No repository field.

+ sass@1.23.0
added 16 packages from 22 contributors and audited 20 packages in 0.973s
found 0 vulnerabilities

Matthews-MacBook-Pro:Chapter10 matthewpark$ 
```

Figure 10.4: Setting up an npm project with npm init

The next step is to install the SASS package in your project folder. We can do this with the **npm install node-sass** command. You can see this demonstrated in the following screenshot:

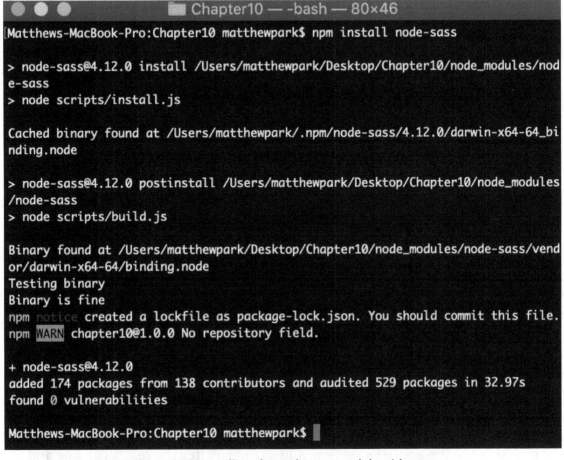

Figure 10.5: Installing the node-sass module with npm

Now, once these installation steps are completed, within your newly created **Chapter10** directory on your desktop, you will find a **package.json** file. The contents will look similar to the following code:

```
{
  "name": "chapter10",
  "version": "1.0.0",
  "description": "HTML5 & CSS3 Workshop Chapter10 Exercises",
  "main": "index.js",
  "scripts": {
    "test": "echo \"Error: no test specified\" && exit 1"
  },
  "author": "Matt Park",
  "license": "ISC",
  "dependencies": {
    "node-sass": "^4.12.0"
  }
}
```

This **package.json** file lives within the root of your project folder and allows other developers to pick up the project and get all the software they need (just the Node SASS module in this case), by running the **npm install** command from the same location as the **package.json** file. You will notice that the generated **package.json** file has created a **scripts** value, with a **test** script line by default. Scripts allow the developer to run commands in the Terminal window and this is where we'll add our own script to compile the SCSS next.

Finally, we'll look at how to compile the SCSS code (a variant of SASS) in CSS code that the web browser can understand to style, we'll need to edit our **package.json** file to add a new script in the **scripts** section. Add a comma after the test script line and add a new line: **"scss": "node-sass --watch scss -o css --output-style expanded"**. This script we've added will use the node sass module to watch the SCSS directory in the project every time an SCSS file is changed (saved), the script will output the compiled CSS file to the CSS directory. The **package.json** file will now look similar to the following code:

```
{
  "name": "chapter10",
  "version": "1.0.0",
  "description": "HTML5 & CSS3 Workshop Chapter10 Exercises",
  "main": "index.js",
  "scripts": {
    "test": "echo \"Error: no test specified\" && exit 1",
```

```
      "scss": "node-sass --watch scss -o css --output-style expanded"
    },
    "author": "Matt Park",
    "license": "ISC",
    "dependencies": {
      "node-sass": "^4.12.0"
    }
}
```

The **--output-style** parameter of the **node-sass** command can have four possible values: **nested** (default), **expanded**, **compact**, and **compressed**. For this chapter, we are going to use **expanded** CSS output, so we can clearly read the output lines of CSS; however, we will demonstrate the difference in the outputs later on in this chapter, once we learn how to write some SCSS.

Now to test this new SCSS script out, inside the project directory, we'll need to create two empty folders named **scss** and **css**. Let's create a new file inside the **scss** folder called **test.scss**. In the **test.scss** file, we can add the following code to test (we'll explain this code in the SCSS *Introduction* section of this chapter):

```
$color-black: #000;
p {
    color: $color-black;
}
```

To test whether this code now compiles, we can run our newly created script in the terminal with the **npm run scss** command. Then, if we save our **test.scss** file, you'll notice it automatically creates the **test.css** file in the **css** directory. See the following screenshot to see it running in Terminal:

```
Matthews-MacBook-Pro:Chapter10 matthewpark$ npm run scss

> chapter10@1.0.0 scss /Users/matthewpark/Desktop/Chapter10
> node-sass --watch scss -o css --output-style expanded

=> changed: /Users/matthewpark/Desktop/Chapter10/scss/test.scss
Rendering Complete, saving .css file...
Wrote CSS to /Users/matthewpark/Desktop/Chapter10/css/test.css
```

Figure 10.6: Running our SCSS script to compile the SCSS code in CSS

You will now see the following compiled CSS code in the **test.css** file:

```
p {
   color: #000;
}
```

You're ready to start using your new setup to code in this chapter. We've now covered how to install Node.js with npm, create our **package.json** projects, and add the node-sass module to its dependencies. We've also covered how to add a new script to our **package.json** file and run the **scss** script to compile our SCSS code in CSS. In the next section, we'll learn about coding SCSS, starting with variables first, which we've just used in our **test.scss** file.

SCSS Introduction

Learning SCSS is an important part of your HTML and CSS journey. It will provide you with invaluable new skills to enable you to achieve more with CSS by writing less code, in theory. It will help your code not only be more maintainable but more transferable between other developers when used correctly. Your SCSS code can be organized and structured in sensible ways that make the code maintainable (we'll discuss this in greater depth in *Chapter 11, Maintainable CSS*). SCSS is a popular and well-adopted scripting language in the CSS preprocessing market and a great place to start learning about the different features CSS **preprocessing** offers.

SCSS is a version of SASS. Compared to regular SASS, SCSS uses syntax formatting similar to that of CSS3, requiring curly braces, semicolons, and other standard CSS coding principles.

If you rename a CSS file from having the **.css** extension to the **.scss** file extension, it will compile without issue (although the output will likely be the same if it's already written in standard CSS), so you can see the syntax is CSS compatible. In the previous section, we learned how to compile a basic SCSS file in CSS using the **npm run scss** command with the script that we created. You'll notice we used a variable in the code, so we'll first look at how we can use variables in SCSS. The syntax for a variable is as follows:

```
$variable-name: value;
```

Now, let's look at a real example of how we can use a variable within a style:

```
$primary-font-color: #000;
h1 {
    color: $primary-font-color;
}
```

At this stage, it may feel like we are writing more to achieve something simple; however, consider how these variables could be used on a larger scale in projects that have hundreds of styles. If you change the primary font color in the future, after coding with a variable in SCSS, then you'll only have one location to update the color reference, and the rest of the styles will inherit the new color through the variable automatically, making the code more maintainable and easier to understand. We'll put this into practice in an exercise.

Exercise 10.01: Using SCSS Variables

In this exercise, we're aiming to put our knowledge of using SCSS variables into practice and create a basic HTML file styled by some CSS compiled from SCSS containing variables for color and font size.

At this point, we are assuming you have already installed Node.js and npm and set up your project (**Chapter10**) folder ready to compile scss. If not, you will need to go back and complete the installation part of this chapter to have your machine and project ready to code this exercise.

We'll start with some starter HTML and CSS and then put the variables in place:

1. Using the **Chapter10** folder, open the file named **Exercise 10.01_Raw.html** from https://packt.live/2p5xjbs and save it as **Exercise 10.01.html**.

2. In the **Chapter10 SCSS** subfolder, the **exercise1.scss** file needs to be created as shown in the https://packt.live/2O6zHXW.

    You can see the code begins with a CSS reset of the default margin and padding for all elements. We then give a base style to the **article** element, with a different background color for every odd element using the **nth-child()** CSS3 selector. We've also created some base font and color styles for the other text elements on the page.

3. Next, navigate in your Terminal window to the **Chapter10** directory; that is, **cd ~/ Desktop/Chapter10**.

   To check you are in the right directory, you can use the **ls** command to list the current path's contents. Check the following figure to see these commands in action:

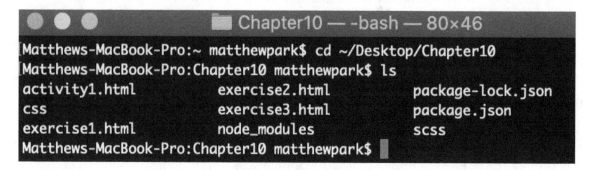

Figure 10.7: Navigating to the directory in Terminal

4. Now we're in the right directory, we can compile the CSS with the SCSS command that we learned about earlier in this chapter. Type the following command at the prompt: **npm run scss**. This command will instruct the node-sass module to watch our SCSS files (including the new **exercise1.scss** file in the **scss** subdirectory) and output the compiled changes to the **exercise1.css** file inside the **css** subfolder upon saving the **scss** file. You can see this in action in the following figure:

```
Matthews-MacBook-Pro:Chapter10 matthewpark$ npm run scss

> chapter10@1.0.0 scss /Users/matthewpark/Desktop/Chapter10
> node-sass --watch scss -o css --output-style expanded

=> changed: /Users/matthewpark/Desktop/Chapter10/scss/exercise1.scss
Rendering Complete, saving .css file...
Wrote CSS to /Users/matthewpark/Desktop/Chapter10/css/exercise1.css
```

Figure 10.8: Running the npm run sass command and then saving the scss file output

At this point, we would expect **`Exercise 10.01.html`** to look like the following figure in the web browser:

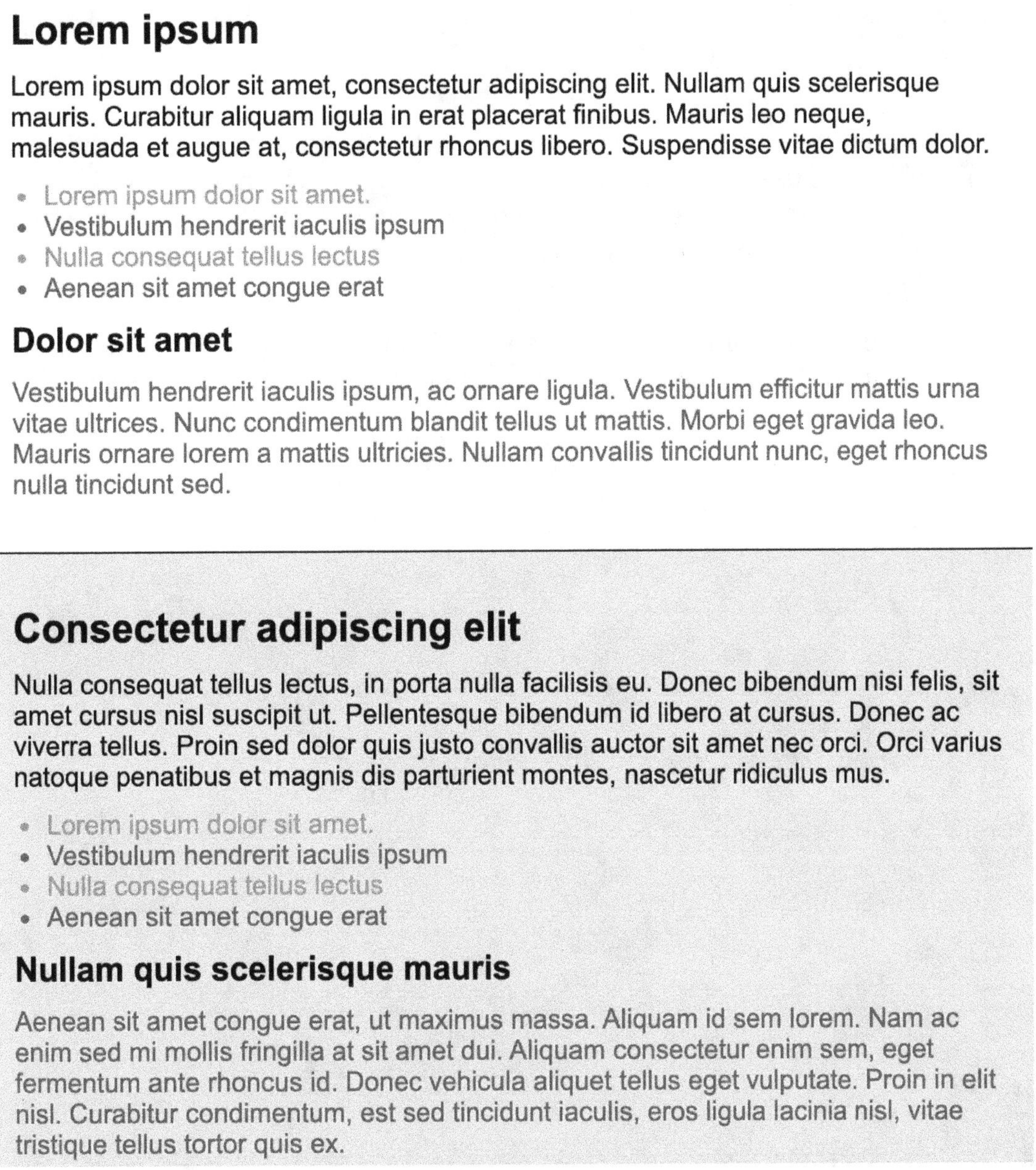

Lorem ipsum

Lorem ipsum dolor sit amet, consectetur adipiscing elit. Nullam quis scelerisque mauris. Curabitur aliquam ligula in erat placerat finibus. Mauris leo neque, malesuada et augue at, consectetur rhoncus libero. Suspendisse vitae dictum dolor.

- Lorem ipsum dolor sit amet.
- Vestibulum hendrerit iaculis ipsum
- Nulla consequat tellus lectus
- Aenean sit amet congue erat

Dolor sit amet

Vestibulum hendrerit iaculis ipsum, ac ornare ligula. Vestibulum efficitur mattis urna vitae ultrices. Nunc condimentum blandit tellus ut mattis. Morbi eget gravida leo. Mauris ornare lorem a mattis ultricies. Nullam convallis tincidunt nunc, eget rhoncus nulla tincidunt sed.

Consectetur adipiscing elit

Nulla consequat tellus lectus, in porta nulla facilisis eu. Donec bibendum nisi felis, sit amet cursus nisl suscipit ut. Pellentesque bibendum id libero at cursus. Donec ac viverra tellus. Proin sed dolor quis justo convallis auctor sit amet nec orci. Orci varius natoque penatibus et magnis dis parturient montes, nascetur ridiculus mus.

- Lorem ipsum dolor sit amet.
- Vestibulum hendrerit iaculis ipsum
- Nulla consequat tellus lectus
- Aenean sit amet congue erat

Nullam quis scelerisque mauris

Aenean sit amet congue erat, ut maximus massa. Aliquam id sem lorem. Nam ac enim sed mi mollis fringilla at sit amet dui. Aliquam consectetur enim sem, eget fermentum ante rhoncus id. Donec vehicula aliquet tellus eget vulputate. Proin in elit nisl. Curabitur condimentum, est sed tincidunt iaculis, eros ligula lacinia nisl, vitae tristique tellus tortor quis ex.

Figure 10.9: Output so far

5. Next, we are going to add our variables to the top of the SCSS files:

```
$color-primary: #000;
$color-secondary: #0000FF;
$color-tertiary: #FF0000;
$fontsize-regular: 16px;
```

We've created three-color variables and one font size variable to use in the next step. Save this in the top of the SCSS file and check your Terminal window to check that it's compiled in CSS. From here, you can see whether there are any compilation errors too, such as a missing semicolon causing a compilation failure, for example.

6. Once the variables are in the file, we can use them within our CSS. Insert the four variables in your CSS to replace any instances of the current values, as shown in the following code:

```
* {
    margin: 0;
    padding: 0;
}
article {
    font-family: Arial, sans-serif;
    padding: 20px;
    background: #DDD;
    border-bottom: 1px solid $color-primary;
}
article:nth-child(odd) {
    background: #FFF;
}
h1 {
    font-size: 24px;
    margin: 10px 0;
    color: $color-primary;
}
h2 {
    font-size: 20px;
```

```
        margin: 10px 0;
        color: $color-primary;
    }
    p {
        margin: 10px 0;
        font-size: $fontsize-regular;
    }
    p.primary {
        color: $color-primary;
    }
    p.secondary {
        color: $color-secondary;
    }
    ul li {
        margin: 0 10px 0 20px;
        font-size: $fontsize-regular;
        color: $color-tertiary;
    }
    ul li:nth-child(even) {
        color: $color-secondary;
    }
```

When you save the SCSS file, check the CSS has compiled in **exercise1.css**. Everything should look the same if you open the HTML file in your web browser.

7. Now we're going to change the variables and observe how the new values compile in CSS. Try assigning new colors to the three color variables and decrease the font size variable too. For example, in the following code, we have updated the variable definitions to new values:

```
$color-primary: #663399;
$color-secondary: #D2691E;
$color-tertiary: #CD5C5C;
$fontsize-regular: 14px;
```

When saving the SCSS file, once compiled in the Terminal window, you will notice in your web browser that if you open/refresh the **Exercise 10.01.html** file, that the colors and font size of the paragraphs and list items have changed. Now check the **exercise1.css** file and you will see how the SCSS has been compiled. In the following figure, you can see the expected output in the browser, followed by the complete SCSS code:

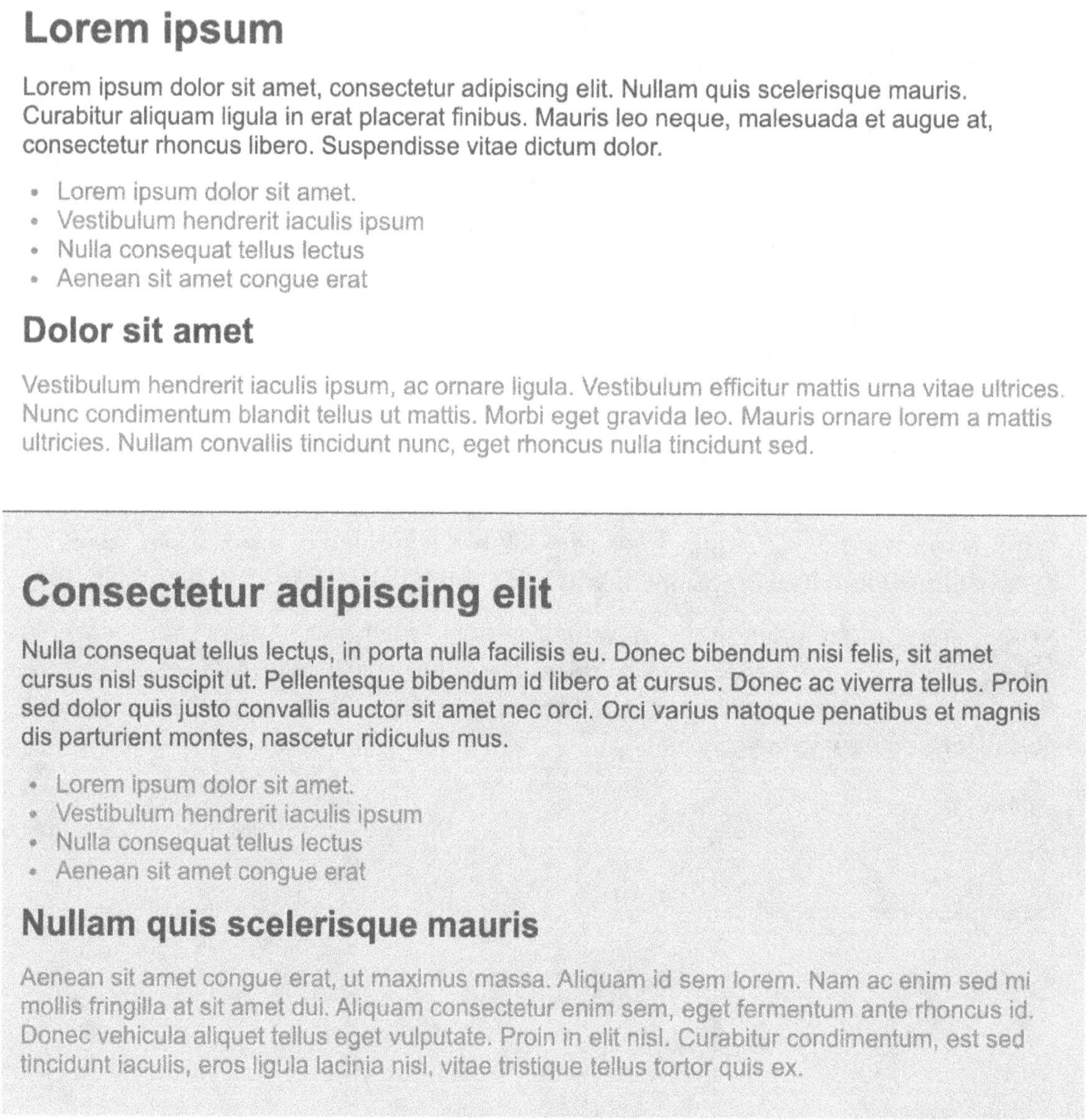

Lorem ipsum

Lorem ipsum dolor sit amet, consectetur adipiscing elit. Nullam quis scelerisque mauris. Curabitur aliquam ligula in erat placerat finibus. Mauris leo neque, malesuada et augue at, consectetur rhoncus libero. Suspendisse vitae dictum dolor.

- Lorem ipsum dolor sit amet.
- Vestibulum hendrerit iaculis ipsum
- Nulla consequat tellus lectus
- Aenean sit amet congue erat

Dolor sit amet

Vestibulum hendrerit iaculis ipsum, ac ornare ligula. Vestibulum efficitur mattis urna vitae ultrices. Nunc condimentum blandit tellus ut mattis. Morbi eget gravida leo. Mauris ornare lorem a mattis ultricies. Nullam convallis tincidunt nunc, eget rhoncus nulla tincidunt sed.

Consectetur adipiscing elit

Nulla consequat tellus lectus, in porta nulla facilisis eu. Donec bibendum nisi felis, sit amet cursus nisl suscipit ut. Pellentesque bibendum id libero at cursus. Donec ac viverra tellus. Proin sed dolor quis justo convallis auctor sit amet nec orci. Orci varius natoque penatibus et magnis dis parturient montes, nascetur ridiculus mus.

- Lorem ipsum dolor sit amet.
- Vestibulum hendrerit iaculis ipsum
- Nulla consequat tellus lectus
- Aenean sit amet congue erat

Nullam quis scelerisque mauris

Aenean sit amet congue erat, ut maximus massa. Aliquam id sem lorem. Nam ac enim sed mi mollis fringilla at sit amet dui. Aliquam consectetur enim sem, eget fermentum ante rhoncus id. Donec vehicula aliquet tellus eget vulputate. Proin in elit nisl. Curabitur condimentum, est sed tincidunt iaculis, eros ligula lacinia nisl, vitae tristique tellus tortor quis ex.

Figure 10.10: The output after variable value changes have been applied

We've now completed the exercise by adding variables in SCSS and then compiling it in CSS, which we then used in our HTML file to display styling in the web browser. You will notice that by using SCSS variables, should we decide to change the primary color on the website in the future, we'd only have to update the variable definition, instead of finding every instance of where the color is used and updating it that way. When you're working with thousands of lines of CSS code, this can be incredibly useful and makes the website code much more maintainable.

Nesting in SCSS

The next topic we're going to cover in SCSS is nesting. This is a very useful feature of SCSS as it prevents repetition in our code, so we don't have to write out the full CSS path for a styling rule each time; we can simply nest it and it gets compiled to the full path. The following is a simple example showing the **article** element (the parent) and the **p** element (the child). You can see the child element is nested inside the parent element:

```
article {
    background: #CCC;
    p {
        color: red
    }
}
```

This would compile to the following CSS code. Upon the compilation of the preprocessed SCSS code, it automatically adds the **article** element tag in front of the **p** element tag styling rule, as shown in the following code:

```
article {
  background: #CCC;
}
article p {
  color: red;
}
```

As you can see, this makes it a lot quicker to code by writing less, but we are still achieving the same output in CSS ultimately. Now, expanding on our example further, in the code, we've now nested an **a** tag within the **p** tag. We have also nested the **hover** state of the **a** tag, which is really useful for keeping the code in the same place, as shown in the following code:

```
article {
    background: #CCC;
    p {
        color: red;
        a {
            color: blue;
            &:hover {
                text-decoration: underline;
            }
        }
    }
}
```

Now, compiling this code in CSS would be done like this:

```
article {
   background: #CCC;
}
article p {
   color: red;
}
article p a {
   color: blue;
}
article p a:hover {
   text-decoration: underline;
}
```

You will notice that, in the succeeding compiled CSS code, the **article** element is added in front of the **p** tag. And also, with the **a** tag, the **article** and **p** tag elements are added before it, to reflect the nesting that is defined in the SCSS code. It's worth noting here that you don't need to nest this deeply normally, unless in exceptional circumstances. The previous example illustrates how nesting works, but three levels deep (for example, **article p a**) should always be the maximum depth, but ideally only one level deep (for example, **article**), or two levels deep (for example, **article p** or **article a**) at most. This will be covered again in *Chapter 11, Maintainable CSS*, where we will explore this topic further.

We can now understand how basic nesting works with SCSS and how it is compiled in CSS. Thinking back to *Chapter 6, Responsive Web Design and Media Queries*, we can incorporate these into nesting to get code together to suit maintainability by keeping the code for each element together in one place. See our example SCSS code with the added media query, which changes the background color of the **article** tag on browser screens of 768 pixels or higher in the following code:

```
article {
    background: #CCC;
    @media (min-width: 768px) {
        background: #FFF;
    }
    p {
        color: red;
        a {
            color: blue;
            &:hover {
                text-decoration: underline;
            }
        }
    }
}
```

We can manage the background color property of the **article** tag for its default (mobile first) and wider screen values in one place thanks to the nested media queries. Once compiled in Terminal, the CSS file will be output with the following contents:

```
article {
  background: #CCC;
}
@media (min-width: 768px) {
  article {
    background: #FFF;
  }
}
article p {
  color: red;
}
```

```
article p a {
    color: blue;
}
article p a:hover {
    text-decoration: underline;
}
```

At this point, now we've written some SCSS, we can demonstrate the difference between the output styles by modifying the **package.json** file and editing the SCSS **node-sass --watch scss -o css --output-style expanded** script. Currently, this is set to expand (as in the preceding output example) to produce clearly readable CSS in our learning. To demonstrate the difference, we will recompile the SCSS again in other output styles:

Nested – This increases the left indent to maintain the visual nesting of the CSS while keeping the closing curly braces on the same line as the same value. See the following code:

```
article {
    background: #CCC; }
    @media (min-width: 768px) {
        article {
            background: #FFF; } }
    article p {
        color: red; }
        article p a {
            color: blue; }
            article p a:hover {
                text-decoration: underline; }
```

Compact – This doesn't have an indent but puts the rule on a single line from start to end, including the curly braces, as shown in the following code:

```
article { background: #CCC; }
@media (min-width: 768px) { article { background: #FFF; } }
article p { color: red; }
article p a { color: blue; }
article p a:hover { text-decoration: underline; }
```

Compressed – This removes most of the whitespace from the CSS output, making the file size as small as possible, which is ideal for a live website where the CSS output doesn't need to be read by a developer, but just needs to be lightweight to minimize the download size for the web browser to read, as shown in the following code:

```
article{background:#CCC}@media (min-width: 768px){article{background:#FFF}}article
p{color:red}article p a{color:blue}article p a:hover{text-decoration:underline}
```

You can use the SCSS script of **node-sass --watch scss -o css --output-style compressed** to achieve this.

Now we've covered nesting in SCSS in some detail, we're going to practice this in the next exercise.

Exercise 10.02: Rewriting Existing CSS with Nested SCSS

In this exercise, we are aiming to take some existing CSS code that we wrote for *Chapter 6, Responsive Web Design and Media Queries, Exercise 6.02, Using Media Queries to Detect Device Orientation*, and convert it into maintainable SCSS, to achieve the same output result in the browser:

1. Get a copy of the code from *Chapter 6, Responsive Web Design and Media Queries, Exercise 6.02, Using Media Queries to Detect Device Orientation*, which you already completed earlier in this book and save the HTML as **Exercise 10.02.html** in your **Chapter10** project folder.

2. Remove the CSS from the style tag in the **<head>** tag of **Exercise 10.02. html** and save this separately in **exercise2.scss** in the **SCSS** subfolder of the **Chapter10** project.

3. Now the CSS has been moved - add the following code within the **<head>** tag of the HTML page:

```
<link href="css/exercise2.css" rel="stylesheet" />
```

4. Next, we'll compile the SCSS we've got so far to check everything is working. Navigate to the project directory in Terminal with the following command:

```
cd ~/Desktop/Chapter10/
```

5. Then, run the build script and save the **exercise2.scss** file to get the **exercise2.css** file generated:

```
npm run scss
```

At this point, we can open **Exercise 10.02.html** in the web browser to check everything has been generated correctly. It should appear as shown in the following figure – loading the compiled CSS:

Home | Nav #1 | Nav #2 | Nav #3

Video Review

Lorem ipsum dolor sit amet, consectetur adipiscing elit. Phasellus scelerisque, sapien at tincidunt finibus, mauris purus tincidunt orci, non cursus lorem lectus ac urna. Ut non porttitor nisl. Morbi id nisi eros.

Donec et purus sit amet odio interdum accumsan eleifend ut sapien. Praesent bibendum turpis non nisl elementum, sed ornare purus semper. In vel sagittis felis.

Suspendisse vitae scelerisque est. Aenean tempus congue lacus vehicula congue. Vivamus congue ligula nec purus malesuada volutpat.

Vestibulum ante ipsum primis in faucibus orci luctus et ultrices posuere cubilia Curae; Curabitur sit amet mattis urna. Morbi id luctus purus, sodales facilisis odio. Orci varius natoque penatibus et magnis dis parturient montes, nascetur ridiculus mus.

Website by Author Name

Figure 10.11: Screenshot of desktop display

6. Now we're going to update the **exercise2.scss** file by using nesting on the **nav** child elements, and the **article** tag and its child elements too. See the following updated SCSS code:

```
* {
  margin: 0;
  padding: 0;
}
header {
    text-align: center;
}
nav {
    padding: 10px;
    ul {
        list-style: none;
        li {
            display: inline-block;
            margin-right: 5px;
            padding-right: 5px;
            border-right: 1px solid black;
            &:last-child {
```

```
                    border-right: 0;
                    padding-right: 0;
                    margin-right: 0;
                }
            a {
                    color: black;
                    text-decoration: none;
                    &:hover {
                        text-decoration: underline;
                    }
                }
            }
        }
    }
article {
    padding: 10px;
    h1, p {
        margin-bottom: 10px;
    }
}
footer {
    padding: 10px;
    background: black;
    text-align: center;
    color: white;
}
@media (orientation: landscape) {
    header {
        display: flex;
        align-items: center;
    }
    nav {
        margin-left: 20px;
    }
    article {
        columns: 100px 2;
    }
}
```

7. The final step in modifying the SCSS is to merge the media queries so they are nested in the appropriate place with the other code so the blocks are grouped together, making them more maintainable. Look at the following updated SCSS code:

```scss
* {
  margin: 0;
  padding: 0;
}
header {
    text-align: center;
    @media (orientation: landscape) {
        display: flex;
        align-items: center;
    }
}
nav {
    padding: 10px;
    @media (orientation: landscape) {
        margin-left: 20px;
    }
    ul {
        list-style: none;
        li {
            display: inline-block;
            margin-right: 5px;
            padding-right: 5px;
            border-right: 1px solid black;
            &:last-child {
                border-right: 0;
                padding-right: 0;
                margin-right: 0;
            }
            a {
                color: black;
                text-decoration: none;
                &:hover {
                    text-decoration: underline;
                }
            }
        }
    }
}
```

```
}
article {
    padding: 10px;
    @media (orientation: landscape) {
        columns: 100px 2;
    }
    h1, p {
        margin-bottom: 10px;
    }
}
footer {
    padding: 10px;
    background: black;
    text-align: center;
    color: white;
}
```

This will produce the following CSS when compiled in **expanded** style output:

```
* {
  margin: 0;
  padding: 0;
}
header {
  text-align: center;
}
@media (orientation: landscape) {
  header {
    display: flex;
    align-items: center;
  }
}
nav {
  padding: 10px;
}
@media (orientation: landscape) {
  nav {
    margin-left: 20px;
  }
}
nav ul {
  list-style: none;
}
```

```
nav ul li {
  display: inline-block;
  margin-right: 5px;
  padding-right: 5px;
  border-right: 1px solid black;
}
nav ul li:last-child {
  border-right: 0;
  padding-right: 0;
  margin-right: 0;
}
nav ul li a {
  color: black;
  text-decoration: none;
}
nav ul li a:hover {
  text-decoration: underline;
}
article {
  padding: 10px;
}
@media (orientation: landscape) {
  article {
    columns: 100px 2;
  }
}
article h1, article p {
  margin-bottom: 10px;
}
footer {
  padding: 10px;
  background: black;
  text-align: center;
  color: white;
}
```

The output in the web browser should still look the same as *Chapter 6, Responsive Web Design and Media Queries, Exercise 6.02, Using Media Queries to Detect Device Orientation*, as shown in *Figure 10.11*.

We are now able to convert an existing CSS script and rework it into maintainable SCSS code using nesting. In the next section, we're going to look at import and control directives in SCSS.

Import, Control Directives, and Mixins in SCSS

In SCSS, we can import other SCSS files so that we can group the styles into their own files, helping to make a project more maintainable (we'll cover this in more detail in *Chapter 11, Maintainable CSS*). If you add an underscore in front of the SCSS filename, then **npm** won't create a CSS file for it, and you can import it into another SCSS file that gets compiled into a CSS file.

The syntax for importing another SCSS file (for example, **'_filename.scss'**) into your main file is **@import 'filename';**. To see this in an example, look at the following code:

_reset.scss

```
* {
  margin: 0;
  padding: 0;
}
ul, li {
  list-style: none;
}
```

build.scss

```
@import 'reset';
header {
  background: #CCC;
  a {
    color: #000;
    &:hover {
      text-decoration: underline;
    }
  }
}
```

This would output with the **_reset.scss** file already compiled and merged into **build.css** as a single file outputted:

```
* {
  margin: 0;
  padding: 0;
}
```

```
ul, li {
  list-style: none;
}
header {
  background: #CCC;
}
header a {
  color: #000;
}
header a:hover {
  text-decoration: underline;
}
```

You can also use control directives within SCSS. These are constructed with an **if** statement using variables; for example, if we wanted a variable called **debug** and we gave it a **true** or **false** value, the syntax would look like this:

```
$debug = true;
@if ($debug) {
    div {
        border: 1px dashed red;
    }
}
```

This code would display a dashed red border on all **div** elements when **$debug** is equal to **true**. If you wanted to give **$debug** a text value instead of a Boolean (**true/false**) one and also utilize the **else if** and **else** statements, then you could write it like so in SCSS:

```
$env: 'test';
div {
  @if ($env == 'dev') {
    border: 1px dotted red;
  } @else if ($env == 'test') {
    border: 1px dotted yellow;
  } @else if ($env == 'live') {
    border: 1px dotted black;
  } @else {
    border: none;
  }
}
```

You can see in the preceding example, we are setting the **$env** variable to a text value this time. With our code example, it will change the border color to red, yellow, or black, depending on the value of the **$env** variable. If **$env** isn't equal to the **'dev'**, **'test'**, or **'live'** values, then the **div** element will have no border applied to it. With SCSS, we can create powerful **if** statements to help us do more by writing less.

Next, we're going to look at mixins with SCSS. These are powerful ways of including reusable code to write less and do more. The new CSS3 **column-count** feature requires browser vendor prefixes to ensure we are catering for the best cross-browser compatibility. Let's look at an example of a simple mixin using the browser vendor prefixes shown in the following code:

```scss
@mixin columns($count) {
  -webkit-column-count: $count;
  -moz-column-count: $count;
  column-count: $count;
}
article {
  @include columns(2);
}
```

The preceding code demonstrates how we can use a mixin in SCSS to write less. Now, every time we need to use the CSS3 feature columns, we could just write **@include columns(x)**, where x represents how many columns we'd like to have. The mixin will automatically output the browser vendor prefix variations of **column-count** in the compiled CSS every time we call the columns mixin in SCSS. You can see an example of the CSS3 column count feature being used in the following screenshot:

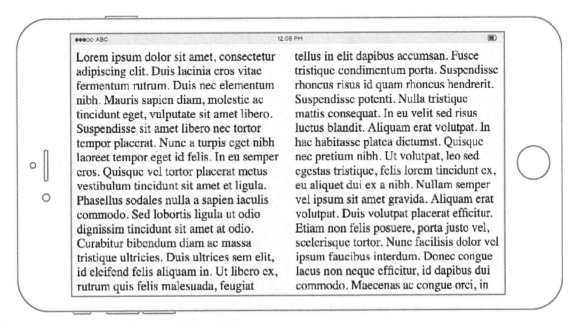

Figure 10.12: Demonstration of column count being used

Now that we have described the basics of control directives and mixins, we will implement them in the following exercise to cement our understanding.

Exercise 10.03: Using SCSS Mixins and Control Directives

The aim of this exercise is to use our knowledge of SCSS mixins and **control directives** to change how the logo on the page is displayed by changing the CSS3 **transform** property:

1. We're going to start by creating **Exercise 10.03.html** in the project folder we created for this chapter. We can use this HTML for the file:

```
<!DOCTYPE html>
<html>
<head>
  <meta name="viewport" content="width=device-width,
    initial-scale=1" />
  <title>Chapter 10: Exercise 10.03</title>
  <link href="css/exercise3.css" rel="stylesheet" />
</head>
<body>
    <div class="transform">
        <img src=
          "https://dummyimage.com/200x100/000/fff&text=Logo"
          alt="" />
    </div>
</body>
</html>
```

2. Next, we're going to create **exercise3.scss** within the SCSS subfolder in our project folder. We can start with the following code:

```
@mixin transform($value) {
  -webkit-transform: rotate($value);
     -ms-transform: rotate($value);
         transform: rotate($value);
}
.transform {
    @include transform(70deg);
    display: inline-block;
    position: relative;
    top: 100px;
    left: 100px;
}
```

3. If you open Terminal and navigate to the project directory, then run the **npm run scss** command as we did in the earlier exercises. Then, once you save the **exercise3.scss** file, it should generate **exercise3.css**. If you view **Exercise 10.03.html**, you should see a rotated logo, as shown in the following figure:

Figure 10.13: Rotated logo file

4. Then we're going to extend the **mixin** transform, reusing what we already have. But to allow us to use a different **transform** property, we're going to achieve this by adding a control directive and another parameter to the **mixin transform** for the property. Change **exercise3.scss** to the following SCSS code and allow the file to recompile to CSS in your Terminal window:

```scss
@mixin transform($property, $value) {
    @if ($property == 'skewY') {
        -webkit-transform: skewY($value);
        -ms-transform: skewY($value);
        transform: skewY($value);
    } @else if ($property == 'scaleY') {
        -webkit-transform: scaleY($value);
        -ms-transform: scaleY($value);
        transform: scaleY($value);
    } @else {
        -webkit-transform: rotate($value);
        -ms-transform: rotate($value);
        transform: rotate($value);
    }
}
.transform {
    @include transform(rotate, 70deg);
    display: inline-block;
    position: relative;
    top: 100px;
    left: 100px;
}
```

As you can see in the preceding code, we've added new properties we can use in the **transform** mixin. It's worth noting, although we've used the browser prefixes for the **transform** property as an example here, as you develop your CSS preprocessing knowledge further, you may want to experiment with a Node.js module called **Autoprefixer**. It can be installed with npm and it does require some more dependencies and a little bit of configuration in your **package.json** file to create the script to run, which is outside of the scope of this book. However, Autoprefixer can automatically add any browser prefixes suggested upon compilation for the last X amount of browser versions, where X is user-defined. Next, we'll write some more HTML elements, so we can demonstrate these in action.

5. We're going to add two additional logos to the HTML, to use our reusable mixin. See the following output HTML:

```
<!DOCTYPE html>
<html>
<head>
  <meta name="viewport" content="width=device-width,
    initial-scale=1" />
  <title>Chapter 10: Exercise 10.03</title>
  <link href="css/exercise3.css" rel="stylesheet" />
</head>
<body>
    <div class="transform transform1">
        <img src=
          "https://dummyimage.com/200x100/000/fff&text=Logo"
          alt="" />
    </div>
    <div class="transform transform2">
        <img src=
          "https://dummyimage.com/200x100/000/fff&text=Logo 2"
          alt="" />
    </div>
    <div class="transform transform3">
        <img src=
          "https://dummyimage.com/200x100/000/fff&text=Logo 3"
          alt="" />
    </div>
</body>
</html>
```

6. We also need to update the SCSS to handle these new classes for **transform** objects 1 to 3. Look at the following updated SCSS code:

```scss
@mixin transform($property, $value) {
    @if ($property == 'skew') {
        -webkit-transform: skew($value);
        -ms-transform: skew($value);
        transform: skew($value);
    } @else if ($property == 'scale') {
        -webkit-transform: scale($value);
        -ms-transform: scale($value);
        transform: scale($value);
    } @else {
        -webkit-transform: rotate($value);
        -ms-transform: rotate($value);
        transform: rotate($value);
    }
}
.transform {
    display: inline-block;
    position: relative;
    top: 100px;
    left: 100px;
}
.transform1 {
    @include transform(rotate, 70deg);
}
.transform2 {
    left: 150px;
    @include transform(skew, 50deg);
}
.transform3 {
    left: 250px;
    @include transform(scale, 1.5);
}
```

7. Once saved and compiled if you view **Exercise 10.03.html** in the web browser again, you should see an output that is the same as the following figure:

Figure 10.14: Logo outputs

The CSS output of the exercise would look as follows:

```
.transform {
  display: inline-block;
  position: relative;
  top: 100px;
  left: 100px;
}
.transform1 {
  -webkit-transform: rotate(70deg);
  -ms-transform: rotate(70deg);
  transform: rotate(70deg);
}
.transform2 {
  left: 150px;
  -webkit-transform: skew(50deg);
  -ms-transform: skew(50deg);
  transform: skew(50deg);
}
.transform3 {
  left: 250px;
  -webkit-transform: scale(1.5);
  -ms-transform: scale(1.5);
  transform: scale(1.5);
}
```

We've been able to create a new mixin and use control directives, allowing the mixin to be reused multiple times for different purposes. Each time we've run the mixin, we've achieved a different **transform** property, and haven't had to type out the browser vendor prefixes each time, as this is done from within the mixin itself. Using a mixin in this way is also good for future maintainability. To explain this better, if we didn't need to use the **-webkit-transform** browser vendor prefix anymore, we could simply remove this output from the mixin, and wouldn't need to go through the entire collection of SCSS files removing every instance; we would just update our mixins. The same would apply if we needed to add a new browser vendor prefix too.

Loops in SCSS

Another way we can use SCSS is to create loops to create multiple CSS classes very quickly and efficiently. This makes creating lots of similar classes more maintainable as well, as opposed to writing them all by hand in normal CSS. There are three types of loops in SCSS: **for**, **each**, and **while**. We'll look at the **for** loop first.

The **for** loop uses the **@for $variable from X to Y {}** syntax, and we can create loops that will output numbers X to Y. Look at the following example:

```
@for $num from 1 through 4 {
    .col-#{$num} {
        column-count: $num
    }
}
```

This SCSS **for** loop will loop through numbers **1** to **4**, outputting the number in the class name and the column count property value too. Take a look at the CSS – this generates the following code:

```
.col-1 {
  column-count: 1;
}
.col-2 {
  column-count: 2;
}
.col-3 {
  column-count: 3;
}
.col-4 {
  column-count: 4;
}
```

As you can see in the preceding CSS code, the SCSS **for** loop code has created 15 lines of CSS code from 4 lines of SCSS code, so it's easier to maintain any future changes, especially if your loop contains a high number of iterations.

The next type of loop is the **each** loop. We can use this to loop through each item in a list using the **@each $item in $list {}** syntax. Look at the following example:

```scss
$fruits: apple pear orange kiwi pineapple melon strawberry;

@each $fruit in $fruits {
    .image-#{$fruit} {
        background: url("images/#{$fruit}.png") no-repeat;
    }
}
```

The SCSS for the **each** loop will loop through the list of fruits, outputting each fruit in the class name and the image path value in the **background** property. Take a look at the CSS – this generates the following code:

```css
.image-apple {
  background: url("images/apple.png") no-repeat;
}
.image-pear {
  background: url("images/pear.png") no-repeat;
}
.image-orange {
  background: url("images/orange.png") no-repeat;
}
.image-kiwi {
  background: url("images/kiwi.png") no-repeat;
}
.image-pineapple {
  background: url("images/pineapple.png") no-repeat;
}
.image-melon {
  background: url("images/melon.png") no-repeat;
}
.image-strawberry {
  background: url("images/strawberry.png") no-repeat;
}
```

As you can see in the preceding CSS code, the SCSS **for** loop code has created 27 lines of CSS code from 6 lines of SCSS code, and just like with the **for** loop, it becomes easier to maintain any future changes.

The next loop type is the **while** loop. The syntax for this is @**while $variable condition value {}**, so the code will loop while a condition is **true**. Take a look at the following example in SCSS:

```
$box: 25;
@while $box > 0 {
    .box-#{$box} {
        width: $box + px;
    }
    $box: $box - 5;
}
```

This SCSS **while** loop will continue to run while the **$box** variable is greater than zero, and within the **while** loop, we are deducting five from the **$box** variable every time the loop runs. Before the loop runs, we assign the **$box** variable with an initial value of 25, so the code will loop 5 times before stopping. Take a look at the resulting CSS shown in the following code:

```
.box-25 {
  width: 25px;
}
.box-20 {
  width: 20px;
}
.box-15 {
  width: 15px;
}
.box-10 {
  width: 10px;
}
.box-5 {
  width: 5px;
}
```

As you can see in the compiled CSS, the loop has output 19 lines, but could easily output an enormous amount of CSS classes if the **$box** variable was higher to start with initially, but still only using 7 lines of SCSS to create it.

All three types of loops serve their own purpose and can be used in different ways. We're now going to put this into practice in our next exercise.

Exercise 10.04: Loops in SCSS

The aim of this exercise is to create a loop in SCSS to output a list of team members with images – by creating the CSS classes for them – for a "Meet the team" web page:

1. Let's start by creating **Exercise 10.04.html** in your **Chapter10** folder. Save it with the following HTML:

```
<!DOCTYPE html>
<html>
<head>
  <meta name="viewport" content="width=device-width,
    initial-scale=1" />
  <title>Chapter 10: Exercise 10.04</title>
  <link href="css/exercise4.css" rel="stylesheet" />
</head>
<body>
    <h1>Meet The Team</h1>
    <div class="team-members">
        <div class="member member-david">
            <h2>David</h2>
            <p>Founder and CEO</p>
        </div>
        <div class="member member-laura">
            <h2>Laura</h2>
            <p>Associate Director</p>
        </div>
        <div class="member member-matt">
            <h2>Matt</h2>
            <p>Head of Project Management</p>
        </div>
        <div class="member member-natalie">
            <h2>Natalie</h2>
            <p>Office Manager</p>
        </div>
        <div class="member member-sarah">
            <h2>Sarah</h2>
            <p>Head of Development</p>
        </div>
        <div class="member member-steve">
            <h2>Steve</h2>
```

```
            <p>FrontEnd Developer</p>
        </div>
        <div class="member member-tahlia">
            <h2>Tahlia</h2>
            <p>BackEnd Developer</p>
        </div>
        <div class="member member-will">
            <h2>Will</h2>
            <p>SEO Specialist</p>
        </div>
    </div>
</body>
</html>
```

2. Next, you'll need to download the images (8 images in total). You can find these here: https://packt.live/2CkSJV7. Save these in a folder called **images** in your **Chapter10** folder.

3. Then, we'll create our **scss/exercise4.scss** file, with the following contents to get started:

```
* {
    margin: 0;
    padding: 0;
}
body {
    font-family: Arial, sans-serif;
}
h1 {
    margin: 20px;
}
.team-members {
    display: grid;
    grid-template-columns: repeat(4, 1fr);
    grid-gap: 20px;
    margin: 20px;
    .member {
        position: relative;
        padding-top: 80%;
        h2,
```

```
        p {
            padding: 0 10px;
            color: #FFF;
            text-shadow: 0px 0px 10px rgba(0, 0, 0, 1);
        }
        p {
            padding-bottom: 10px;
        }
    }
}
```

4. Now we need to add the loop code to our SCSS file to output the images to our grid items. Look at the following code, which will be appended to the `.member` selector before it closes:

```
$members: david, laura, matt, natalie, sarah, steve, tahlia, will;
@each $member in $members {
    &-#{$member} {
        background: url(../images/#{$member}.jpg) no-repeat;
        background-size: 100% auto;
    }
}
```

As you can see in this code, we are looping through a list of names for each team member, and outputting these as a selector in this format: `.member-<name>`, with the background image selected.

5. Compile your SCSS file by going to the **Chapter10** directory in Terminal, then run the **npm run scss** command, and then save your SCSS file to generate **css/exercise4.css**. The output CSS in the **css/exercise4.css** file can be found in the https://packt.live/2O3DGok link.

You can see how the **each** loop in SCSS has iterated through the list of names to create all the classes and background image paths.

6. Open **Exercise 10.04.html** in your web browser and look at your web page. It should look like the following figure:

Meet The Team

Figure 10.15: Output for loop in SCSS

In this exercise, we have learned how to put the SCSS **each** loop into action to quickly create new styles that are not only maintainable but also save writing more and copying and pasting similar code over and over again.

We've now covered SCSS in all the core areas, including how to compile it on your computer; writing variables; nesting selectors; and using file imports, control directives, mixins, and loops. You can use SCSS to save time by writing less and achieving more, and can also use SCSS to produce more maintainable CSS code for your web pages. We'll now look at applying your knowledge of SCSS in the activity that follows by updating your video store home page to use SCSS.

Activity 10.01: Converting the Video Store Home Page into SCSS

For this activity, we're going to take our responsive video store home page code from *Chapter 6, Responsive Web Design and Media Queries, Activity 6.01, Refactoring the Video Store Product Cards into a Responsive Web Page,* convert the CSS into preprocessed SCSS, and then compile it into minified CSS code:

1. Get a copy of your code from *Chapter 6, Responsive Web Design and Media Queries, Activity 6.01, Refactoring the Video Store Product Cards into a Responsive Web Page,* and save it as a new file within your **Chapter10** project folder called **Activity 10.01.html**.

2. Remove the CSS code from within the **<style>** tags and save this as **activity1. scss** within your **SCSS** subfolder in the project. Add a link to the **activity1.css** file in your document, from the **CSS** subfolder of your project.

3. Edit the SCSS code to add color variables, utilize nesting, and move the media queries to inside the element tags in the appropriate places, instead of having them at the end of the file.

4. Edit your **package.json** in your project to have the SCSS script output in compressed format.

5. Using Terminal, compile your SCSS code into a minified CSS code.

6. The end result should look the same as *Chapter 6, Responsive Web Design and Media Queries, Activity 6.01, Refactoring the Video Store Product Cards into a Responsive Web Page,* but we are now coding it with SCSS and outputting this into compiled and compressed (minified) CSS. Look at the following expected output figure for reference:

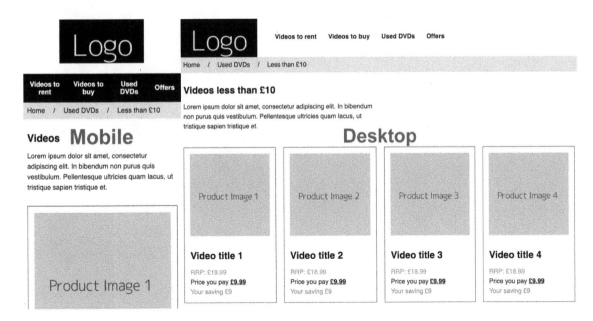

Figure 10.16 Output of mobile and desktop layouts

> **Note**
>
> The solution to this activity can be found in page 624.

Summary

In this chapter, we introduced CSS preprocessors, developing an understanding of how we configure the software to run the compilation of these on our machine with Node.js, npm, and with the node-sass module. We've covered how to create a project **package. json** file and write a script command, which we can run to compile the CSS.

We've learned about the SCSS preprocessed scripting language and its different output styles, and how we can write SCSS code to create variables; use nesting, mixins, and control directives; and import other SCSS files.

Using this knowledge, you should now be able to write and compile SCSS code to create CSS for your website projects.

In the next chapter, we will learn about what it means to create more maintainable CSS, looking at ways to implement this and how it can benefit developers working on a project.

11

Maintainable CSS

Overview

By the end of this chapter, you will be able to write CSS code using the BEM approach and explain semantic CSS; describe componentization and rule grouping and apply these principles while writing CSS code; write maintainable SCSS with sensible file and folder structures; and implement CSS best practices to create better maintainable code. This chapter introduces the concept of maintainable CSS in terms of what it looks like and how we go about creating it. With the knowledge that's given in this chapter, you will be able to create more manageable CSS codebases in SCSS and make your web projects more future-ready for changes and other developers to pick up easily.

Introduction to Maintainable CSS

In the previous chapters, we learned about a wide variety of different aspects of HTML5 and CSS3, including everything you need to know as a beginner to write HTML5 and CSS3 yourself and develop your first website, but how can we help keep a website maintainable? How do we help prevent the common pitfalls of managing large codebases in CSS? How do we write code that allows a team of developers to work on a website project at the same time and that's also friendly to future developers?

In this chapter, we're going to explore what maintainable CSS looks like, how to write it, and why we use it. Let's start by understanding why we would want to write maintainable CSS.

Say you are working with a large **codebase** that includes thousands of lines of CSS and the client wants to change a button style that's used throughout the website. Easy, right? Well if we had written maintainable CSS for this project, then we could change this button style in one of the SCSS files that the button component lived in, recompile the CSS, and then the job would be done. But if we didn't write maintainable CSS, updating this button style could mean updating many references in different places to ensure the button is changed globally across the website. This could involve writing overrides to force the button style to change and could end up being very messy and rather unmanageable. We could end up with even more lines of CSS overall for the browser to download, some lines of which probably wouldn't even be required anymore because they're related to the old button style. However, we wouldn't necessarily know where to remove it from if the CSS hasn't been maintained well, causing the codebase to become very bloated with old styles over time.

Good CSS coding practice when working in a team on a website development project is to ensure that any developer can pick up the code where you left off, with all the developers maintaining the same CSS standards to ensure that the codebase makes sense and that modifications can be made quickly, safely, and efficiently to minimize **code debt** (which, in short, means minimizing unnecessary code). We can understand how projects become unmaintainable if we consider reasons such as budget limitations or other pressures such as tight deadlines, which can make some developers write code that causes CSS code debt, or worse – unmaintainable CSS code.

Block, Element, and Modifier

The first stage of writing maintainable CSS is to write and use meaningful elements identifying rules. A maintainable approach for this is to use **Block**, **Element**, and **Modifier (BEM)**. BEM is a semantic approach to identifying elements and applying CSS rules to them.

Semantic CSS means naming elements according to what they are, rather than what they look like. For example, using a class of `.article-box` instead of `.blue-box` on an HTML `div` element would allow the color property of the `div` to be changed later and it would still make sense semantically, given that the color wasn't associated with the selector name in the first place. Therefore, this helps make the code more maintainable in the future.

Let's explain each block that makes up BEM.

Block

This is the main entity and is meaningful on its own. This could be a header (`.header`), navigation (`.nav`), field (`.field`), article (`.article`), and so on.

Element

This would be a child of a block, based on a semantic meaning with block, and it wouldn't make sense by itself. Examples include header logo (`.header__logo`), navigation item (`.nav__item`), field label (`.field__label`), and article paragraph (`.article__paragraph`).

Modifier

This is a variation in a block or element and can be used to change its appearance or behavior. Examples include a header in a blue theme variation (`.header--blue`), a navigation item that's highlighted (`.nav__item--highlighted`), a field with a checkbox input with the state "checked" (`.field__input--checked`), and a button with a state of success (`.button--success`).

So, what does BEM look like when written to CSS? The syntax is as follows. To declare a block, you would write just the block name:

```
.block {}
```

To modify the appearance or behavior of the block, you would add a **modifier**:

```
.block--modifier {}
```

To style an element of a block, you would use the following syntax:

```
.block__element {}
```

To modify the appearance or behavior of an element, you would add a modifier:

```
.block__element--modifier {}
```

All of these CSS identifiers refer to the HTML classes that have been assigned and the styles that are inherited this way. Let's look at the HTML syntax of these rules:

```html
<div class="block block--modifier">
    <div class="block__element block__element--modifier">
        <!-- content here -->
    </div>
</div>
```

Now, let's look at an example of BEM in action. First, we'll take a simple navigation structure and apply BEM to it. Let's start with the HTML:

```html
<nav class="nav">
    <ul class="nav__list">
        <li class="nav__item">
            <a href="#" class="nav__link">Home</a>
        </li>
        <li class="nav__item">
            <a href="#" class="nav__link">About</a>
        </li>
        <li class="nav__item">
            <a href="#" class="nav__link">Contact</a>
        </li>
    </ul>
</nav>
```

You will notice how we haven't created "grandchild" selectors such as .nav__item__ link as this becomes unnecessarily messy. An element only needs to semantically be related to one block to be maintainable. At this point, it's worth noting that you can create a new block within a block to ensure the code is more semantically relevant.

Now, let's create the CSS for our simple navigation structure:

```css
.nav {
    background: black;
}
.nav__list {
    display: flex;
}
.nav__item {
    list-style: none;
    flex: 1 1 auto;
    border-right: 1px solid white;
}
```

```
.nav__item:last-child {
    border-right: none;
}
.nav__link {
    display: block;
    color: white;
    padding: 10px;
    text-align: center;
    text-decoration: none;
}
.nav__link:hover {
    text-decoration: underline;
}
```

Here, you can see what the CSS for the BEM marked up navigation looks like. The following screenshot shows how this would be displayed in the browser:

Figure 11.1: Navigation bar

Now, let's say we wanted to change an instance of the navigation and make the contact navigation link stand out more. First, we'd give the link a modifier class:

```
<a href="#" class="nav__link nav__link--contact">Contact</a>
```

Then, we can add a new CSS rule to make the link stand out more:

```
.nav__link--contact {
    color: red;
    font-weight: bold;
}
```

This would update the navigation to appear as follows:

Figure 11.2: Highlighted contact navigation link

If we wanted to change the appearance of all the links, for instance, use different colors and fonts for navigation that appears on a different part of the website, then we could apply a modifier to the block itself, rather than the individual element. The HTML could be changed for `.nav` by adding a new modifier called **theme2**, like so:

```
<nav class="nav nav--theme2">
```

We would then create a new CSS rule to apply the **theme2** styling to the navigation bar. This would look something like this:

```
.nav--theme2 {
    background: lightgray;
}
.nav--theme2 .nav__link {
    background: lightgray;
    color: black;
    font-family: Arial;
}
```

The following screenshot shows the end result of the **theme2** modifier being applied, the `.nav` styles being inherited, and the modifier changing/extending the styles to give a different appearance:

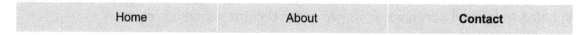

Home	About	Contact

Figure 11.3: Navigation with theme2 modifier styles applied

Next, we'll look at putting BEM markup into action by practicing its use in an exercise.

Exercise 11.01: Using BEM Markup

The aim of this exercise is to get you writing maintainable BEM code, starting with some existing HTML and CSS markup, and transforming this into a more maintainable semantic BEM-based markup. Let's get started:

1. Start your exercise with the HTML code in **Exercise 11.01_Raw.html** from https://packt.live/2X7JOQh. Save this as **Exercise 11.01.html** in a new folder called **Chapter11**.

2. Next, we're going to create the **exercise1.css** file within a subfolder called **css** in our **Chapter11** project folder. Save the CSS extract to your file from https://packt.live/32CC6yZ. This is shown in the following code:

3. When you access **Exercise 11.01.html** in your web browser now, you should see a web page that looks as follows:

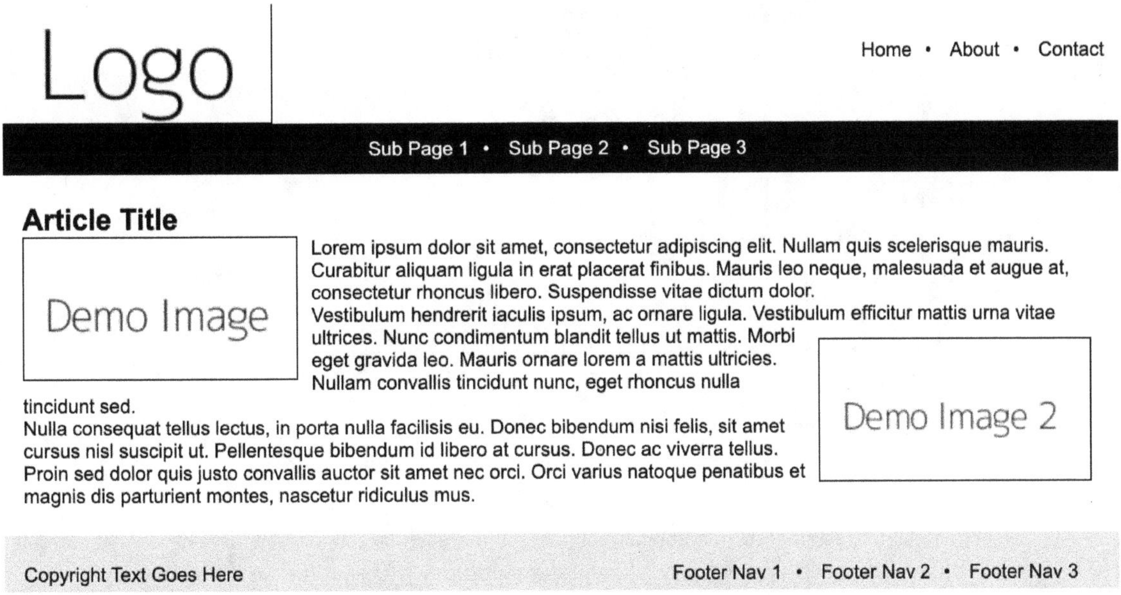
Logo
Home • About • Contact
Sub Page 1 • Sub Page 2 • Sub Page 3
Article Title
Demo Image
Lorem ipsum dolor sit amet, consectetur adipiscing elit. Nullam quis scelerisque mauris. Curabitur aliquam ligula in erat placerat finibus. Mauris leo neque, malesuada et augue at, consectetur rhoncus libero. Suspendisse vitae dictum dolor.
Vestibulum hendrerit iaculis ipsum, ac ornare ligula. Vestibulum efficitur mattis urna vitae ultrices. Nunc condimentum blandit tellus ut mattis. Morbi eget gravida leo. Mauris ornare lorem a mattis ultricies. Nullam convallis tincidunt nunc, eget rhoncus nulla tincidunt sed.
Demo Image 2
Nulla consequat tellus lectus, in porta nulla facilisis eu. Donec bibendum nisi felis, sit amet cursus nisl suscipit ut. Pellentesque bibendum id libero at cursus. Donec ac viverra tellus. Proin sed dolor quis justo convallis auctor sit amet nec orci. Orci varius natoque penatibus et magnis dis parturient montes, nascetur ridiculus mus.
Copyright Text Goes Here
Footer Nav 1 • Footer Nav 2 • Footer Nav 3

Figure 11.4: Web browser screenshot

4. Next, we'll be updating the HTML references in **`Exercise 11.01.html`** so that it uses BEM. The following is the updated HTML code:

```
<!DOCTYPE html>
<html>
<head>
  <meta name="viewport" content="width=device-width,
    initial-scale=1" />
  <title>Chapter 11: Exercise 11.01</title>
  <link href="css/exercise1.css" rel="stylesheet" />
</head>
<body>
    <header class="header">
        <div class="logocontainer">
            <img class="logocontainer__img" src="https://
              dummyimage.com/200x100/fff/000&text=Logo" alt="Logo" />
        </div>
        <nav class="nav nav--header">
            <ul class="nav__list">
                <li class="nav__item">
                    <a class="nav__link" href="#">Home</a>
                </li>
                <li class="nav__item">
                    <a class="nav__link" href="#">About</a>
                </li>
```

```
            <li class="nav__item">
                <a class="nav__link" href="#">Contact</a>
            </li>
        </ul>
    </nav>
</header>
<nav class="nav nav--subnav">
    <ul class="nav__list">
        <li class="nav__item">
            <a class="nav__link" href="#">Sub Page 1</a>
        </li>
        <li class="nav__item">
            <a class="nav__link" href="#">Sub Page 2</a>
        </li>
        <li class="nav__item">
            <a class="nav__link" href="#">Sub Page 3</a>
        </li>
    </ul>
</nav>
<article class="article">
    <h1 class="article__title">Article Title</h1>
    <p class="article__text"><img class="article__image"
      src=
      "https://dummyimage.com/200x100/fff/000&text=
      Demo+Image" alt="Demo Image" /> Lorem ipsum dolor
      sit amet, consectetur adipiscing elit. Nullam quis
      scelerisque mauris. Curabitur aliquam ligula in erat
      placerat finibus. Mauris leo neque, malesuada et
      augue at, consectetur rhoncus libero.
      Suspendisse vitae dictum dolor.
    </p>
    <p class="article__text">Vestibulum hendrerit iaculis
      ipsum, ac ornare ligula. Vestibulum efficitur mattis
      urna vitae ultrices. Nunc condimentum blandit tellus
      ut mattis. <img class=
      "article__image article__image--right" src=
      "https://dummyimage.com/200x100/fff/000&text=
      Demo+Image+2" alt="Demo Image 2" />  Morbi eget
      gravida leo. Mauris ornare lorem a mattis ultricies.
      Nullam convallis tincidunt nunc, eget rhoncus nulla
      tincidunt sed.
    </p>
```

```
        <p class="article__text">Nulla consequat tellus lectus,
            in porta nulla facilisis eu. Donec bibendum nisi
            felis, sit amet cursus nisl suscipit ut. Pellentesque
            bibendum id libero at cursus. Donec ac viverra
            tellus. Proin sed dolor quis justo convallis auctor
            sit amet nec orci. Orci varius natoque penatibus et
            magnis dis parturient montes, nascetur ridiculus
            mus.
        </p>
    </article>
    <footer class="footer">
        <p class="footer__copyright">Copyright Text Goes Here</p>
        <nav class="nav nav--footer">
        <ul class="nav__list">
            <li class="nav__item">
                <a class="nav__link" href="#">Footer Nav 1</a>
            </li>
            <li class="nav__item">
                <a class="nav__link" href="#">Footer Nav 2</a>
            </li>
            <li class="nav__item">
                <a class="nav__link" href="#">Footer Nav 3</a>
            </li>
        </ul>
        </nav>
    </footer>
</body>
</html>
```

As you can see, all the navigation instances share the same `.nav` styles initially, but each unique instance has a modifier class (for instance, `.nav--header`) to tweak its appearance for that instance. We've added a modifier class to each of the reused **nav** blocks (for instance, `.nav--header`), so that we can reuse the `.nav` block as a reusable component. It's default styling rules (from `.nav`) can be inherited on the header (`.nav--header`), sub navigation (`.nav--subnav`), and footer (`.nav--footer`).

5. Now, we'll update the **exercise1.css** file so that it matches the new BEM HTML markup. The following is the updated CSS:

```css
* {
    margin: 0;
    padding: 0;
}


body {
    font-family: Arial, sans-serif;
    font-size: 14px;
    color: #000;
}


.header {
  display: flex;
  align-items: center;
}
.logocontainer {
  flex: 0 0 auto;
}
.logocontainer__img {
  width: 100%;
  height: auto;
  display: block;
  border: 1px solid #000;
}
.nav--header {
  flex: 1 1 auto;
  text-align: right;
  padding-right: 20px;
}
.nav__list {
  list-style: none;
}
.nav__item {
  display: inline-block;
  padding-right: 10px;
}
```

```css
.nav__item:after {
    content: '\02022';
    padding-left: 10px;
}
.nav__item:last-child {
    padding-right: 0;
}
.nav__item:last-child:after {
    display: none;
}
.nav__link {
  color: #000;
  text-decoration: none;
}
.nav__link:hover {
    text-decoration: underline;
}
.nav--subnav {
    text-align: center;
    background: #000;
    padding: 10px 0;
}
.nav--subnav .nav__item {
    color: #FFF;
}
.nav--subnav a {
    color: #FFF;
}
.article {
    margin: 20px;
}
.article__image {
    float: left;
    margin: 0 10px 10px 0;
    border: 1px solid #000;
}
.article__image--right {
    float: right;
    margin: 10px 0 0 10px;
}
```

```
.footer {
    background: #DDD;
    clear: both;
    display: flex;
    padding: 20px;
}
.footer__copyright {
    flex: 1 1 auto;
}
.nav--footer {
    flex: 0 0 auto;
    text-align: right;
}
```

You will notice that when using BEM in CSS, not much nesting is required, as we only need to nest when modifiers are used at a parent level.

6. If you reload the **Exercise 11.01.html** file in your web browser, it should appear as follows. If this isn't the case, then you will need to recheck the HTML and CSS modifications that were made when you migrated this exercise into BEM markup:

Logo

Home • About • Contact

Sub Page 1 • Sub Page 2 • Sub Page 3

Article Title

Demo Image

Lorem ipsum dolor sit amet, consectetur adipiscing elit. Nullam quis scelerisque mauris. Curabitur aliquam ligula in erat placerat finibus. Mauris leo neque, malesuada et augue at, consectetur rhoncus libero. Suspendisse vitae dictum dolor.
Vestibulum hendrerit iaculis ipsum, ac ornare ligula. Vestibulum efficitur mattis urna vitae ultrices. Nunc condimentum blandit tellus ut mattis. Morbi eget gravida leo. Mauris ornare lorem a mattis ultricies. Nullam convallis tincidunt nunc, eget rhoncus nulla tincidunt sed.

Demo Image 2

Nulla consequat tellus lectus, in porta nulla facilisis eu. Donec bibendum nisi felis, sit amet cursus nisl suscipit ut. Pellentesque bibendum id libero at cursus. Donec ac viverra tellus. Proin sed dolor quis justo convallis auctor sit amet nec orci. Orci varius natoque penatibus et magnis dis parturient montes, nascetur ridiculus mus.

Copyright Text Goes Here

Footer Nav 1 • Footer Nav 2 • Footer Nav 3

Figure 11.5: Completed web browser screenshot

We've now completed this exercise and you should have a good understanding of what BEM markup is and how we can take existing HTML and CSS markup and convert it into the more maintainable BEM markup standard. Next, we're going to review how we can use BEM in SCSS.

Using BEM Markup with SCSS

Developing our knowledge further from *Chapter 10, Preprocessors and Tooling*, we're going to take a look at how we can write BEM in our SCSS code so that we can compile it into more maintainable CSS code. We can nest BEM elements inside the block with SCSS. For example, if we had the **.nav__list** element, then we could nest it inside a **.nav** block using the **&** syntax, as shown in the following example:

```
.nav {
    &__list {
        /* Styles Here */
    }
}
```

Let's look at the sample CSS we had earlier in this chapter:

```
.nav {
    background: black;
}
.nav__list {
    display: flex;
}
.nav__item {
    list-style: none;
    flex: 1 1 auto;
    border-right: 1px solid white;
}
.nav__item:last-child {
    border-right: none;
}
.nav__link {
    display: block;
    color: white;
    padding: 10px;
    text-align: center;
    text-decoration: none;
}
```

```scss
.nav__link:hover {
    text-decoration: underline;
}
```

To achieve the preceding CSS output, we would write the following SCSS:

```scss
.nav {
    background: black;

    &__list {
        display: flex;
    }
    &__item {
        list-style: none;
        flex: 1 1 auto;
        border-right: 1px solid white;

        &:last-child {
            border-right: none;
        }
    }
    &__link {
        display: block;
        color: white;
        padding: 10px;
        text-align: center;
        text-decoration: none;

        &:hover {
            text-decoration: underline;
        }
    }
}
```

The SCSS is 9 lines and 146 characters more compact than the original CSS, so it's quicker to write and easier to maintain as it's still using BEM. As we discussed in *Chapter 10, Preprocessors and Tooling*, we can get our **npm run scss** command to output our CSS in compressed format too, thereby making the production website version super compact but keeping the SCSS file easily maintainable.

Next, we're going to look at adding the modifier class to the SCSS file. For example, if we had a modifier class of **.nav__link--contact**, then we could add this nested part of the **&__link** part of the SCSS, as follows:

```
.nav {
    &__link {
        &--contact {
          color: red;
          font-weight: bold;
        }
    }
}
```

The preceding SCSS code would construct the **.nav__link--contact** class and apply the bold format and red color to the text. If we added a modifier class to the block called **.nav--theme2**, and if we wanted to change the appearance of an element within the modified block, then we could nest the block modifier within the element SCSS, as shown in the following code:

```
.nav {
    &__link {
        .nav--theme2 & {
            background: lightgray;
            color: black;
            font-family: Arial;
        }
    }
}
```

As you can see in the preceding code, we are changing the font, font color, and background color of **.nav__link** if it's preceded by the **.nav--theme2** block modifier class. We're going to put this into practice in the next exercise by updating our CSS and making it into SCSS.

Exercise 11.02: Applying SCSS to BEM

The aim of this exercise is to put writing SCSS in BEM markup into practice so that we can make our code even more maintainable. We'll use the HTML and CSS code we wrote in *Exercise 11.01, Using BEM Markup*, as the basis to begin this exercise and then we'll rewrite the CSS so that it can be compiled from SCSS instead. Let's get started:

1. Take a copy of **Exercise 11.01.html** and save this as **Exercise 11.02.html** in your **Chapter11** project folder.

2. Update the **<title>** and CSS link to the CSS file in the **<head>** of **Exercise 11.02.html** to point to the **exercise2.css** file that we'll compile in the following steps:

```
<link href="css/exercise2.css" rel="stylesheet" />
```

3. Create a new subfolder called **scss** and save the **exercise2.scss** file with the contents of **css/exercise1.css** inside it.

4. Now, we'll modify **exercise2.scss** to convert the existing CSS into SCSS. The following is the resulting SCSS code:

```scss
* {
    margin: 0;
    padding: 0;
}
body {
    font-family: Arial, sans-serif;
    font-size: 14px;
    color: #000;
}
.header {
  display: flex;
  align-items: center;
}
.logocontainer {
    flex: 0 0 auto;
    &__img {
        width: 100%;
        height: auto;
        display: block;
        border: 1px solid #000;
    }
}
.nav {
```

```
&--header {
    flex: 1 1 auto;
    text-align: right;
    padding-right: 20px;
}
&--subnav {
    text-align: center;
    background: #000;
    padding: 10px 0;
}
&--footer {
    flex: 0 0 auto;
    text-align: right;
}
&__list {
    list-style: none;
}
&__item {
    display: inline-block;
    padding-right: 10px;
    &:after {
        content: '\02022';
        padding-left: 10px;
    }
    &:last-child {
        padding-right: 0;
    }
    &:last-child:after {
        display: none;
    }
    .nav--subnav & {
        color: #FFF;
    }
}
&__link {
    color: #000;
    text-decoration: none;
    &:hover {
        text-decoration: underline;
    }
    .nav--subnav & {
        color: #FFF;
```

```scss
                }
            }
        }
    .article {
        margin: 20px;
        &__image {
            float: left;
            margin: 0 10px 10px 0;
            border: 1px solid #000;
            &--right {
                float: right;
                margin: 10px 0 0 10px;
            }
        }
    }
    .footer {
        background: #DDD;
        clear: both;
        display: flex;
        padding: 20px;
        &__copyright {
            flex: 1 1 auto;
        }
    }
```

In the preceding SCSS code, inside every block are the elements and some modifier classes. This will create the same CSS output as **exercise1.css** does when it's compiled into **exercise2.css**. The original CSS file was 112 lines; we've written 109 lines of SCSS to replace it. When output into **exercise2.css** in a compressed output style, it's only 1 long line, which is super optimized for use on a production website. However, you've still got the SCSS file so that you can make edits, which makes this very maintainable.

5. Now, we'll need to compile the SCSS, just as we did in *Chapter 10, Preprocessors and Tooling*. In the **Chapter11** project folder, we'll need to add our **package.json** file with the following contents:

```json
{
    "name": "chapter11",
    "version": "1.0.0",
    "description": "HTML5 & CSS3 Workshop Chapter11 Exercises",
    "main": "index.js",
    "scripts": {
```

```
        "test": "echo \"Error: no test specified\" && exit 1",
        "scss": "node-sass --watch scss -o css --output-style
          compressed"
    },
    "author": "Matt Park",
    "license": "ISC",
    "dependencies": {
        "node-sass": "^4.12.0"

    }

}
```

6. Once this file is in place, we can navigate to the **Chapter11** folder in the Terminal and run the **npm install** command to ensure we've got the node-sass dependencies ready to compile next.

7. Now, we're going to compile the **exercise2.scss** file by running our **npm run scss** command. This will output the contents of **scss/exercise2.scss to css/exercise2.css** upon saving the **.scss** file.

8. Check the **Exercise 11.02.html** file in your web browser to check that the web page is still working as expected, as shown in the following screenshot:

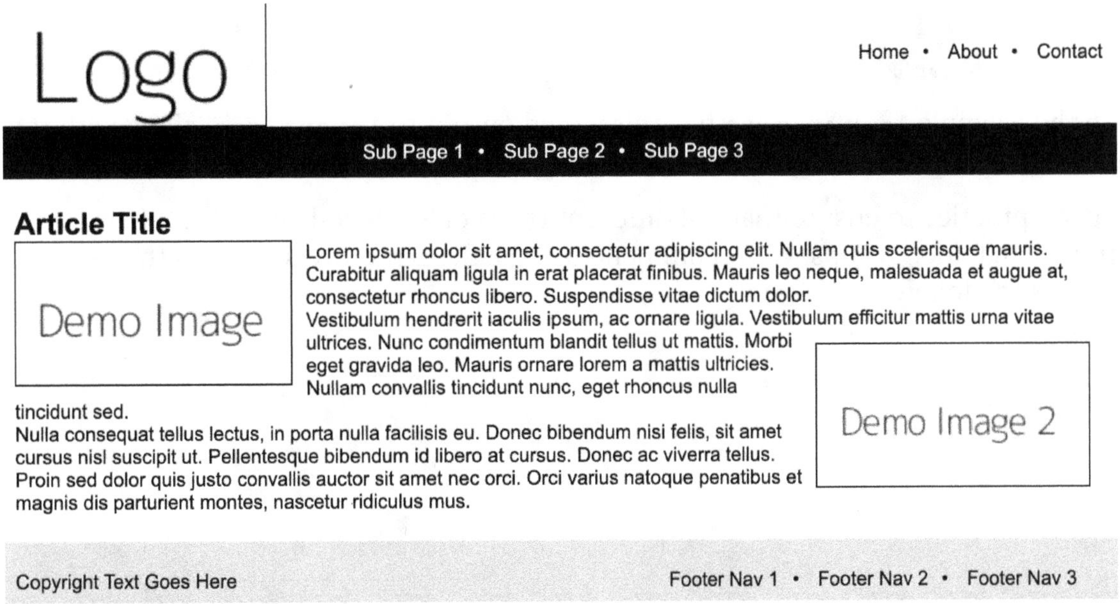

Figure 11.6: Browser output

Now that we've completed this exercise, you should be able to write SCSS in the maintainable BEM markup standard and understand how CSS can be converted into nested SCSS and vice versa. In the next section of this chapter, we'll look at structuring our SCSS files into an even more maintainable format.

Structuring Your SCSS into Maintainable Files

In this section, we're going to explore how to make our SCSS more maintainable by splitting the code into separate files and folders, which is very useful when managing large code bases of CSS.

It is recommended to group SCSS rules together into blocks or components. For example, the navigation would have its own SCSS file, and all of those SCSS files would be merged into a single CSS file output when compiled.

Each individual SCSS file would have an underscore in front of the filename to prevent it from being individually compiled. This means we can create the **_navigation.scss** file and add our navigation styles to it. Inside our main file (**main.scss**), which is going to be output into CSS, we can list all the components we are going to import, like so:

```
/* Layout File */
@import '_header';
@import '_navigation';
@import '_footer';
```

With the example **layout.scss** file, this would output to **layout.css**, along with the contents of the **_header**, **_navigation**, and **_footer** SCSS files merged in.

It's good practice to ensure that the different types of SCSS code belong in their appropriate folder groups. We recommend using a file structure similar to the one shown in the following diagram:

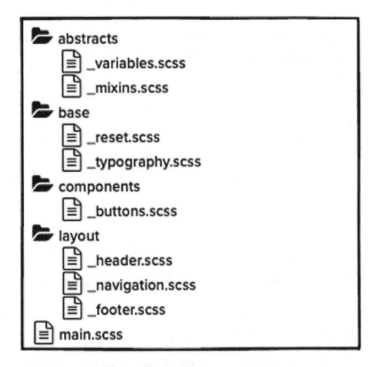

Figure 11.7: Suggested basic layout for SCSS files and folders

As you can see, we have the following folders:

- **abstracts**: This is where the SCSS files that don't output any CSS belong; for instance, variables, mixins, and other helper tools.

- **base**: This is where the reset SCSS files and general "base" rules such as typography belong.

- **components**: This is where the different components or modules live. This could be buttons, dropdowns, and other reusable components throughout the website.

- **layout**: This is where key structural shared blocks are located; for example, the website header, navigation, and footer.

We also have **main.scss**, which is where all the mentioned folders would be imported. We can achieve this with a wildcard folder import so that as we add new files to the existing folders, they are imported without changing the **main.scss** file. See the suggested contents of **main.scss** in the following code, based on the folders in the preceding diagram :

```
@import "abstracts/_variables";
@import "abstracts/_mixins";
@import "base/_reset";
@import "base/_typography";
@import "components/_buttons";
@import "layout/_header";
@import "layout/_navigation";
@import "layout/_footer";
```

This code will get all the other SCSS code from these files and output via a single file called **main.css**.

Exercise 11.03: Using Structured SCSS Files

The aim of this exercise is to create SCSS file structures using a maintainable approach. We'll use the SCSS code we developed for *Exercise 11.02, Applying SCSS to BEM*, and split it into the appropriate files and folders in order to create a maintainable code base for the website to grow on. Let's get started:

1. Take a copy of **Exercise 11.02.html** from your **Chapter11** project folder and save it as **Exercise 11.03.html**.

2. Update **<title>** and the CSS link to the CSS file in the **<head>** of **Exercise 11.03.html** to point to the **exercise3.css** file that we'll compile in the next step:

```
<link href="css/exercise3.css" rel="stylesheet" />
```

3. Save the **scss/exercise3.scss** file with the contents of **scss/exercise2.scss** inside it.

4. Take a copy of the code from **scss/exercise3.scss** and create the following subfiles (note the new subfolders):

   scss/exercise3/base/_reset.scss:

```
* {
    margin: 0;
    padding: 0;
}
```

scss/exercise3/base/_typography.scss:

```scss
body {
    font-family: Arial, sans-serif;
    font-size: 14px;
    color: #000;
}
```

scss/exercise3/components/_article.scss:

```scss
.article {
    margin: 20px;
    &__image {
        float: left;
        margin: 0 10px 10px 0;
        border: 1px solid #000;
        &--right {
            float: right;
            margin: 10px 0 0 10px;
        }
    }
}
```

scss/exercise3/layout/_header.scss:

```scss
.header {
  display: flex;
  align-items: center;
}
.logocontainer {
    flex: 0 0 auto;
    &__img {
        width: 100%;
        height: auto;
        display: block;
        border: 1px solid #000;
    }
}
```

You also have to create a **_navigation.scss** file. The complete code for the **scss/exercise3/layout/_navigation.scss** file can be found at https://packt.live/36XoXnB.

scss/exercise3/layout/_footer.scss:

```
.footer {
    background: #DDD;
    clear: both;
    display: flex;
    padding: 20px;
    &__copyright {
        flex: 1 1 auto;
    }
}
```

5. Now, we'll update the **scss/exercise3.scss** file with the following contents:

```
@import 'exercise3/base/_reset';
@import 'exercise3/base/_typography';
@import 'exercise3/layout/_header';
@import 'exercise3/layout/_navigation';
@import 'exercise3/layout/_footer';
@import 'exercise3/components/_article';
```

This will import all the SCSS subfiles we just created.

6. Now, we're going to compile the **exercise3.scss** file by running our **npm run scss** command. This will output the contents of **scss/exercise3.scss** to **css/exercise3.css** upon saving the **.scss** file.

7. Check the **Exercise 11.03.html** file in the web browser at this point to check that the web page is still working as expected. This is shown in the following screenshot:

Home • About • Contact

Sub Page 1 • Sub Page 2 • Sub Page 3

Article Title

Demo Image

Lorem ipsum dolor sit amet, consectetur adipiscing elit. Nullam quis scelerisque mauris. Curabitur aliquam ligula in erat placerat finibus. Mauris leo neque, malesuada et augue at, consectetur rhoncus libero. Suspendisse vitae dictum dolor. Vestibulum hendrerit iaculis ipsum, ac ornare ligula. Vestibulum efficitur mattis urna vitae ultrices. Nunc condimentum blandit tellus ut mattis. Morbi eget gravida leo. Mauris ornare lorem a mattis ultricies. Nullam convallis tincidunt nunc, eget rhoncus nulla tincidunt sed.

Demo Image 2

Nulla consequat tellus lectus, in porta nulla facilisis eu. Donec bibendum nisi felis, sit amet cursus nisl suscipit ut. Pellentesque bibendum id libero at cursus. Donec ac viverra tellus. Proin sed dolor quis justo convallis auctor sit amet nec orci. Orci varius natoque penatibus et magnis dis parturient montes, nascetur ridiculus mus.

Copyright Text Goes Here

Footer Nav 1 • Footer Nav 2 • Footer Nav 3

Figure 11.8: Browser output

8. Now, we're going to add a new SCSS file for variables so that we can create some color variables to use in our exercise web page. Create the following file:

scss/exercise3/abstracts/_variables.scss

```
$primary-color: #004275;
$secondary-color: #ffd421;
$tertiary-color: #00e1ea;
```

9. We'll go through our SCSS files and replace the colors we've already defined for the new color variables, as follows:
   Replace any "**#000**" values for **$primary-color** variable
   Replace any "**#FFF**" values for **$secondary-color** variable
   Replace any "**#DDD**" values for **$tertiary-color** variable
   Save the files after you've made the changes.

10. Now, we just need to add a new line to the top of **scss/exercise3.scss** and save the file:

```
@import 'exercise3/abstracts/_variables';
```

Note that it's very important to ensure that this is placed at the top of the **scss/exercise3.scss** file. This ensures that the variables are defined before we run the other SCSS code.

11. Then, we're going to compile the **exercise3.scss** file again by running our **npm run scss** command and saving the **.scss** file to make it generate the CSS for **css/exercise3.css**.

12. Check the **Exercise 11.03.html** file in the web browser again to ensure that the web page is still working as expected with the new color variables in use. This is shown in the following screenshot:

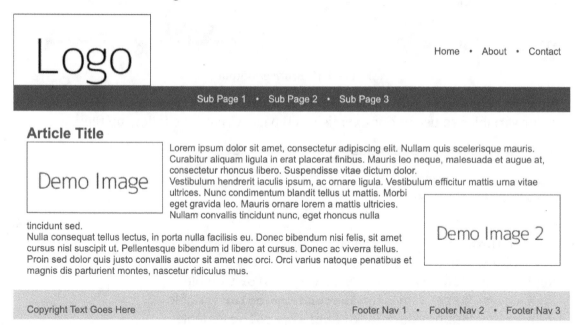

Figure 11.9: Browser output after new variable colors are added

We've now got a maintainable SCSS file structure with variables, reset styles, base typography styles, layout styles for the header, footer, navigation, and component styling for an article. This will be much more manageable as the website grows with components, pages, and styling rules. If we hadn't applied these best practices, it would be harder to manage the website project as it grows with more and more styles.

Since we can group the rules into components and different blocks, if we remove a certain component from the website in the future, cleaning up the styles would be as simple as removing the SCSS file for that component and removing the import line from the main SCSS file as well. This clean add/remove approach to new CSS files for blocks and components greatly reduces the amount of technical debt created over time as the project grows in size.

Good Practices for Maintainable CSS

There are several other good practices to follow in addition to using BEM with SCSS in structured folders and files. There are six points to consider when creating maintainable CSS:

1. We can make use of the extend functionality in SCSS with the **%class-name** syntax. We can define a class for extending its use by using **@extend class-name;** in many places to save writing the same code again and again. See the following example in SCSS:

```scss
%font-normal {
    font-family: Arial, sans-serf;
    font-size: 14px;
    color: #000;
}
body {
    @extend %font-normal;
    background: #CCC;
}
input[type=text] {
    @extend %font-normal;
    border: 1px solid #000;
}
```

This generates the following CSS:

```css
body,
input[type=text] {
    font-family: Arial, sans-serf;
    font-size: 14px;
    color: #000
}
```

```
body {
    background: #CCC
}
input[type=text] {
    border: 1px solid #000
}
```

As you can see, wherever we use the extend class, it groups together and outputs in one go in the CSS. However, it's easier to maintain in SCSS as we still edit the code in its actual rule location, and so we don't have to create multiple rules for the same code – it's all handled upon the compilation of it.

2. It's worth making sure you avoid "deep nesting," meaning that it shouldn't be necessary to nest more than three levels deep as this makes it much harder to override styles in the code, thus reducing its maintainability. Not only this, but longer CSS rule selectors have a performance hit in rendering the CSS.

3. We should always remain consistent with using quotation marks in our SCSS (and CSS) and not mixing between single and dark quotation marks. When using SCSS, it's preferred to use single quotation marks, but the important point of ensuring consistency and maintainability is to not use both in your SCSS files.

4. Our SCSS rules should follow a sensible order so that they can easily be located by another developer or even yourself a few days later. They should follow a logical order while we're using BEM since we can store our SCSS in an appropriate subfolder to help group rules in logical places. However, within each file, we can order our rules in a way that just makes sense to read. Within our SCSS selectors, we can also group our properties together logically. For example, width, height, margin, and padding can be underneath each other. Typography-related properties can be grouped together (for instance, font-size, color, and font-family). Take a look at the following example code. You can see that the properties are randomly ordered:

```
.unordered-selector {
    width: 100px;
    font-family: Arial;
    background: black;
    color: white;
    padding: 5px;
    text-decoration: underline;
    height: 100px;
    border: 1px solid white;
    font-size: 20px;
}
```

With the following ordered example, you can see that the properties follow a grouping order, that is, of the type of properties, thus resulting in more maintainable CSS:

```css
.ordered-selector {
    width: 100px;
    height: 100px;
    padding: 5px;
    font-family: Arial;
    font-size: 20px;
    color: white;
    text-decoration: underline;
    background: black;
    border: 1px solid white;
}
```

5. It's worth understanding when we should use the `!important` flag in CSS. The short answer is, very rarely. Let's understand why:

   When CSS gets out of control with the deep nesting of rules (for instance, beyond three nesting levels deep), it's often very easy for a new developer to come into the website project and add an `!important` flag onto the end of various CSS properties so that they can use a shorter CSS selector (for instance, one or two nesting levels deep) in order to override a value with something else.

   You can imagine that, after a while, this becomes very messy, and soon becomes difficult to maintain, with various CSS selectors trying to override each other with the `!important` flag. Use this very sparingly to avoid a maintenance disaster.

6. Avoid using ID CSS selectors with your styling rules. These are mainly used in JavaScript and are too specific and not reusable. You should use BEM classes in your CSS instead of IDs to make the CSS more identifiable, reusable, and maintainable.

We've now covered various aspects of writing maintainable CSS, including understanding what is meant by the phrase semantic CSS, writing CSS, SCSS with the BEM approach, understanding componentization and rule grouping, writing maintainable SCSS with sensible file and folder structures, and finally, following good CSS practices to generally create more maintainable code. We'll now look at applying our knowledge of maintainable CSS to our activity. We'll do this by updating our video store home page with the new techniques we've learned about.

Activity 11.01: Making Our Video Store Web Page Maintainable

The aim of this activity is to take our existing video store home page (from *Chapter 10, Preprocessors and Tooling, Activity 10.01, Converting a Video Store Home Page into SCSS*) and refactor the code so that it uses BEM semantic markup with suitable SCSS file structuring to create a more maintainable web page. Let's get started:

1. Take a copy of the *Chapter 10, Preprocessors and Tooling, Activity 10.01, Converting a Video Store Home Page into SCSS*, HTML file and save it in your **Chapter11** project folder under **Activity 11.01.html**.

2. Take a copy of *Chapter 10, Preprocessors and Tooling, Activity 10.01, Converting a Video Store Home Page into SCSS*, SCSS file and save it in your **Chapter11** project folder under **scss/activity1.scss**.

3. Edit your **Activity 11.01.html** file so that it uses BEM semantic markup.

4. Edit your **activity1.scss** to follow the updated BEM semantic markup.

5. Split the **activity1.scss** file's SCSS code into suitable subfolders and files, and then import these into the **activity1.scss** file.

6. Compile the SCSS in your Terminal in order to compile your **activity1.css** file.

7. Test the web page in your browser to ensure it's loading as expected, as shown in the following screenshot:

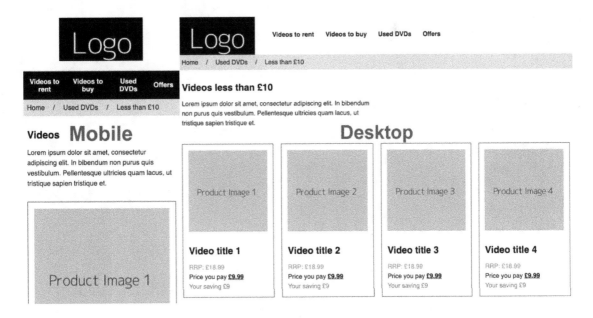

Figure 11.10: Browser output on mobile and desktop

> **Note**
>
> The complete solution can be found in page 629.

Summary

In this chapter, we covered various aspects of creating and using maintainable CSS in our work. First, we got BEM markup for our HTML and CSS and then we looked at how SCSS can work with BEM and how to separate its folders and files to enable better CSS codebase management at a larger scale. We also reviewed additional good practices to help us write maintainable CSS. You should now have an idea of what maintainable CSS code looks like and how you can write your CSS to follow the same standards.

In the next chapter of this book, we will learn about web components and how APIs, when combined, can create powerful features on a web page.

12

Web Components

Overview

By the end of this chapter, you will be able to create custom elements to use on a web page; encapsulate the structure and styles of a custom element with the Shadow DOM; create a simple-modal using HTML templates; and create and share your own web component. This chapter introduces the three technologies used to create web components – custom elements, the Shadow DOM, and HTML templates – and how they can be combined to make reusable components.

Introduction

In the previous chapter, we learned about the techniques we can use to ensure we are writing in a well-supported standard. In this chapter, we will use many of these techniques to create web components – bits of UI that we can safely share across multiple web apps and sites.

The idea behind **web components** is that you, as a developer, are able to create a custom HTML element that is reusable.

To facilitate the reusability of the component, we need to be able to encapsulate the functionality or behavior of the component so that it doesn't pollute the rest of our code and, in turn, is not polluted by outside influences.

For example, we may want to create a web component that handles notifications or alert messages on a web page. To do this, we would want the colors for these messages to be consistent in all cases; perhaps we would add a red background for an error alert, a yellow background for an information alert, and green background for a successful action alert.

Let's assume that these alerts are styled with the following CSS styles:

```
<style>
.error {
  background: red;
}
.info {
  background: yellow;
}
.success {
  background: green;
}
</style>
```

If we added these alerts to a page not completely under our control, for example, a WordPress blog or large site using a UI library, it would be easy, for example, to have another error class set that obscures text by setting a style later in the HTML document:

```
.error {
  color: red;
}
```

Now, our error alert would appear as a red block with no visible text, which could be a real problem for our users. The encapsulation of a component, in this case, would solve this problem and give us more confidence in our alert component.

If you want to reuse a complex behavior or UI widget before using web components, you would have to:

1. Copy and paste a block of HTML.

2. Add the related CSS to the head element of the HTML document.

3. Add JavaScript near the bottom of the HTML document.

4. Make sure the changes have had no adverse effects on the rest of your HTML document.

5. Make sure the rest of the document does not have any adverse effects on the added component.

This is not a great developer experience and it is prone to errors.

The great benefit of web components is that a developer can add a custom element to their web app or HTML document and everything that's needed for the component to work will be encapsulated or contained within that element. This provides a better developer experience, can lead to cleaner code and better-controlled interfaces for your components, and make the interaction between components easier.

We will look at the benefits of web components in more detail later in this chapter but first, we will look at the three technologies we need to make web components. To create web components, we will be working with a combination of three new technologies in HTML5, as follows:

- **Custom elements**

- **The Shadow DOM**

- **HTML templates**

We will look at each of these technologies separately, starting with custom elements, before we look at how combining them makes web components a reality for the modern web.

Custom Elements

We've looked at a lot of the elements provided by the HTML5 standard in previous chapters; for example, the **p** element defines a paragraph within the body of an HTML document, the **a** element represents an anchor or link, and the **video** element defines a video source that we can embed in our HTML document.

HTML5 also tolerates non-standard elements; you could add a tag such as `<something></something>` in a HTML5 document and it would pass a validation check. However, this element would not have any semantic meaning for a browser. With a custom element, we can do more.

Custom elements let you create your own element types that you can then use on a web page. A custom element is a lot like one of those standard HTML elements; they use the same syntax (a name surrounded by angle brackets) and the main difference is that they are not defined in the HTML5 standard, so we have to register them to be able to use them.

To create a custom element, we will need to add it to the custom element registry, which we do with JavaScript. In JavaScript, this is defined as the `CustomElementRegistry` object. We can access the `CustomElementRegisty` object through a property called `customElements`, which is defined in the global `window` object.

> **Note**
>
> The global object (that is, in the case of the browser, the window object) contains the DOM document and all the browser APIs that are available in JavaScript. Whenever a JavaScript script runs in the browser, it has access to this window object.

The define Method

The main method of `customElements` that we want to use is the `define` method as it allows us to register a new custom element. We call the define method with the `customElements.define()` JavaScript code and we can pass arguments into the method between the brackets. We need to pass the method a unique name, which is the name for the custom element, and a JavaScript class that defines its behavior.

> **Note**
>
> The `class` keyword is used to create a JavaScript object with a custom set of methods and properties. We can create an instance of a class using the `new` keyword, and each instance will have the methods and properties defined in the class.
>
> A class can inherit from another class using the `extends` keyword. This means that the new class will inherit the methods and properties of the class it extends.

Let's look at an example:

```
<main-headline />
<script>
    customElements.define("main-headline",
        class MainHeadline extends HTMLElement {
        }
    );
</script>
```

The **define** method takes a string as its first argument. This is the name you wish to register your new custom element with. In this example, we have named our custom element, **main-headline**. Any instances of the **main-headline** element in the HTML document will, therefore, be registered as this custom element.

The second argument of the **define** method is the element's constructor. This a JavaScript class that extends **HTMLElement** and defines the behavior of our custom element. In our example, the behavior adds nothing to **HTMLElement**, but we will look at some possibilities for extending the behavior of **HTMLElement** and other built-in elements later in this chapter.

While a JavaScript class can have any name that is valid for a JavaScript variable – the first character must be a letter or underscore and all the other characters can be an alphanumeric character or underscore – there is a convention for class names to use a capitalized first letter and capital letters for the first letter of each word, for instance, **MainHeadline**.

There is an optional third argument for the **define** method that takes an **options** object. Currently, the **options** object has one possible property, that is, the **extends** property. The **extends** property lets us specify a built-in element from which our custom element will inherit.

We will look at extending built-in HTML elements in more detail later in this chapter.

Naming Conventions

For a custom element name to be valid, it must have a hyphen somewhere in the name. For example, **main-headline** is a valid custom element name, but **mainheadline** is not.

There is no restriction on how many hyphens appear in the name, which means, for example, **sub-sub-headline** is a valid custom element name as well.

If we were to provide a non-valid name to the **define** method it would throw an error, like so:

```
<!DOCTYPE html>
<html lang="en">
    <head>
        <meta charset="utf-8">
        <meta name="viewport" content="width=device-width,
            initial-scale=1, shrink-to-fit=no">
        <title>Naming convention</title>
    </head>
    <body>
        <mainheadline />
        <script>
            customElements.define("mainheadline",
                class MainHeadline extends HTMLElement {
                }
            );
        </script>
    </body>
</html>
```

This code would result in our browser throwing an error that we can view in the console tab of the developer tools. In Chrome, we can access the developer tools with the keyboard shortcuts *Ctrl* + *Shift* + I (PC) or *Cmd* + *Opt* + I (Mac).

For example, in the **Console** tab of the Chrome developer tools, we would see a message similar to the one shown in the following screenshot. The message, highlighted in red, says **DOMException: Failed to execute 'define' on 'CustomElementRegistry': 'mainheadline' is not a valid custom element name**:

Figure 12.1: Error message for an invalid custom element name

Unique Names

A custom element name can only be registered once. For example, we cannot define two different **main-headline** elements. If you try to define a custom element when a custom element with the same name has already been defined, the browser will throw an error.

In code, this would look something like the following example, where we have called the **customElements.define** method twice with the name **main-headline**:

```html
<!DOCTYPE html>
<html lang="en">
    <head>
        <meta charset="utf-8">
        <meta name="viewport" content="width=device-width,
          initial-scale=1, shrink-to-fit=no">
        <title>Naming convention</title>
    </head>
    <body>
        <main-headline />
        <script>
            customElements.define("main-headline",
                class MainHeadline extends HTMLElement {
                }
            );
            customElements.define("main-headline",
                class OtherMainHeadline extends HTMLElement {
                }
            );
        </script>
    </body>
</html>
```

In the **Console** tab of the Chrome developer tools, we will see a message similar to the one shown in the following screenshot. The message, highlighted in red, says **DOMException: Failed to execute 'define' on 'CustomElementRegistry': this name has already been used with this registry**:

Figure 12.2: Error message for an already used custom element name

Extends HTMLElement

If the constructor for our custom element does not extend **HTMLElement**, the browser will once again throw an error. For example, let's say we try and extend **SVGElement** instead of **HTMLElement**:

```
<sub-headline></sub-headline>
<script>
     customElements.define("sub-headline",
         class SubHeadline extends SVGElement {
         }
     );
</script>
```

This code would, again, result in an error being thrown, with a message similar to the one shown in the following screenshot. The message, highlighted in red, says **TypeError: Illegal constructor**:

Figure 12.3: Error message for constructor not inheriting from HTMLElement

The constructor can extend a subclass of **HTMLElement** rather than extend **HTMLElement**. For example, **HTMLParagraphElement** extends **HTMLElement**, and that means our constructor can extend **HTMLParagraphElement** rather than **HTMLElement**.

We will see more examples of extending a subclass of **HTMLElement** when we look at extending built-in elements later in this chapter.

So far, we've seen how we can define a custom element and pass it a class constructor that extends **HTMLElement** to define its behavior. Now, we'll try creating our own custom element.

Exercise 12.01: Creating a Custom Element

In this exercise, we are going to create a custom element and add it to a web page. We will create a **blog-headline** custom element for a blog post.

Here are the steps we will follow:

1. Firstly, we want to create a new directory. We will call this **Chapter12**.

2. In **Chapter12**, we will create an **Exercise 12.01.html** file. Open that file and copy this HTML code in to create a web page:

```
<!DOCTYPE html>
<html lang="en">
    <head>
        <meta charset="UTF-8">
        <title>Exercise 12.01: Creating a Custom Element</title>
    </head>
    <body>
        <main class="blog-posts">
            <article>
                <!-- add main-heading here -->
            </article>
        </main>
    </body>
</html>
```

3. We want to replace the **<!-- add main heading here -->** comment with our custom element. When we register our custom element, we will call it **blog-headline**. We will start by adding an instance of the **blog-headline** element to our web page:

```
<blog-headline />
```

4. We now need to define our **blog-headline** custom element and add it to the custom elements registry. We will do this by adding a script element after the closing tag of the **main** element. We want to add this after the **main** element as we can be sure the content in the **main** element has loaded when the script is executed:

```
        </main>
        <script>
            // register blog-headline
        </script>
    </body>
```

5. Next, we need to create our custom element's class constructor, which will extend **HTMLElement**:

```
<script>
    class BlogHeadline extends HTMLElement {}
</script>
```

6. We will then register our custom element constructor using the **define** method on the global **customElements** object:

```
<script>
    class BlogHeadline extends HTMLElement {}
    window.customElements.define("blog-headline", BlogHeadline);
</script>
```

7. So far, we have registered our **blog-headline** element but it doesn't do anything. In order to see our **blog-headline** element appear in a web page, we will add the text "Headline" as text content to the **blog-headline** element. We need to add a constructor method to our **BlogHeadline** class:

```
<script>
    class BlogHeadline extends HTMLElement {
        constructor() {
            super();
        }
    }
    window.customElements.define("blog-headline", BlogHeadline);
</script>
```

The constructor is a special method that is called when an instance of our custom element is created. We have to call the super function here as that will construct the parent class, that is, the **HTMLElement** class, that our **BlogHeadline** class extends. If we do not call super in the constructor, our **BlogHeadline** class will not have properly extended **HTMLElement**, which will cause an **Illegal constructor** error.

8. In our constructor, the **this** keyword refers to the scope of our class, which refers to an instance of the custom element. This means we can add text to the element with the **textContent** property:

```
<script>
    class BlogHeadline extends HTMLElement {
        constructor() {
            super();
            this.textContent = "Headline";
        }
    }
    window.customElements.define("blog-headline", BlogHeadline);
</script>
```

9. Finally, to show you that we can create multiple instances of our **blog-headline** element, we will duplicate the **article** element with the **blog-headline** element in it as shown in the following code:

```
<article>
    <blog-headline />
</article>
<article>
    <blog-headline />
</article>
```

If you now right-click on the filename in VSCode on the left-hand side of the screen and select **Open In Default Browser**, you will see the result of this code that is shown in the following image. We have created a custom element, the **blog-headline** element, which currently adds the word "Headline" wherever it is added to the HTML document. In the next few exercises, we will expand upon the functionality of our **blog-headline** element to make it more useful:

Headline
Headline

Figure 12.4: The blog-headline custom element in action

By adding text to our blog-headline element, we have had a glimpse at how we can add behavior to our custom elements. So far, this custom element doesn't do much, but we will learn about techniques for making more complex custom elements and for adding styles, behavior, and customized instances of the element to make them more reusable.

Next, we will delve deeper into how we can customize the behavior and functionality of a custom element.

Behavior of a Custom Element

Like the standard elements in HTML5, our custom elements can have content (text or child elements) and can provide an interface for modifying the element via attributes. As each custom element has a JavaScript class associated with it, we can enhance the element with a JavaScript API, where we can define properties and methods that add functionality to our custom element.

We can add text to an instance of a custom element and add attributes in the same way we would with the built-in elements. For example, we could have a custom element, the **styled-text** element, with some text content and options for a dark, highlight, or light theme via the **theme** attribute:

```
<styled-text theme="dark">The castle stood tall.</styled-text>
```

We can then refer to the attribute from within our custom element's JavaScript class using the **getAttribute**, **setAttribute**, and **hasAttribute** methods. In JavaScript, we are able to control the flow of our code using if/else statements and switch statements.

> **Note**
>
> In JavaScript, we can use the **if**, **else if**, and **else** statements to control which lines of code are executed. **if/else** statements will check a condition that can be either true or false and will execute the code it encompasses if that condition is met. For example, the if(true) statement would always run, whereas if(false) would never run. An **else** statement will run when the condition of a preceding **if** statement is not met.

To style the text according to the theme attribute, we could define the custom element styled text. Here, we use the **setStyle** method to change the background color and color styles of the element depending on the value of the theme attribute. The code is as follows:

Example 12.01.html

```
23              setStyle(theme) {
24                  if (theme === "highlight") {
25                      this.style.backgroundColor = "yellow";
26                      this.style.color = "black";
27                  }
28                  else if (theme === "light") {
29                      this.style.backgroundColor = "white";
30                      this.style.color = "black";
31                  }
32                  else if (theme === "dark") {
33                      this.style.backgroundColor = "black";
34                      this.style.color = "white";
35                  }
36              }
```

The complete code for this example is available at: https://packt.live/2NMEWfn

If we apply each of the themes to the same text, we get the following output. The different theme attribute values, that is, highlight, **light**, and **dark**, each style the text differently:

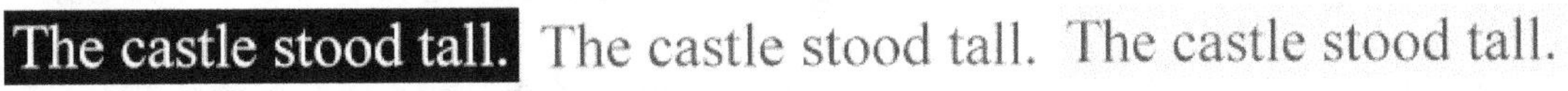

The castle stood tall. The castle stood tall. The castle stood tall.

Figure 12.5: A custom styled text element

In the next exercise, we will use a similar technique to that shown in the preceding example to work with multiple attributes on a custom element. We will set attributes on a blog-headline element to change the appearance of the headline.

As well as attributes with a value such as **"light"** or **"dark"** and **"dog"** or **"cat"**, we can use **boolean** attributes. A **boolean** attribute is an attribute that is either present or absent from the element. You do not have to set a value for this attribute. The fact is that if it is present it is true and if it is absent it is false. As an example, we can set our blog-headline to either be a new post or not use an attribute.

For example, the following code would set the first custom element as an old post and the second element as a new post. This is done by using the **newpost** attribute:

```
<blog-headline>Old post</blog-headline>
<blog-headline newpost>New post</blog-headline>
```

We can put these techniques for adding attributes to a custom element and using those attributes to control the behavior of a custom element into practice with an exercise.

Exercise 12.02: Adding and Using Custom Elements with Attributes

In this exercise, we will expand upon the functionality of the blog-headline element that we created in our previous exercise.

We will add text content to the **blog-headline** element so that we have different text for each blog-headline instance.

We will then modify the blog-headline with attributes:

- We will add an attribute for **type** that will add an icon to the heading depending on the type of blog post.

- We will also add a **newpost** attribute, which will take a **boolean** value. If the **newpost** attribute exists on the **blog-headline** element, we will add an icon of a clock to the **blog-headline** element.

The steps are as follows:

1. We start by creating an **Exercise 12.02.html** file in the **Chapter12** directory.

2. Our starting point is the code resulting from **Exercise 12.01.html**. You can copy the following code into **Exercise 12.02.html** or make a copy of **Exercise 12.01.html** and rename that file. We have changed the title to *Exercise 12.02, Custom Element with Attributes*, but nothing else has changed:

```
<head>
    <meta charset="UTF-8">
        <title>Exercise 12.02: Custom Element with Attributes
        </title>
</head>
```

3. We will start by developing our articles. We will replace the two closed **blog-headline** elements with proper text content. For this exercise, we will add the text **"Blog Post About Kittens"** to the **blog-heading** element in the first article and we will add the text "Blog Post About Puppies" to the **blog-heading** element in the second article:

```
<main class="blog-posts">
    <article>
        <blog-headline>Blog Post About Kittens
        </blog-headline>
    </article>
    <article>
        <blog-headline>Blog Post About Puppies
        </blog-headline>
    </article>
</main>
```

4. Next, we will add a **type** attribute to each **blog-headline** element and set the value to either "**dogs**" or "**cats**", depending on the content of the article:

```
<main class="blog-posts">
    <article>
        <blog-headline type="cats">Blog Post About
          Kittens</blog-headline>
    </article>
    <article>
        <blog-headline type="dogs">Blog Post About
          Puppies</blog-headline>
    </article>
</main>
```

5. We can now update the **BlogHeadline** class to handle the different values for the type attribute. We will keep the constructor but remove the line that sets the text content to "**Headline**". The text content will now be set when the element is added to the web page:

```
<script>
    class BlogHeadline extends HTMLElement {
        constructor() {
            super();
        }
    }
    window.customElements.define("blog-headline",
        BlogHeadline);
</script>
```

6. Next, we add a getter and setter to handle the type attribute. Getters and setters are special methods that can control how a property or attribute is updated. Rather than a property being set directly with the value it is given, we can do checks to make sure the property is valid or change the value with a method. The code is as follows:

```
<script>
    class BlogHeadline extends HTMLElement {
        constructor() {
            super();
        }
        get type() {
            return this.getAttribute("type");
        }
        set type(newValue) {
            this.setAttribute("type", newValue);
        }
    }
    window.customElements.define("blog-headline",
        BlogHeadline);
</script>
```

When the type attribute is **set**, we want to add an icon to the headline. Depending on whether the type attribute value is dogs or cats, the icon will either be a dog or a cat. We will use the Unicode values 🐱 and 🐶 to create our cat and dog icons, respectively. You can try out other Unicode characters too; search online for a Unicode version of a symbol that you can copy into the code in place of the cat or dog.

7. Next, we will create a **setIcon** method. This block of code will be called whenever we **set** the type of the custom element. It will check whether we have set the type attribute to cats or dogs and will change the text content of the custom element accordingly:

```
class BlogHeadline extends HTMLElement {
    constructor() {
            super();
    }
    get type() {
        return this.getAttribute("type");
    }
    set type(newValue) {
        this.setAttribute("type", newValue);
        this.setIcon(this.type);
    }
    setIcon(type) {
        if (type === "cats") {
            this.textContent = "🐱 " +
                this.textContent + " 🐱";
        }
        else if (type === "dogs") {
            this.textContent = "🐶 " +
                this.textContent + " 🐶";
        }
    }
}
```

8. Next, we will check for the type attribute in the constructor using **hasAttribute** to make sure the icon gets set when the custom element is instantiated:

```
constructor() {
        super();
    if (this.hasAttribute("type")) {
        this.setIcon(this.type);
    }
}
```

9. Now, we will add another attribute, that is, the **newpost** attribute. We will add this to the **blog-headline** element for the first article, that is, the one about kittens. This attribute will be a **boolean**, so we are only interested in whether it exists on the element. We add it to the **blog-headline** element:

```
<main class="blog-posts">
    <article>
        <blog-headline type="cats" newpost>Blog Post
            About Kittens</blog-headline>
    </article>
    <article>
        <blog-headline type="dogs">Blog Post
            About Puppies</blog-headline>
    </article>
</main>
```

10. To handle the **newpost** attribute in our **BlogHeadline** class, we add a check for **newpost** at the end of the constructor.

```
constructor() {
        super();
    if (this.hasAttribute("type")) {
        this.setIcon(this.type);
    }
    this.newpost = this.hasAttribute("newpost");
}
```

11. Next, we will add getters and setters for the **newpost** attribute. We can add these just above the **setIcon** method:

```
get newpost() {
    return this.hasAttribute("newpost");
}
set newpost(newValue) {
    if (newValue) {
        this.setAttribute("newpost", "");
    }
    else {
        this.removeAttribute("newpost");
    }
    this.setIsNewPostIcon(this.newpost);
}
```

12. Next, we want to create a **setIsNewPostIcon** method to handle the **newpost** attribute value:

```
setIsNewPostIcon(isNewPost) {
    if (isNewPost) {
        this.classList.add("new-post");
    }
    else {
        this.classList.remove("new-post");
    }
}
```

The **setIsNewPostIcon** method will add a class called **"newpost"** to the **blog-headline** element if the **newpost** attribute exists on the element. We have a boolean (true or false) value called **isNewPost**, which is the argument for the **setIsNewPostIcon** method. This **isNewPost** value is set to true or false based on whether the **this.hasAttribute('type')** check is true or false. The class will add a clock icon to elements with the **newpost** attribute via CSS.

13. Finally, we will add some CSS to the head element of our HTML document to style the new-post class:

```
<style>
    .new-post::before {
        font-weight: bold;
        content: "⏰ NEW! ⏰";
    }
</style>
```

If you now right-click on the filename in VSCode on the left-hand side of the screen and select **Open In Default Browser**, you will see the following image that shows what this would look like in the browser:

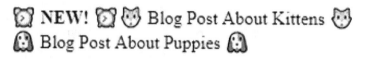

Figure 12.6: blog-headline version 2 with extra icons!

There are a couple of problems with the **blog-headline** element that we have created.

One problem that may cause us issues is that the element is not encapsulated; it could influence and be influenced by the rest of the web page. For example, we have added the **new-post** class to the **head** element, which means it can be used throughout the HTML document. We want our web components to be more contained.

We will look at the Shadow DOM later in this chapter and learn how we can use that technology to protect a component from outside influences.

Equally problematic is the fact that we are only checking attributes in the constructor of our **BlogHeadline** class. This means that if we change attributes with JavaScript, they will not be reflected in the **blog-headline** instances on the page.

To handle the second problem and many other issues, custom elements give us some life cycle methods, which we will look at in the next section.

Custom Element Life Cycle

A custom element has several life cycle methods that can be used to call blocks of code at the right time in the custom elements life cycle.

The constructor of a custom element is called when the element is instantiated, but there are several ways to create new instances of an element for a web page: adding them in an HTML document or creating them with JavaScript using **document. createElement**. This means we can't guarantee that a custom element has been connected to the web page when the constructor is called. This is why custom elements provide a **connectedCallback** method.

The **connectedCallback** method is invoked when a custom element is added into the HTML document. It can happen more than once, for example, if an element is connected and then disconnected and then reconnected to the document.

The **disconnectedCallback** method is invoked when the element is disconnected from the HTML document.

The **adoptedCallback** method is invoked if the custom element moves to a new document. An example of this callback being invoked is when we move a custom element between an HTML document and an iframe.

The **attributeChangedCallback** method is invoked when an element's attributes change, that is, whether the value changes or the attribute is added or removed. The callback receives three arguments:

- **name**: The name of the attribute
- **oldValue**: The old value of the attribute before the change
- **newValue**: The new value of the attribute after the change

Not all changes to attributes will trigger **attributeChangedCallback**; rather, we, as developers, are responsible for maintaining a whitelist of attributes we wish to observe. We do this with the static **observedAttributes** method.

For example, if I wanted to know when the type attribute changed in our previous exercise, I would set the array returned by the **observedAttributes** method to include "type". I would then handle the change to the type attribute with **attributeChangedCallback** like so:

```
static get observedAttributes() {
    return ["type"];
}
attributeChangedCallback(name, oldValue, newValue) {
    if (name === "type") {
        // handle changes to the value of the type attribute

    }
}
```

Something else to consider is that **oldValue** and **newValue** may not necessarily be different. The callback is invoked because the attribute has been set, but depending on what we want to do when the value changes, we may want to check whether the value has actually changed. We can do this by comparing the old and new value and escaping the function as early as possible if the two values are the same:

```
attributeChangedCallback(name, oldValue, newValue) {
    if (oldValue === newValue) { return; }
```

In the next exercise, we will make changes to the **blog-headline** element. This time, we will use the life cycle methods to improve our custom element. We will add some functionality to trigger the different life cycle methods so that we can experience them for ourselves in a better way.

Exercise 12.03: Custom Element Life Cycle

We will start from where we left off in the previous exercise. In this exercise, we will make further improvements to the **blog-headline** element and we will add some functionality so that we can test the life cycle of our custom element by adding and removing the element and changing attributes dynamically.

Here are the steps to follow:

1. We will start by creating an **Exercise 12.03.html** file in the **Chapter12** directory.

2. Our starting point is the code resulting from **Exercise 12.02.html**, so we will make a copy of **Exercise 12.02.html** and rename that file to **Exercise 12.03. html**. We will change the title to *Exercise 12.03: Custom Element Life Cycle*.

3. Next, we will add a **test-ui** style to give the UI a bit more whitespace between items:

    ```css
    .test-ui {
        margin-top: 2rem;
    }
    ```

4. We want to know when the **type** attribute and the **newpost** attribute have changed. To do this, we will add them to the array that's returned by the **observedAttributes** function:

    ```js
    static get observedAttributes() {
        return ["type", "newpost"];
    }
    ```

5. Next, we will add **attributeChangedCallback** to where we can handle any changes to the values of the **type** attribute and the **newpost** attribute. We'll check that the old value and the new value are actually different and then handle changes to **type** or **newpost** with the **setIcon** and **setIsNewPostIcon** methods, respectively:

    ```js
    attributeChangedCallback(name, oldValue, newValue) {
        if (oldValue === newValue) { return; }
        if (name === "type") {
            this.setIcon(this.type);
        }
        else if (name === "newpost") {
            this.setIsNewPostIcon(this.newpost);
        }
    }
    ```

6. We are going to add a getter for the **_heading** property so that we can access it as the variable heading:

    ```js
    get heading() {
        return this._heading;
    }
    ```

7. Now that we are responding to changes to the blog-headline elements' attributes, we don't need to do this in the constructor of our **BlogHeading** class. For now, we will simplify **constructor** so that it calls **super** and stores the initial text content of the **heading**:

```
constructor() {
    super();
    this._heading = this.textContent;
}
```

8. At this point, we have made our custom element respond to attribute changes. We can test this by adding a UI to allow us to change the attributes dynamically. To do this, we will add the following HTML beneath the **main** element:

```
<div class="test-ui">
    <button id="swap-type-1">Swap type attribute
    </button>
    <button id="swap-newpost-1">Swap newpost attribute
    </button>
</div>
```

9. We want these two buttons to swap the attribute values of the first **blog-headline** element when they are clicked. To do that, we will need a reference to the buttons and the **blog-headline** element, and we will add the following to the bottom of our script element:

```
const headline = document.querySelector("blog-headline");
const swapTypeButton = document.getElementById("swap-type-1");
const swapNewpostButton =
  document.getElementById("swap-newpost-1");
swapTypeButton.addEventListener("click", function() {
    const type = headline.getAttribute("type");
    headline.setAttribute("type", type ==
      "cats" ? "dogs" : "cats");
});
swapNewpostButton.addEventListener("click", function() {
    const newpost = headline.hasAttribute("newpost");
    if (newpost) {
        headline.removeAttribute("newpost");
    }
    else {
        headline.setAttribute("newpost", "");
    }
});
```

With those buttons, we can test whether the attribute value changes are reflected in our **blog-headline** element. For example, if we click the **Swap type attribute** button when the blog-headline has cats, it will swap to dogs and the icon should change to dogs.

So far, the code we have written will result in the following output:

⏰ **NEW!** ⏰ 🐱 Blog Post About Kittens 🐱

🐶 Blog Post About Puppies 🐶

Swap type attribute   Swap newpost attribute

Figure 12.7: Custom headline elements with buttons to update attributes

By clicking the left button, we can change the type attribute of the top **blog-headline** element, as shown in the following image:

⏰ **NEW!** ⏰ 🐶 Blog Post About Kittens 🐶

🐶 Blog Post About Puppies 🐶

Swap type attribute   Swap newpost attribute

Figure 12.8: Swapping the type attribute value from cats to dogs

By clicking the right button, we can toggle the **newpost** attribute of the top **blog-headline** element and remove the new-post class attribute, as shown in the following image:

🐶 Blog Post About Kittens 🐶

🐶 Blog Post About Puppies 🐶

Swap type attribute   Swap newpost attribute

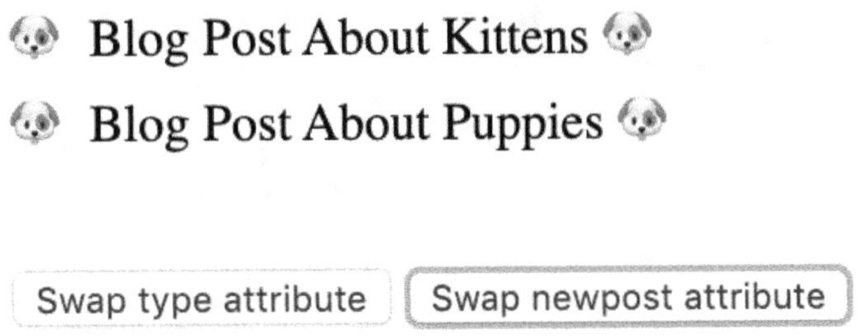

Figure 12.9: Top blog-headline element without the new-post class

10. We will add another button to connect and disconnect the blog-headline element from the HTML document. First, we will add the button to **div.test-ui**:

```
<div class="test-ui">
    <button id="swap-type-1">Swap type attribute
    </button>
    <button id="swap-newpost-1">Swap newpost attribute
    </button>
    <button id="toggle-connect-1">Toggle connection
    </button>
</div>
```

11. Next, we want to get a reference to the button. We also want to get a reference to the parent element that hosts the headline. This will make it easier to add and remove the headline element from the DOM:

```
const headline = document.querySelector("blog-headline");
const swapTypeButton = document.getElementById("swap-type-1");
const swapNewpostButton = document.getElementById("swap-newpost-1");
const toggleConnectionButton = document.getElementById("toggle-
    connect-1");
const headlineParent = headline.parentElement;
```

12. When the **Toggle Connection** button is clicked, it will remove or add the headline element to the DOM:

```
toggleConnectionButton.addEventListener("click", function() {
    if (headline.parentElement) {
        headlineParent.removeChild(headline);
    }
    else {
        headlineParent.insertBefore(headline, headlineParent.
            firstChild);
    }
});
```

13. Using the console.log method, we can output a message to the JavaScript console. We will add this to **connectedCallback** and **disconnectedCallback** so that we can see when these methods have been triggered. We can see the message in the console tab of the developer tools (see *Chapter 1, Introduction to HTML and CSS*, for an introduction to the developer tools):

```
connectedCallback() {
    console.log("Connected");
    this.setIcon(this.type);
}
disconnectedCallback() {
    console.log("Disconnected");
}
```

If you now right-click on the filename in VSCode on the left-hand side of the screen and select **Open In Default Browser**, you will see the results of the exercise that is shown in the following image. The buttons let us update the first headline by changing the element's attributes and connecting and disconnecting the element:

Figure 12.10: Custom headline elements with buttons to update the attributes and elements

If we open the Chrome developer tools, we will see the messages **Connected** and **Disconnected** logged in the console each time we click the Toggle connection button, as shown in the following screenshot:

Console	Elements	Sources	Network	Performance	Memory	Application	Security	Redux	Audits		
top		▼	⊙	Filter		Default levels ▼					⚙

```
Connected                                                            exercise 3:43
Disconnected                                                         exercise 3:48
Connected                                                            exercise 3:43
Disconnected                                                         exercise 3:48
```

Figure 12.11: Disconnected and Connected messages logged in the console

So far, we have looked at the life cycle of a custom element. Now, we will look at how we can extend built-in elements such as the anchor (**<a />**) element.

Extending a Built-in Element

Our custom element does not have to directly extend **HTMLElement** as long as it extends another built-in element that is a subclass of **HTMLElement**. This means we can customize the behavior of existing elements such as the **p** element via **HTMLParagraphElement** or the **a** element via **HTMLAnchorElement**.

We extend the custom element with the third optional argument of the **customElements.define** method. We pass the argument an options object with an **extends** property. The value we give the extends property is the name of the built-in element we wish to extend. For example, if we want to create a custom **h1** element that restricts the size of the text content, we would set the **extends** property to "**h1**":

```
window.customElements.define("short-headline",
   ShortHeadline, { extends: "h1"});
```

The **ShortHeadline** constructor would now extend **HTMLHeadingElement** instead of **HTMLElement**:

```
class ShortHeadline extends HTMLHeadingElement {
     //... functionality for short headlines
}
```

To use this custom element, we would not create a short-headline element; instead, we would use the **is** attribute on an **h1** element and set the value to our short-headline custom element:

```
<h1 is="short-headline">Headline</h1>
```

If we extend an element, we can use all the properties and attributes available to that element. For example, we could extend the anchor tag based on its attributes. We could add information about a link, such as whether it opens in a new browser window.

In the next exercise, we will look at how we can extend the anchor element to provide this information.

Exercise 12.04: Custom Element Extending HTMLAnchorElement

In this exercise, we are going to create a new custom element that extends the anchor element and provides information about a link before a user clicks it.

We are going to take advantage of the target attribute of the anchor element. The target attribute can be set to **_self**, **_blank**, **_parent**, or **_top**. The default behavior, **_self**, is for a URL that's navigated to from the anchor element to load in the current browser context. When the target is set to **_blank**, the URL will launch in a new tab or window (depending on your browser's configuration). The other two options relate to iframes, that is, launching in the parent or top-level browser context. They will both act the same as **_self** if there is no parent context.

Here are the steps:

1. We start by creating an **Exercise 12.04.html** file in the **Chapter12** directory.

2. We'll start with the following code, which creates a web page with a main element that has an unordered list with two list items. Each of these list items has a link to a URL on the Packt website. The first link opens in the current window, whereas the second link will open the URL in a new tab or window. Copy this code into your **Exercise 12.04.html** file and save it:

```html
<!DOCTYPE html>
<html lang="en">
    <head>
        <title>Exercise 12.04: Extending HTMLAnchorElement</title>
    </head>
    <body>
        <main class="blog-post">
            <ul class="links">
                <li><a href="https://www.packtpub.com/
                    web-development">Link that opens in current
                    window</a></li>
                <li><a href=
                    "https://www.packtpub.com/free-learning"
                    target="_blank">Link that opens in a new
                    window.</a>
                </li>
            </ul>
        </main>
        <script>

        </script>
    </body>
</html>
```

3. In the **script** element, we are going to define a custom element and create a constructor that extends the anchor element with **HTMLAnchorElement**:

```
<script>
  class AnchorInfo extends HTMLAnchorElement {
      constructor() {
          super();
          this._originalText = this.textContent;
      }
  }
  window.customElements.define("anchor-info",
     AnchorInfo, {extends: "a"});
</script>
```

4. We will check the target attribute of the custom element and see if it opens in a new window, that is, if the target attribute has a value of **_blank** and add an appropriate icon, ⊞, if it is. To do this, we will add an **observedAttributes** method to observe the "target" attribute:

```
static get observedAttributes() {
    return ["target"];
}
```

5. We then need to handle changes to the **target** attribute using **attributeChangedCallback**. When the value of the **target** attribute changes, we will trigger a method called **checkTarget**. This method will set a boolean value of **opensInNewTab**:

```
attributeChangedCallback(name, oldValue, newValue) {
    if (oldValue === newValue) { return; }
    switch (name) {
        case "target":
            this.checkTarget();
            break;
    }
}
checkTarget() {
    this.opensInNewTab = this.target === "_blank";
}
```

6. We will call a render method that checks the value of **opensInNewTab** and renders the component accordingly:

```
checkTarget() {
    this.opensInNewTab = this.target === "_blank";
    this.render();
}
render() {
    const targetState = this.opensInNewTab ? "
    " : "";
    this.textContent = this._originalText +
        targetState
}
```

7. Finally, to make our anchor elements behave as the anchor-info custom element, we need to add the **is** attribute with the anchor-info value to them:

```
<ul class="links">
    <li><a is="anchor-info" href="https://www.packtpub.com/
    web-development">Link that opens in current
    window</a></li>
    <li><a is="anchor-info"
    href="https://www.packtpub.com/free-
    learning" target="_blank">Link that opens in a new
    window.</a></li>
</ul>
```

The second link on the page opens in a new window. If you now right-click on the filename in VSCode on the left-hand side of the screen and select **Open In Default Browser**, you will see it will have the external link icon added, as shown in the following screenshot:

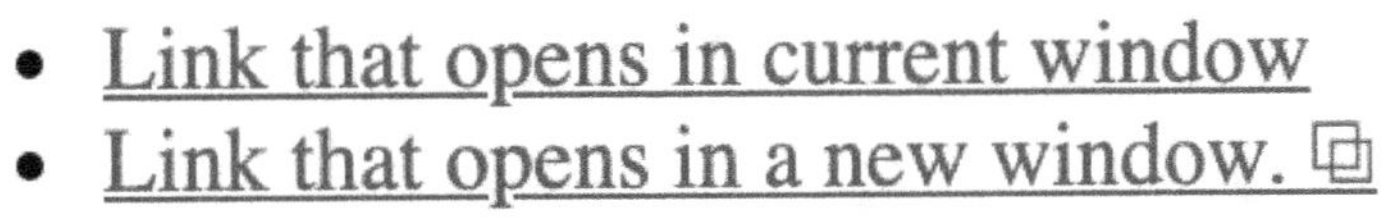

- Link that opens in current window
- Link that opens in a new window. ⧉

Figure 12.12: Custom anchor-info element that extends the anchor element

We've looked at creating custom elements but up until now, we haven't been able to encapsulate the element. We will look at the Shadow DOM and see how it can help protect our custom element from its context and vice versa.

Shadow DOM

The **Shadow DOM** is a feature that lets us control access to parts of the DOM.

If we think of our HTML document as the non-shadow, or light, DOM then all the objects and styles are accessible from the root of the document by traversing the DOM tree.

Even if we were to include a third-party script or stylesheet on our web page, we could still change the style or behavior with our own code. A user can even add their own stylesheets in many browsers and run scripts on your page through dev tools or via extensions.

The Shadow DOM lets us protect parts of our code from these outside influences, which is vital if we want our web components to work across various web pages or apps where we can't possibly know what the context is.

The Shadow DOM is an HTML fragment or DOM tree that is hidden from the rest of the light DOM. It needs to be attached to a shadow host, which is a node in the visible DOM. As with all DOM trees, the Shadow DOM will have a Shadow root from which the rest of the tree branches.

Attaching a Shadow DOM

To make use of the Shadow DOM, we need to use the **attachShadow** method on an element via JavaScript. When we call the **attachShadow** method, we can pass an options object with a mode property. There are two options available for the mode property: "open" and "closed".

Here's an example of attaching a shadow DOM in open mode to a div element with the "host" ID attribute. When we attach the shadow DOM in open mode, we are able to access the shadow root of the shadow DOM, which means we can access that DOM from the outside:

```
<div id="host"></div>
<script>
  const hostElement = document.getElementById("host");
  const openShadowDOM = hostElement.attachShadow({ mode: "open" });
  const shadowRoot = hostElement.shadowRoot;
  console.log(shadowRoot); // will return the shadow root
</script>
```

If we have a reference to the shadow root, we can manipulate the shadow DOM, which means we can append elements to it or remove or change elements within the DOM.

For example, we can append a paragraph to the shadow DOM via the root:

```
<div id="host"></div>
<script>
  const hostElement = document.getElementById("host");
  hostElement.attachShadow({ mode: "open" });
  const shadowRoot = hostElement.shadowRoot;

  const paragraph = document.createElement("p");
  paragraph.textContent = "Lorem ipsum, etcetera"
  shadowRoot.appendChild(paragraph);
</script>
```

The open mode does offer a degree of encapsulation. For example, if we try to apply a style to the head element of a web page that applies to all the paragraph elements, it will not apply to a paragraph within the Shadow DOM. We have a single point of access via the shadow root.

We can see this via the result of the following code, which is shown in the screenshot following this code:

Example 12.02.html

```
16          <script>
17          const hostElement = document.getElementById("host");
18            hostElement.attachShadow({ mode: "open" });
19            const shadowRoot = hostElement.shadowRoot;
20
21            const paragraph = document.createElement("p");
22            paragraph.textContent = "Paragraph in the Shadow DOM";
23          shadowRoot.appendChild(paragraph);
24          </script>
```

The complete code for this example is available at: https://packt.live/2Nr61po

In the following screenshot, we can see that the two paragraphs either side of our host div have been styled according to a style rule targeted at the p element selector. However, despite the shadow DOM having a paragraph element, it has not received the style.

This is an example of the encapsulation the Shadow DOM achieves, even in open mode.

We have included the dev tools representation of the elements in the following screenshot as this shows the shadow root (**#shadow-root** and the mode in brackets). We will explore the topic of using the dev tools to inspect the Shadow DOM further in the next section:

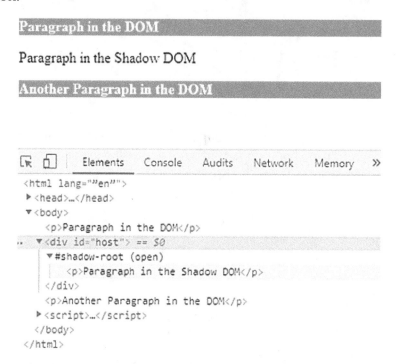

Figure 12.13: Shadow DOM with a style not applied to it

By changing the mode to closed, we change the output of the shadow root. It will now return null, which means we cannot access the Shadow DOM from the rest of the DOM through the **shadowRoot** property of an element:

```
<div id="host"></div>
<script>
 const hostElement = document.getElementById("host");
 hostElement.attachShadow({ mode: "closed" });
 const shadowRoot = hostElement.shadowRoot;
 console.log(shadowRoot); // will return null
</script>
```

This does not mean, however, that you cannot get a reference to the shadow root. **attachShadow** will still return a reference to **shadowRoot**. The idea is that you will use this inside a custom element or a web component and therefore will have a reference to the shadow root without it being accessible via an element in the main HTML document.

Inspecting the Shadow DOM

Most browsers come with a set of tools designed to help web developers when they are working on a web page. In the next exercise, we will use the Chrome dev tools to inspect the Shadow DOM of the browser's in-built input element.

It can be useful to inspect the Shadow DOM of in-built elements as it will give us a better idea of what these elements are doing and how to build our own web components.

To inspect the Shadow DOM of browser elements in the Chrome dev tools, we need to know a little bit about the dev tools and we need to set them up so that they let us inspect the Shadow DOM.

In Chrome, we can access the dev tools with the keyboard shortcuts *Ctrl* + *Shift* + *I* (on PC) and *Cmd* + *Opt* + *I* (on Mac). On most websites, you can also access the context menu by right-clicking and then choosing Inspect from the menu options.

The dev tools will appear like so:

Figure 12.14: The developer tools in Chrome

The Chrome dev tools provide a vast array of tools for web development, most of which are beyond the scope of this chapter. Across the top panel in the preceding screenshot, we can see a set of tabs, including **Elements**, **Console**, **Network**, and so on. We will be focusing on the **Elements** tab here.

The Elements tab gives you access to the HTML document of your web page. You can hover over elements in the Elements display and they will be highlighted on the web page. You can also edit these elements; change the text or attributes to immediately see what effect those changes will have on your web page.

It is also possible to inspect the Shadow DOM via the Elements tab. To do this, we may need to change the settings of our dev tools. On the right of the top panel, next to the cross that closes the dev tools, there are three vertical dots that open the controls for the dev tools. Click this and select the **Settings** option. Alternatively, you can press the F1 key. This will bring up the Settings pane that's shown in the following screenshot:

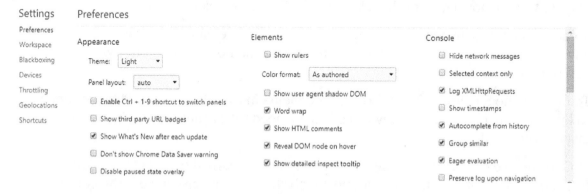

Figure 12.15: Settings pane of Chrome dev tools

There are a great many settings available for you to change, but we want to find the Elements heading and toggle **Show user agent shadow DOM** so that it is checked. We can now close the settings pane and return to the Elements tab.

We should now have access to the shadow DOM of the browser. You can determine a shadow DOM because the shadow root is marked in the Elements panel. For example, given an **input** element with the type set to **number**, we will see the results shown in the following screenshot in the Elements panel:

```
<input type="number" name="tel">
```

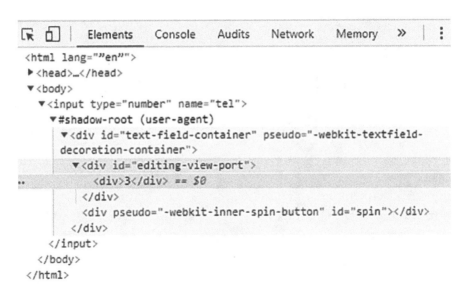

Figure 12.16: Inspecting the Shadow DOM of a number input

In the preceding screenshot, we can see that the **input** element actually attaches a Shadow DOM. Attached to the shadow root, we can see a container div (with an ID of **'text-field-container'**) hosting a viewport and a spin button that lets you increment the number value of the input.

In the next exercise, we will go back to a custom element we worked on in *Exercise 12.01, Creating a Custom Element, Exercise 12.02, Custom Element with Attributes*, and *Exercise 12.03, Custom Element Life Cycle*, and look at how we can use the Shadow DOM with that custom element to encapsulate its structure and styles.

Exercise 12.05: Shadow DOM with a Custom Element

In this exercise, we will encapsulate the structure and styles of a custom element with the Shadow DOM.

Here are the steps:

1. Start by creating an **Exercise 12.05.html** file in the **Chapter12** directory.

2. We'll start with code based on the blog-headline custom element we worked on in *Exercise 12.01, Creating a Custom Element, Exercise 12.02, Custom Element with Attributes*, and *Exercise 12.03, Custom Element Life Cycle*. Copy and save the following code:

```html
<!DOCTYPE html>
<html lang="en">
    <head>
        <meta charset="UTF-8">
        <title>Exercise 12.05: Shadow DOM</title>
        <style>
            .new-post::before {
                font-weight: bold;
                content: "⏰ NEW! ⏰";
            }
        </style>
    </head>
    <body>
        <main class="blog-posts">
            <article>
                <blog-headline type="cats" newpost>Blog Post
                    About Kittens</blog-headline>
            </article>
            <article>
                <blog-headline type="dogs">Blog Post About Puppies
                </blog-headline>
            </article>
        </main>
        <script>
            class BlogHeadline extends HTMLElement {
                static get observedAttributes() {
                    return ["type", "newpost"];
                }
                constructor() {
                    super();
                    this._heading = this.textContent;
```

```
  }
  connectedCallback() {
    this.setIcon(this.type);
  }
  disconnectedCallback() {
      console.log("Disconnected");
  }
  attributeChangedCallback(name, oldValue, newValue) {
      if (oldValue === newValue) { return; }
      switch(name) {
          case "type":
              this.setIcon(this.type);
              break;
          case "newpost":
              this.setIsNewPostIcon(this.newpost);
              break;
      }
  }
  get heading() {
      return this._heading;
  }
  get type() {
      return this.getAttribute("type");
  }
  set type(newValue) {
      this.setAttribute("type", newValue);
      this.setIcon(this.type);
  }
  get newpost() {
      return this.hasAttribute("newpost");
  }
  set newpost(newValue) {
      if (newValue) {
          this.setAttribute("newpost", "");
      }
      else {
          this.removeAttribute("newpost");
      }
      this.setIsNewPostIcon(this.newpost);
  }
  setIcon(type) {
      switch(type) {
```

```
                            case "cats":
                                this.textContent = "🐱 " +
                                    this.heading + " 🐱";
                                break;
                            case "dogs":
                                this.textContent = "🐶 " +
                                    this.heading + " 🐶";
                                break;
                        }
                    }
                    setIsNewPostIcon(isNewPost) {
                        if (isNewPost) {
                            this.classList.add("new-post");
                        }
                        else {
                            this.classList.remove("new-post");
                        }
                    }
                }

            window.customElements.define("blog-headline",
                BlogHeadline);
        </script>
    </body>
</html>
```

3. Here, we can see the effects of using the Shadow DOM. We are going to make the **BlogHeadline** element wrap its text content in an **h1** element, which we'll create in the constructor and keep as a property called **headingElement**:

```
constructor() {
    super();
    this._heading = this.textContent;
    this.textContent = "";
    this.headingElement = document.
        createElement("h1");
    this.headingElement.textContent =
        this._heading;
    this.appendChild(this.headingElement);
}
```

4. We need to update the **setIcon** and **setIsNewPostIcon** methods in order to update **headingElement**:

```
setIcon(type) {
    if (type === "cats") {
        this.headingElement.textContent = "🐱 " +
            this.heading + " 🐱";
    }
    else if (type === "dogs") {
        this.headingElement.textContent = "🐶 " +
            this.heading + " 🐶";
    }
}
setIsNewPostIcon(isNewPost) {
    if (isNewPost) {
        this.headingElement.classList.add
            ("new-post");
    }
    else {
        this.headingElement.classList.remove
            ("new-post");
    }
}
```

5. Next, we'll add a normal **h1** element and set a bold new style for h1 elements:

```
        h1 {
            font-family: Arial, Helvetica, sans-serif;
            color: white;
            background:lightskyblue;
            padding: 16px;
        }
    </style>
</head>
<body>
    <main class="blog-posts">
        <article>
            <h1>Normal H1 Heading</h1>
        </article>
```

If we look at the results of the code so far, we will see that the h1 style affects both our normal **h1** element and the **blog-headline** custom elements, as shown in the following output:

Figure 12.17: h1 style affecting custom elements

6. To prevent the **h1** style leaking into our custom elements from the document, we will attach a shadow DOM to the custom element. We do this in the **BlogHeadline** constructor and then append the heading element to **shadowRoot** instead of the custom element itself:

```
constructor() {
    super();
    this._heading = this.textContent;
    this.textContent = "";
    this.attachShadow({ mode: "open" });
    this.headingElement = document.
        createElement("h1");
    this.headingElement.textContent =
        this._heading;
    this.shadowRoot.appendChild(this.
        headingElement);
}
```

The result will be the encapsulation of our blog-headline's DOM from the surrounding document, which we can see in the following screenshot:

Figure 12.18: Encapsulated blog-headline custom elements

7. By moving to a Shadow DOM, we have caused an issue. The **blog-headline** element no longer gets the style from adding the **new-post** class. We need to move that style out of the **head** of the document and make it part of the Shadow DOM. We will do this by recreating the style element in the constructor:

```
this.styleElement = document.
  createElement("style");
this.styleElement.innerText = `.new-post:
  :after { font-weight: bold; content:
  "⏰ NEW! ⏰"; margin-left: 8px; }`;

this.shadowRoot.appendChild
  (this.styleElement);
```

8. Finally, we have an encapsulated **blog-headline** custom element. If you now right-click on the filename in VSCode on the left-hand side of the screen and select **Open In Default Browser**, you will see the result is as shown in the following screenshot:

Figure 12.19: Encapsulated blog-headline custom elements

We've seen how we can encapsulate the structure and style of our custom elements using the Shadow DOM, but using DOM manipulation in JavaScript to work with the Shadow DOM is not as easy as working with HTML.

In the next section, we will see how we can improve the developer experience of working with web components using HTML templates. We will look at how we can use templates to create versatile HTML structures that we can use in an encapsulated component.

HTML Templates

HTML templates let you create flexible templates in HTML that you can use in multiple places. If you have an HTML structure that is used several times on a web page but with different content or data, you can create a **template** element that defines that structure without immediately showing it to the user on the web page. We can reference the template to create multiple copies of that HTML structure.

For example, we could create a very simple template for a styled button with the following code, which creates a template with an ID attribute of **ok-button-template**. The markup is simply a button element with the **btn** and **ok-btn** classes applied to it. The button is styled with a purple background color, white text, and a hover state that darkens the background:

```
<template id="ok-button-template">
    <style>
        .btn {-webkit-appearance: none;
            appearance: none;
            background-color: #3700B3;
            border: none;
            border-radius: 2px;
            color: white;
            cursor: pointer;
            min-width: 64px;
            outline: none;
            padding: 4px 8px;}
        .btn:hover {background-color: #6200EE;}
    </style>
    <button class="btn ok-btn">OK</button>
</template>
```

If we created this template in a web page, we wouldn't see it in the page. To use it, we have to attach it to the HTML document. We can easily do that by getting a reference to the template and appending it to the body of the HTML document, like so:

```
<script>
    const buttonTemplate = document.getElementById
        ("ok-button-template");
    document.body.appendChild(buttonTemplate.content);
</script>
```

This code will get the content of our template and append it to the HTML document. The result is a button on the web page, as shown in the following screenshot:

Figure 12.20: OK button template in the web page

We may come across a problem with adding templates using this method. If we tried to do the same again to create two buttons, we would only see one button. Templates are meant to be reusable, so what is going on here? The answer is: we need to clone the content of the template each time we want to use it. For example, we can create multiple buttons with **cloneNode**:

```
<script>
    const buttonTemplate = document.getElementById
        ("ok-button-template");
    document.body.appendChild(buttonTemplate.content.cloneNode(true));
    document.body.appendChild(buttonTemplate.content.cloneNode(true));
</script>
```

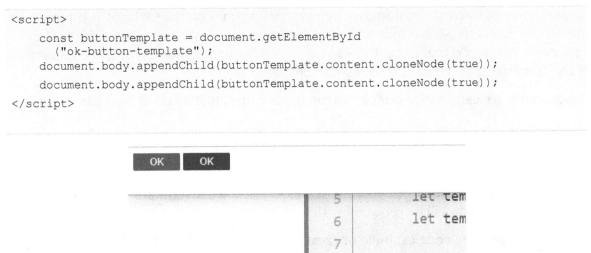

Figure 12.21: Multiple OK buttons cloned from the template

HTML templates are particularly useful when paired with the Shadow DOM in web components because we can create a structure and attach it to the shadow DOM and minimize JavaScript DOM manipulation.

A minimal example of using **ok-button-template** and combining it with a custom element could look like this:

```
<script>
    class StyledButton extends HTMLElement {
        constructor() {
            super();
            this.attachShadow({ mode: "open" });
        }
        connectedCallback() {
            const buttonTemplate = document.getElementById
                ("ok-button-template");
            const node = document.importNode
                (buttonTemplate.content, true);
            this.shadowRoot.appendChild(node);
        }
    }
    window.customElements.define("styled-button", StyledButton);
</script>
```

We've attached a shadow DOM again and this time, when the custom element is connected to the DOM, we use **importNode** to import the node from one document into a subdocument (in this case, the Shadow DOM of our component). Then, we can create instances of our **OK** button in the HTML document with the **<styled-button />** custom element.

As useful as it is to create multiple instances of a button with the **OK** label, it is perhaps even more useful to be able to use a template but vary the content when we create new instances. We can do that using the **slot** element in our template. This lets us set areas of the template that we want to be able to change.

For example, we can change our **OK** button to have any label by creating a label slot:

```html
<template id="slotted-button-template">
    <button class="btn primary-btn"><slot>Label</slot></button>
</template>
```

Here, we have set a default placeholder as a label but if we instantiate the button with text content, the slot will be replaced with that text content.

Here is the complete code example of creating a default version of the button and two labeled versions:

Example 12.03.html

```html
<template id="styled-button-template">
    <style>
        .btn {
            -webkit-appearance: none;
            appearance: none;
            background-color: #3700B3;
            border: none;
            border-radius: 2px;
            color: white;
            cursor: pointer;
            min-width: 64px;
            outline: none;
            padding: 4px 8px;
        }

        .btn:hover {
            background-color: #6200EE;
        }
    </style>
    <button class="btn primary-btn"><slot>Label</slot></button>
</template>
```

The complete code for this example is available at: https://packt.live/2WQjwC8

The output is as follows:

Figure 12.22: Multiple styled-button elements with different labels

You can also create named slots in a template by adding the name attribute to a **slot** element, which can then be used to target multiple slots with different content. For example, a template could have named slots for a heading and content:

```
<template id="article-template">
    <article>
        <h1><slot name="heading">Heading goes here…</slot></h1>
        <slot name="content">Content goes here…</slot>
    </article>
</template>
```

If we used this template in a custom element called **short-article**, we would then populate it:

```
<short-article>
    <span slot="heading">HTML and CSS</span>
    <p slot="content">HTML and CSS are the foundations of a web
        page.</p>
</short-article>
```

We now have a reusable custom element; we can provide different content to each instance and because they are HTML elements, we can provide children that are also custom elements.

To see the power of HTML templates in action, we are going to try them out on a slightly more complex structure with slots in the next exercise.

Exercise 12.06: Templates

In this exercise, we will use HTML templates to create a modal, which we will turn into a component called **simple-modal**.

Here are the steps:

1. We will start by creating an **Exercise 12.06.html** file in the **Chapter12** directory.

2. Our starting point will be the following web page with a simple-modal element already attached. Copy and save it in **Exercise 12.06.html**:

```
<!DOCTYPE html>
<html lang="en">
    <head>
        <meta charset="UTF-8">
        <title>Exercise 12.06: HTML Template</title>
        <style>
            html, body { margin: 0; padding: 0; }
        </style>
    </head>
    <body>
    <simple-modal></simple-modal>
    </body>
</html>
```

3. Beneath the body element, we will create a template for **simple-modal** called **simple-modal-template**, and we will create an HTML structure with elements for **container**, **overlay**, and **modal**. The modal will have three parts (a header, body, and footer):

```
<template id="simple-modal-template">
<section class="modal-container">
    <div class="overlay"></div>
    <div class="modal">
        <header class="header">
            <h1><!-- heading here --></h1>
        </header>
        <div class="body">
            <!-- content here -->
        </div>
        <footer class="footer">
            <button class="btn">OK</button>
        </footer>
```

```
        </div>
    </section>
</template>
```

4. Next, we will style the modal by adding a **style** element block in the template. We will make the overlay a fixed position, semi-transparent rectangle that takes up the whole window. We will center the content of the modal container and give the modal some **padding** and a white background. We will use the same button style that we used previously in this chapter:

```
<style>
    .modal-container {
        display: flex;
        align-items: center;
        justify-content: center;
        height: 100vh;
    }
    .overlay {
        position: fixed;
        background: rgba(0, 0, 0, 0.5);
        top: 0;
        bottom: 0;
        left: 0;
        right: 0;
    }
    .modal {
min-width: 250px;
        position: relative;
        background: white;
        padding: 16px;
 box-shadow: 0 4px 16px rgba(0, 0, 0, 0.8);
    }
    .btn {
        -webkit-appearance: none;
        appearance: none;
        background-color: #3700B3;
        border: none;
        border-radius: 2px;
        color: white;
        cursor: pointer;
        min-width: 64px;
        outline: none;
        padding: 4px 8px;
```

```
        }
        .btn:hover {
            background-color: #6200EE;
        }
    </style>
```

5. Next, we are going to add the **script** element and create our custom element definition. We add this block of code at the end of the web page after the template with the simple-modal-template ID has been defined. We attach the Shadow DOM and attach the template to it:

```
<script>
    customElements.define("simple-modal",
        class SimpleModal extends HTMLElement {
            constructor() {
                super();
                this.attachShadow({ mode: "open" });
            }
            connectedCallback() {
                const tmpl = document.getElementById
                    ("simple-modal-template");
                const node = document.importNode(tmpl.content, true);
                this.shadowRoot.appendChild(node);
            }
        });
</script>
```

So far, we have created our modal and an instance of it appears on our web page, like so:

Figure 12.23: Modal with no content

6. Our next task is to define slots in our template so that we can provide content when we instantiate the element. We will set up heading and content slots to replace comments in the **h1** and **body** of the template:

```
<template id="simple-modal-template">
 <section class="modal-container">
    <div class="overlay"></div>
    <div class="modal">
        <header class="header">
            <h1><slot name="heading"></slot></h1>
        </header>
        <div class="body">
            <slot name="content"></slot>
        </div>
        <footer class="footer">
            <button class="btn">OK</button>
        </footer>
    </div>
 </section>
</template>
```

7. Finally, we can populate the modal when we instantiate it:

```
<simple-modal>
    <span slot="heading">You've Opened a Modal!</span>
     <p slot="content">
         Finished with the modal? Click OK.
     </p>
</simple-modal>
```

If you now right-click on the filename in VSCode on the left-hand side of the screen and select **Open In Default Browser**, you will see the result of our component as shown in the following screenshot:

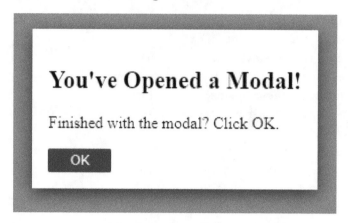

Figure 12.24: Modal with content

We have now looked at the three technologies that are vital for making web components available in the browser. It is time to put all three together and create a web component.

Creating a Web Component

By combining custom elements with a Shadow DOM and HTML templates, we have a convenient set of tools for making a reusable component – a web component – that is encapsulated from the document it appears in.

We are going to create our own reusable components in the upcoming activity. This component will add an avatar to a web page with a default placeholder image that can be replaced by our own avatar image.

Activity 12.01: Creating a Profile

We've been tasked with creating a profile component for the *Films On Demand* website.

The profile will be a web component. It will include the name and email of the user and an avatar, which will show a default placeholder image (a silhouette). We will be able to replace the image with a profile image of the user.

1. We are creating a reusable component that can be used throughout the site and we want to make use of the techniques we've learned about to encapsulate the component.

2. We want to create a template with slots for the profile image, name, and email address. We'll give the template the ID **fod-profile-template**. We will call the custom element for our component **fod-profile**.

The avatar placeholder can be found at https://packt.live/2K0jcLT. We want to create a profile component similar to the following screenshot:

Name: John Doe

E-mail: john.doe@gmail.com

Figure 12.25: Design for the fod-profile component

3. To connect the **fod-profile** custom element to our template with the ID **fod-profile-template** and to attach the **shadowDom**, we will copy and paste the following script at the end of the web page:

```
<script>
    customElements.define("fod-profile",
        class FODProfile extends HTMLElement {
            constructor() {
                super();
                this.attachShadow({ mode: "open" });
            }
            connectedCallback() {
                const tmpl = document.getElementById
                    ("fod-profile-template");
                const node = document.importNode(tmpl.content,
                    true);
                this.shadowRoot.appendChild(node);
            }
        });
</script>
```

Note

The solution to this activity can be found on page 640.

In this activity, we've learned how to create a reusable web component, but there are still some improvements we can make to the shareable nature of our component.

Sharing a Web Component

The great benefit of web components is that they are reusable and because of the protections that the Shadow DOM and templates afford, we can be fairly certain that they can be used on different websites with ease. This means we can share components and use third-party components with more confidence.

To make our component shareable, we really want to make it something that is self-contained in a single file that can be attached to a web page and then used as a component.

To do this, we can put any scripts in an external file; for example, working from the simple-modal component we created in *Exercise 12.06*, *Template*, we can create an external script file called **simple-modal.js** that wraps the custom element definition in an immediately executing function:

```
(function() {
    customElements.define("simple-modal", class SimpleModal extends
        HTMLElement {
        constructor() {
            super();
            this.attachShadow({ mode: "open" });
        }
        connectedCallback() {
            const tmpl = document.getElementById
                ("simple-modal-template");
            const node = document.importNode(tmpl.content, true);
            this.shadowRoot.appendChild(node);
        }
    });
})();
```

We still need to include the HTML template for the component in this file. The easiest way to do this is to create the template dynamically and use a template literal to populate it. We then add the template to the document body so that we can use it:

```
const template = document.createElement("template");
    template.id = "simple-modal-template";
    template.innerHTML = `
    <style>
        .modal-container {
            display: flex;
            align-items: center;
```

```
        justify-content: center;
        height: 100vh;
    }
    .overlay {
        position: fixed;
        background: rgba(0, 0, 0, 0.5);
        top: 0;
        bottom: 0;
        left: 0;
        right: 0;
    }
    .modal {
        min-width: 250px;
        position: relative;
        background: white;
        padding: 16px;
        box-shadow: 0 4px 16px rgba(0, 0, 0, 0.8);
    }
    .btn {
        -webkit-appearance: none;
        appearance: none;
        background-color: #3700B3;
        border: none;
        border-radius: 2px;
        color: white;
        cursor: pointer;
        min-width: 64px;
        outline: none;
        padding: 4px 8px;
    }
    .btn:hover {
        background-color: #6200EE;
    }
</style>
<section class="modal-container">
    <div class="overlay"></div>
    <div class="modal">
        <header class="header">
            <h1><slot name="heading"></slot></h1>
        </header>
        <div class="body">
            <slot name="content"></slot>
```

```
            </div>
            <footer class="footer">
                <button class="btn">OK</button>
            </footer>
        </div>
    </section>
    `;
document.body.appendChild(template);
```

The template and custom element definition can now be contained in one file, which is easy to add to a web page. For example, to use simple-modal.js, we would add the file and create an instance of the simple-modal element:

```
<body>
    <simple-modal>
        <span slot="heading">You've Opened a Modal!</span>
        <p slot="content">
            Finished with the modal? Click OK.
        </p>
    </simple-modal>
</body>
<script src="simple-modal.js"></script>
```

Summary

In this chapter, through multiple exercises and activities, we have looked at the features that have been added to HTML5 that allow us to create web components. We have learned how to create a custom HTML element and how to define the behavior of that custom element. We have also learned about the Shadow DOM and how we can use it to encapsulate our custom elements; in other words, we have learned how to keep our custom elements safe from outside influences and prevented them, in turn, from polluting the rest of a web page. Finally, we have learned how to create HTML templates that make our custom elements more flexible and allows us to reuse components in more situations.

Combining all of these features of HTML5, we have applied our new knowledge to create a modal and a blog-headline element, and we have learned how to create web components that can interact with one another to make reusable, versatile UI components that can be used across multiple projects.

In the next chapter, we will be looking at new web technologies to see exciting and experimental features you may want to work with.

13

The Future of HTML and CSS

Overview

By the end of this chapter, you will be able to identify the compatibility of features with the browser using the caniuse website; create a CSS paint worklet with the CSS Paint API; create a paint worklet with mouse input; and apply the progressive enhancement to a web page. In this chapter, we will first introduce some methods for keeping up with changes to the web platform. We will then apply these techniques to look at some experimental browser features, such as the CSS Paint API, which is a new web technology defined by the CSS Houdini task force. We will look at what the CSS Paint API can do, and we will also see how we can use a development version of the Chrome browser to try out cutting-edge features before they are shipped to the general public.

Introduction

Throughout the previous chapters, you've learned a lot about the current state of HTML and CSS. Through the course of this chapter, we will attempt to cast our eyes to the future. We will look into the crystal ball of the web and see where these technologies are headed.

The nature of this chapter is going to be somewhat speculative, but we are going to look at some technology that is available through web browsers at the moment and may become available throughout the web ecosystem in the future.

The web moves fast, and keeping up with changes is a key skill to learn for those interested in technology and developing for the web. We hope to provide some really useful strategies and resources that can help you to do just that; to keep up with changes, to stay ahead of new web technologies, and to see what exciting and experimental features you may want to work with.

Before we experiment with the CSS Paint API and see what it allows web developers to do, we will consider some useful ways to help us, as developers, keep up with the ever-changing environment of the web.

Keeping up with the Web

Keeping up with changes in web technology and the web as a platform is one of the toughest challenges you will face as a web developer or designer.

Design trends change and technologies evolve all the time. All of the browser vendors are continuously improving their browsers on desktop and mobile, bringing out new features and improving user and developer experience with new tools and capabilities.

Several browsers are made available to developers during their development cycles, and these can be useful to a developer as they let us use experimental features and see what is changing in the next release. We can keep an eye on upcoming changes and see what effect they may have on our websites, and we can act accordingly rather than having to react in an unplanned manner when the browser updates.

Chrome Canary

In this section, we will look at Chrome Canary, the nightly build of Google's Chrome web browser. There are versions of Chrome Canary available for Windows (32-bit and 64-bit), Mac, and Android, all of which can be downloaded from the Chrome Canary website (https://packt.live/36DQASx).

Chrome Canary is a nightly build of the Chrome browser that may be unstable, but often contains experimental code and features that haven't been tested as thoroughly as they would be before being released in Google Chrome. This means there can be security issues, and such browsers should be used to assist development only. In other words, don't use them to store personal information such as credit card details and passwords.

Once you've downloaded and installed Chrome Canary, you can use it. It is very recognizable as it looks similar to the Chrome browser, though it does have a lot more yellow in the app's icon:

Figure 13.1: Chrome Canary (left) and Chrome (right)

You can also see from the About Google Chrome page that Chrome Canary will be running at least a few major versions ahead of Google Chrome.

In the following screenshot, you can see that Google Chrome is running major version 76:

Figure 13.2: About page for Google Chrome official build

In the following screenshot, you can see that Google Chrome Canary is running major version 78:

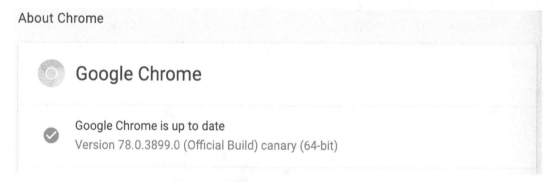

Figure 13.3: About page for Chrome Canary official build

We will look at some features that are only available in Chrome Canary later in this chapter when we look at the CSS Paint API and CSS Custom Properties API.

Nightly builds are also available for Firefox (from Firefox Nightly), Safari (Safari dev), and Edge (Edge Nightly). These browsers offer regular updates and are continuously improving the features they offer.

Experimental Flags

As well as new developments being available in the **nightly** builds of a browser, features are often released in the official build of the browser but behind a feature flag. This allows a developer to opt into a feature so that they can experiment or prototype a solution based on a cutting-edge browser feature while protecting the general public from a feature that is still being worked on and has not been standardized yet.

You can enable and disable flags in the Chrome browser by visiting **chrome://flags/** (to visit the page, type **chrome://flags/** in your browser's address bar and hit Enter). This page provides a list of available experiments that the current version of the browser is running and gives you the option to enable or disable these experiments.

The following screenshot shows you what the **chrome://flags/** page looks like. You can search for features and set them to disabled, enabled, or default. As the warning suggests, these are experimental features and should be treated as such. For example, there is always the potential that they can introduce an undetected security issue:

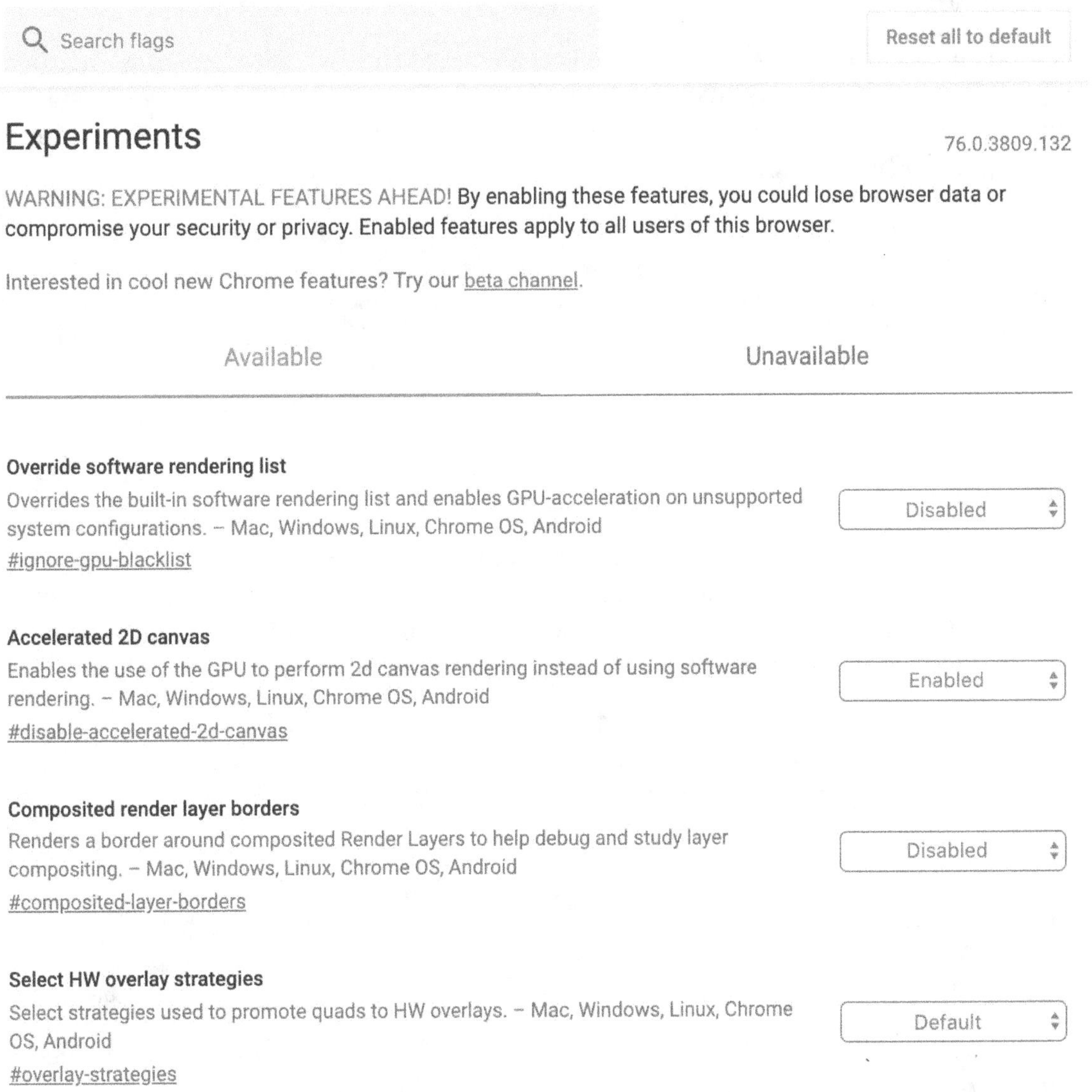

Q Search flags

Reset all to default

Experiments

76.0.3809.132

WARNING: EXPERIMENTAL FEATURES AHEAD! By enabling these features, you could lose browser data or compromise your security or privacy. Enabled features apply to all users of this browser.

Interested in cool new Chrome features? Try our beta channel.

Available	Unavailable

Override software rendering list

Overrides the built-in software rendering list and enables GPU-acceleration on unsupported system configurations. – Mac, Windows, Linux, Chrome OS, Android

#ignore-gpu-blacklist

Disabled

Accelerated 2D canvas

Enables the use of the GPU to perform 2d canvas rendering instead of using software rendering. – Mac, Windows, Linux, Chrome OS, Android

#disable-accelerated-2d-canvas

Enabled

Composited render layer borders

Renders a border around composited Render Layers to help debug and study layer compositing. – Mac, Windows, Linux, Chrome OS, Android

#composited-layer-borders

Disabled

Select HW overlay strategies

Select strategies used to promote quads to HW overlays. – Mac, Windows, Linux, Chrome OS, Android

#overlay-strategies

Default

Figure 13.4: Chrome experimental flags page (chrome://flags)

These experiments relate to all aspects of the browser and can include experimental dev tools features, aspects of browser performance, and experimental web platform features such as HTML, CSS, and JavaScript. For example, in the past, this flag has included experiments relating to the new CSS grid layout and flexbox layout features, both of which we covered in *Chapter 2, Structure and Layout*.

Browser Vendor Status

While nightly builds of browsers can give you a chance to work with the latest features during their development and while they are still in an experimental state, browser vendors also provide useful resources for keeping up with the latest features and future roadmap of their browsers.

For the Chrome browser, you can keep up to date with current and future developments on the Chrome status website (https://packt.live/2qvpagS) and see what new features are being launched and how they are useful to web developers, The Chrome dev team's updates blog (https://packt.live/2PVR1Br) is also a great resource.

In the case of Google Chrome, information specifically about web dev tools improvements can also be accessed via the dev tools' "What's New" panel.

Caniuse

The browser landscape is a complicated one. Sometimes, a new feature will be accepted as standard and will be implemented in most or all browsers very quickly, but this is not always the case. Some features can appear in one browser and go through many iterations before a standard is agreed upon and they are implemented elsewhere.

For example, CSS **Flexbox**, which you learned about in *Chapter 2, Structure and Layout*, first appeared way back in 2009 and went through several experimental versions before a recommendation for the CSS standards was agreed upon. The browser support is now quite good, but it has taken 10 years to get there.

We, as developers, often have to work to browser specifications (that is, a list of browsers the page must support), and we may need to know whether a browser feature is available for all of the browsers included in the specification. There are several sources for this information, and one of the most palatable is the caniuse website.

If we need to know if a browser feature is available in Edge we can check the **caniuse** website, and it will provide a breakdown of supported browsers.

For example, to check support for the CSS grid layout, which lets us create complex layouts and page structures without adding a lot of unwanted markup to a web page, we would visit https://packt.live/2JY3z7I. The result is a map of browser support where green means full support, red means no support, and a lighter green means some support that is either behind a vendor prefix or is not the standardized version of the feature.

At the time of writing, the results for the CSS grid layout are as follows. As you can see, most of the current browsers support this feature with global support at around 91% with no prefix. IE 11 support has an older version of the specification, so it is supported with some caveats:

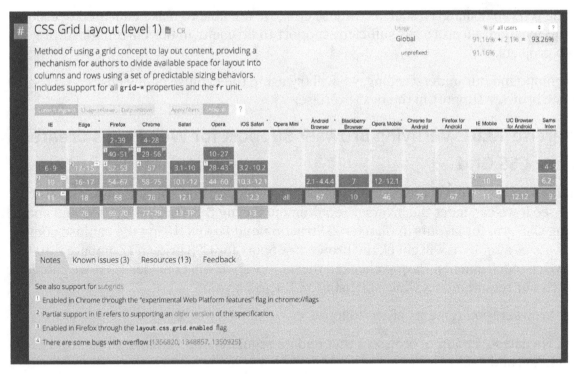

Figure 13.5: CSS Grid support

If we were to do the same for CSS grid layout level 2, that is, the CSS subgrid feature, which we can find at https://packt.live/34F3629, the story would be very different. There is very little support for this feature at the moment, with the one glimmer of green being a future version of the Mozilla Firefox browser. The results, at the time of writing, are shown in the following screenshot:

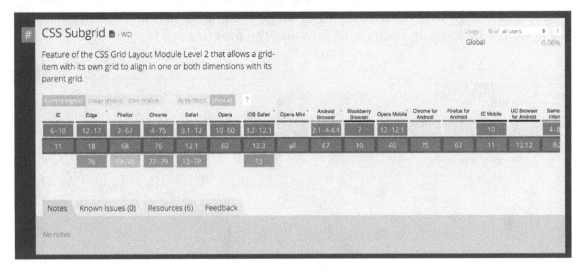

Figure 13.6: CSS Subgrid support

Using reference material such as caniuse.com, we are able to make educated decisions about whether a feature has sufficient support to be useful in our current or future web projects.

To compound our understanding, we will try using the caniuse.com website as a tool to check browser support in the next exercise.

Exercise 13.01: Verifying Browser Support for Web Pages Created Using CSS Grid

In this exercise, we will create a web page using CSS grid and check browser support and see if we can meet the browser requirements using this feature. You learned about using CSS grid for layouts in *Chapter 2, Structure and Layout*. Using the caniuse.com website as a tool, we will check the browser support for CSS grid and compare it to the browser requirements. Having checked the support, we will consider what we can do to match our requirements while still using a CSS grid layout.

Our browser requirements are as follows:

- The latest version of browsers that update regularly, for example, Chrome, Firefox, Edge, and Safari

- Internet Explorer support from version 11

The steps are as follows:

1. Firstly, create a folder named **Chapter13** and then under it create a file called **Exercise 13.01.html**. Save this file.

2. Add the following code to the **Exercise 13.01.html** file. This code creates a simple web page with minimal styles to set box-sizing for all the elements, which will set the padding and margin to zero and the font to **Arial**:

```html
<!DOCTYPE html>
<html lang="en">
    <head>
        <meta charset="utf-8">
        <title>Exercise 13.01: Browser support</title>
        <style>
         body {
                padding: 0;
                margin: 0;
                box-sizing: border-box;
```

```
            font-family: Arial, Helvetica, sans-serif;
        }
        *, *::after, *::before {
            box-sizing: inherit;
        }
    </style>
</head>
<body>

</body>
</html>
```

3. We are going to use the CSS grid to create a layout like so:

```
<div class="container grid">
    <header class="header">
        <h1>Browser support</h1>
    </header>
    <nav class="nav">
        <ul>
            <li class="current">Article 1</li>
            <li><a href="#">Article 2</a></li>
        </ul>
    </nav>
    <main class="main">
        <h2>Article 1</h2>
        <p>
        Lorem ipsum dolor sit amet, consectetur adipiscing
        elit. Duis sit amet porttitor dolor. Nunc
        sodales sodales risus. Donec vitae ex tempor
        leo blandit egestas sed sed odio. Vivamus
        nisi ligula, pharetra vel nisl sed, aliquam
        varius tellus. Maecenas vel semper eros,
        a pellentesque massa. Nullam rhoncus elit metus,
        sed rutrum ipsum malesuada sit amet.
        Maecenas nibh metus, fringilla vitae
        vulputate varius, consectetur nec ipsum.
        Suspendisse vitae fermentum felis, scelerisque
        imperdiet quam. Duis posuere maximus ex,
        tincidunt hendrerit dolor commodo id.
        </p>
```

```
        <p>
        Aenean id laoreet ligula. Ut blandit odio arcu.
        Sed ex felis, auctor eget lobortis quis, iaculis
        in enim. Cras vehicula blandit odio. Aenean at
        mperdiet ex, sed lobortis dolor. Vivamus vehicula
        consectetur sem faucibus mollis. Pellentesque ac
        enim a velit ullamcorper varius in et dolor.
        Nam ultricies, urna at luctus feugiat, ante
        nisl maximus sapien, in rutrum dui dui at
        dolor. Fusce eu lorem ipsum.
        </p>
    </main>
    <section class="advertisement">
        <h2>Advertise here</h2>
        <p>Want your product to be noticed?</p>
        <p>Advertise here!</p>
    </section>
    <footer class="footer">
        <p>Add Copyright info here</p>
    </footer>
</div>
```

The following screenshot shows the result of adding this markup to the **body** element of **Exercise 13.01.html**:

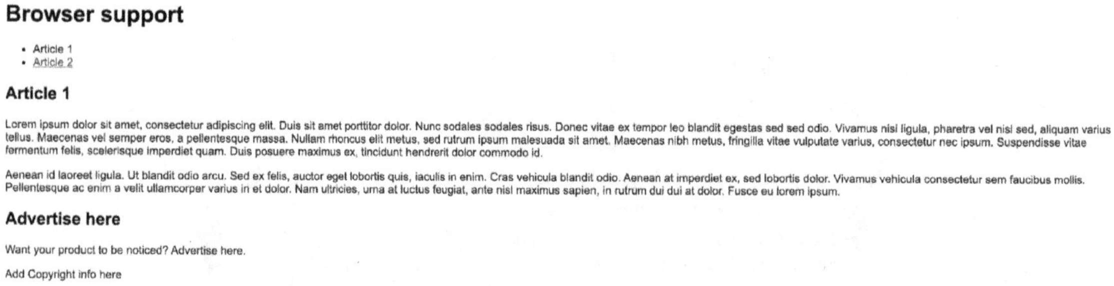

Browser support

- Article 1
- Article 2

Article 1

Lorem ipsum dolor sit amet, consectetur adipiscing elit. Duis sit amet porttitor dolor. Nunc sodales sodales risus. Donec vitae ex tempor leo blandit egestas sed sed odio. Vivamus nisl ligula, pharetra vel nisl sed, aliquam varius tellus. Maecenas vel semper eros, a pellentesque massa. Nullam rhoncus elit metus, sed rutrum ipsum malesuada sit amet. Maecenas nibh metus, fringilla vitae vulputate varius, consectetur nec ipsum. Suspendisse vitae fermentum felis, scelerisque imperdiet quam. Duis posuere maximus ex, tincidunt hendrerit dolor commodo id.

Aenean id laoreet ligula. Ut blandit odio arcu. Sed ex felis, auctor eget lobortis quis, iaculis in enim. Cras vehicula blandit odio. Aenean at imperdiet ex, sed lobortis dolor. Vivamus vehicula consectetur sem faucibus mollis. Pellentesque ac enim a velit ullamcorper varius in et dolor. Nam ultricies, urna at luctus feugiat, ante nisl maximus sapien, in rutrum dui dui at dolor. Fusce eu lorem ipsum.

Advertise here

Want your product to be noticed? Advertise here.

Add Copyright info here

Figure 13.7: Markup without styles

The next step is to add markup for our layout that we can then style using CSS and apply a grid layout. The markup includes a container div on which we can apply the display, that is, the grid CSS property. Within the container, we create five elements – a header element with the page heading, a **nav** element with the navigation for the page, the main element for an article, a section for an advertisement, and a footer element that may contain legal information and site navigation. We can add some placeholder content to fill out the page.

4. We can now use the CSS grid to layout the elements of our page. We set the grid class to use **display: grid** and set the element's height to the entire viewport (**100 vh**). Here, we will make use of the **grid-template-areas** property to set out our layout with named areas. We'll create a map of what our layout will look like; for example, the header section spans two columns of the first row, starting in the first column:

```
<style>
  .grid {
        display: grid;
        height: 100vh;
        grid-template-rows: 100px 1fr 100px;
        grid-template-columns: 100px 1fr 200px;
        grid-template-areas:
            "header header advert"
            "nav main advert"
            "footer footer footer";
  }
  .grid .footer {
        grid-area: footer;
  }
  .grid .header {
        grid-area: header;
  }
  .grid .main {
        grid-area: main;
  }
  .grid .nav {
        grid-area: nav;
  }
  .grid .advertisement {
        grid-area: advert;
  }
</style>
```

If you now right-click on the filename in VSCode on the left-hand side of the screen and select **Open In Default Browser**, you will see the following screenshot that shows the result of applying the CSS grid layout:

Browser support

Advertise here

Want your product to be noticed?

Advertise here!

- Article 1
- Article 2 **Article 1**

Lorem ipsum dolor sit amet, consectetur adipiscing elit. Duis sit amet porttitor dolor. Nunc sodales sodales risus. Donec vitae ex tempor leo blandit egestas sed sed odio. Vivamus nisi ligula, pharetra vel nisl sed, aliquam varius tellus. Maecenas vel semper eros, a pellentesque massa. Nullam rhoncus elit metus, sed rutrum ipsum malesuada sit amet. Maecenas nibh metus, fringilla vitae vulputate varius, consectetur nec ipsum. Suspendisse vitae fermentum felis, scelerisque imperdiet quam. Duis posuere maximus ex, tincidunt hendrerit dolor commodo id.

Aenean id laoreet ligula. Ut blandit odio arcu. Sed ex felis, auctor eget lobortis quis, iaculis in enim. Cras vehicula blandit odio. Aenean at imperdiet ex, sed lobortis dolor. Vivamus vehicula consectetur sem faucibus mollis. Pellentesque ac enim a velit ullamcorper varius in et dolor. Nam ultricies, urna at luctus feugiat, ante nisl maximus sapien, in rutrum dui dui at dolor. Fusce eu lorem ipsum.

Add Copyright info here

Figure 13.8: Grid layout applied

The layout now takes up the whole **viewport**. The header spans two columns, with the advert spanning two rows on the right. The footer spans the whole of the bottom of the page. The article expands responsively and takes up the middle section of the page.

5. Next, we will update the styles of the page to give each section a bit more differentiation. We will add some borders to distinguish between sections, style the nav, and use a background color to define the advertisement and footer elements. We'll also apply some **padding** to give the content a bit of breathing space:

```
.header {
    border-bottom: 1px solid gray;
}
.nav {
    border-right: 1px solid gray;
}
.nav ul {
    list-style-type: none;
```

```css
    margin: 0;
    padding:0;
    width: 100%;
}
.nav ul li {
    width: 100%;
    margin: 0;
    padding: 16px 8px;
}
.current {
    background: lightgray;
    border-bottom: 1px solid gray;
}
.main {
    display: block;
    padding: 16px;
    overflow:auto;
}
.advertisement {
    padding: 16px;
    border-left: 1px solid gray;
    background: greenyellow;
}
.footer {
    background: black;
    color: white;
    padding: 0 1rem;
}
```

The following screenshot shows the result of these changes:

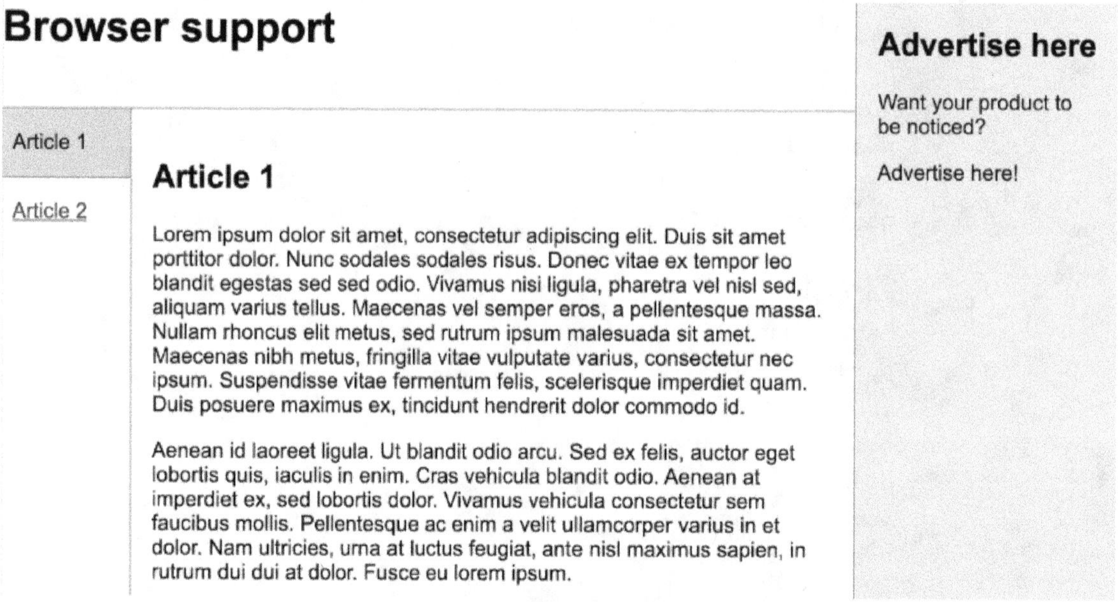

Browser support

Advertise here

Want your product to
be noticed?

Advertise here!

Article 1

Article 2

Article 1

Lorem ipsum dolor sit amet, consectetur adipiscing elit. Duis sit amet
porttitor dolor. Nunc sodales sodales risus. Donec vitae ex tempor leo
blandit egestas sed sed odio. Vivamus nisi ligula, pharetra vel nisl sed,
aliquam varius tellus. Maecenas vel semper eros, a pellentesque massa.
Nullam rhoncus elit metus, sed rutrum ipsum malesuada sit amet.
Maecenas nibh metus, fringilla vitae vulputate varius, consectetur nec
ipsum. Suspendisse vitae fermentum felis, scelerisque imperdiet quam.
Duis posuere maximus ex, tincidunt hendrerit dolor commodo id.

Aenean id laoreet ligula. Ut blandit odio arcu. Sed ex felis, auctor eget
lobortis quis, iaculis in enim. Cras vehicula blandit odio. Aenean at
imperdiet ex, sed lobortis dolor. Vivamus vehicula consectetur sem
faucibus mollis. Pellentesque ac enim a velit ullamcorper varius in et
dolor. Nam ultricies, urna at luctus feugiat, ante nisl maximus sapien, in
rutrum dui dui at dolor. Fusce eu lorem ipsum.

Figure 13.9: Further styles applied to differentiate between sections of the layout

6. Our web page now has a layout defined with CSS grid template areas. CSS grids
    mean we don't have to use additional markup to control our layout and keep style
    and content separate. However, we have browser requirements to meet and the
    CSS grid, being relatively new, may give us some concerns. In this step, we will
    check support using caniuse.com.

    To do so, we simply visit https://packt.live/2JY3z7I in a browser. We navigate to
    that page and then we will see the support table for the CSS grid.

    We can see the result in *Figure* 13.5. It shows the caniuse.com support table for the
    CSS grid.

    Looking at the support table, we can see a lot of green, which means support for
    the CSS grid in most major browsers is really good. There are a few red blocks
    for Opera Mini and Blackberry Browser, but neither of those is included in
    our requirements.

    The one area that may concern us is IE, which only seems to have partial support
    for the CSS grid.

7. Hover the mouse over the IE 11 block so that we can see more details of what partial support means.

The following screenshot shows details of IE 11's partial support for the CSS grid:

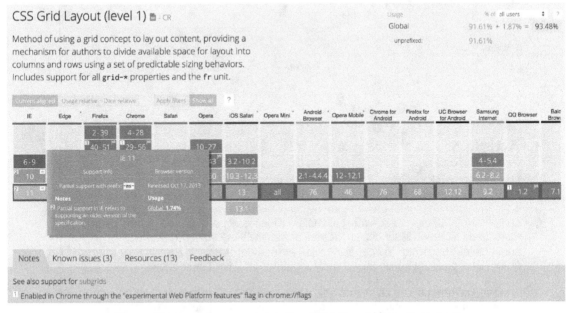

Figure 13.10: caniuse.com details of IE 11's grid layout support

The concerns we have are that the browser supports an older version of the CSS grid specification and it also expects the vendor prefix, that is, **-ms-**, to be used.

Checking the page in IE 11 shows us the issues that caniuse has flagged. In the following screenshot, we can see the result of opening the page in IE 11:

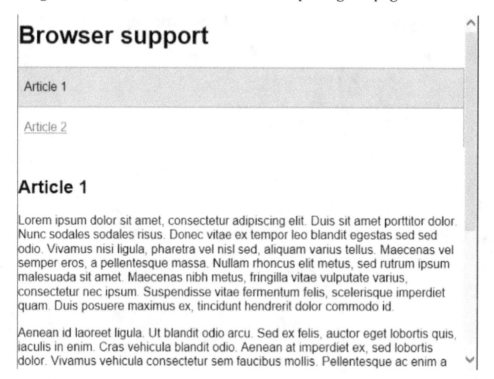

Browser support

Article 1

Article 2

Article 1

Lorem ipsum dolor sit amet, consectetur adipiscing elit. Duis sit amet porttitor dolor. Nunc sodales sodales risus. Donec vitae ex tempor leo blandit egestas sed sed odio. Vivamus nisi ligula, pharetra vel nisl sed, aliquam varius tellus. Maecenas vel semper eros, a pellentesque massa. Nullam rhoncus elit metus, sed rutrum ipsum malesuada sit amet. Maecenas nibh metus, fringilla vitae vulputate varius, consectetur nec ipsum. Suspendisse vitae fermentum felis, scelerisque imperdiet quam. Duis posuere maximus ex, tincidunt hendrerit dolor commodo id.

Aenean id laoreet ligula. Ut blandit odio arcu. Sed ex felis, auctor eget lobortis quis, iaculis in enim. Cras vehicula blandit odio. Aenean at imperdiet ex, sed lobortis dolor. Vivamus vehicula consectetur sem faucibus mollis. Pellentesque ac enim a

Figure 13.11: Result in IE 11

The result is that none of our layout code is working. We just get each element appearing one after the other.

We aren't meeting our browser requirements at the moment but we now know that IE 11 can support some version of grid layout. To meet our browser requirements, we need to add some IE-specific CSS to solve the issues in IE 11. With CSS, the last rule that the browser recognizes will be the one that's been applied so that we can support both IE and modern browser implementations of the CSS grid.

8. Firstly, we apply the **-ms-** prefix for the grid and the row and column sizing. Instead of **grid-template-rows**, we apply the sizing to **-ms-grid-rows** and instead of **grid-template-columns**, we use **-ms-grid-columns**:

```
.grid {
    display: -ms-grid;
    display: grid;
    height: 100vh;
    -ms-grid-rows: 100px 1fr 100px;
    grid-template-rows: 100px 1fr 100px;
    -ms-grid-columns: 100px 1fr 200px;
```

```
        grid-template-columns: 100px 1fr 200px;
        grid-template-areas:
            "header header advert"
            "nav main advert"
            "footer footer footer";
    }
```

9. Because the IE version of grid does not support named grid areas and the **grid-template-areas** property, we need to set out each of our template areas with **-ms-grid-row** and **-ms-grid-column** to set the starting row and column position and then spans to control the amount of space the elements take up. Here is the code to do this:

```
.grid .footer {
    -ms-grid-row: 3;
    grid-row: 3;
    -ms-grid-column: 1;
    grid-column: 1;
    -ms-grid-column-span: 3;
    grid-area: footer;
}
.grid .header {
    -ms-grid-row: 1;
    grid-row: 1;
    -ms-grid-column: 1;
    grid-column: 1;
    -ms-grid-column-span: 2;
    grid-area: header;
}
.grid .main {
    -ms-grid-row: 2;
    grid-row: 2;
    -ms-grid-column: 2;
    grid-column: 2;
    grid-area: main;
}

.grid .nav {
    -ms-grid-row: 2;
    grid-row: 2;
    -ms-grid-column: 1;
    grid-column: 1;
    grid-area: nav;
```

```
        }
   .grid .advertisement {
       -ms-grid-row: 1;
        grid-row: 1;
       -ms-grid-column: 3;
        grid-column: 3;
       -ms-grid-row-span: 2;
        grid-area: advert;
   }
```

10. The result is a CSS grid layout that works in all modern browsers and IE back to version 10, which more than meets our browser requirements.

    The following screenshot shows the resulting layout in IE 11:

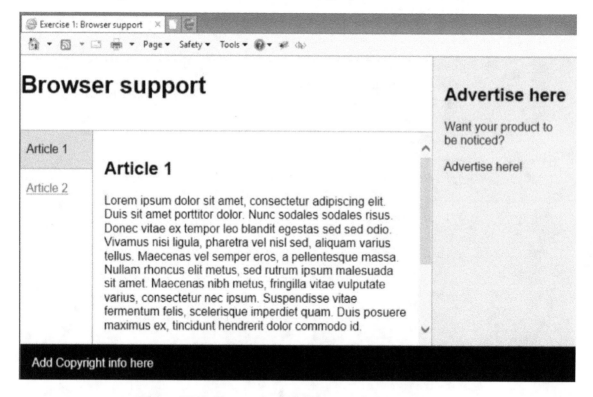

Figure 13.12: The result of making use of IE 11's CSS grid

As the preceding screenshot shows, we've solved the issues with our layout in IE 11 and now meet our browser support requirements.

By keeping track of the browser features we use on a web page and using tools such as caniuse.com, we can improve browser support and meet the requirements of our users and clients.

In the next part of this chapter, we will look at the CSS Paint API, which is a proposed specification that was created by the CSS Houdini Taskforce and designed to give developers more access to the CSS rendering pipeline and the CSSOM.

We will look at how we can use the technology of today by combining some of the techniques we've discussed in this chapter. We will also see how we can use progressive enhancement to provide a good user experience for those users who don't have access to the technology because it is not available in the browser they are using.

First, we will provide an explanation of the role of the CSS Houdini Taskforce and what it means for the future of web development.

CSS Houdini

The CSS **Technical Architecture Group (TAG)** Houdini Task Force (more prominently known as CSS Houdini) has been tasked with opening up the black box systems that make up browser rendering by providing APIs for developers to work with and for developers to change the behaviors of processes such as layout and paint.

The CSS Houdini group creates draft specifications to enhance CSS by giving developers more access to the render pipelines of the browser. We will look at two of these APIs: the CSS Paint API, which gives us greater control of the painting aspects of CSS such as background colors, gradients, and masks, and the CSS Properties and Values API, which you can use to register and define custom properties as being of a CSS type (such as a color or length). We will see how the two APIs can be used in unison.

By giving the web developer low-level access to the CSS render pipeline for layout, composition, and paint processes, developers are better able to control these processes and create more optimal scripts for rendering at the CSSOM level.

CSS Paint API

The first API from CSS Houdini to see the light of day, with an actual browser implementation, is the Paint API, which first appeared in Chrome 65 and is now available in Opera with the intent for Firefox, Safari, and Edge to implement it as well.

The Paint API gives us a way to control a 2D rendering context that we can programmatically draw to using JavaScript. This is a lot like the Canvas element we learned about in *Chapter 7, Media – Audio, Video, and Canvas.*

There are a couple of significant differences between the Paint API and the 2D rendering context of the Canvas:

- The main difference is that the Paint API allows us to use the 2D image with any CSS property that accepts an image as a value. For example, we can create a 2D image and then use it as the value for a **background-image**, **border-image**, or **mask-image**.

- The Paint API also makes use of a worklet, which is a bit like a worker, that is, a thread running parallel to the main browser thread. Essentially, this means that the worklet won't block the browser from doing other processes. The worklet has very restricted functionality, which means it can be heavily optimized for rendering.

To create a paint worklet, we will need to follow a few steps:

1. Firstly, we will create a simple JavaScript class with a **paint** method. The paint method will receive a 2D rendering context, a geometry object with the dimensions of the area in which we can paint, and, optionally, a set of properties.

2. We need to create this class in a separate file.

3. To be able to use the paint worklet, we must register it, which we will do with the **registerPaint** function. This function is only made available when we add the file to our web page as a paint worklet module. The **registerPaint** function takes two arguments: a unique key that you will use to identify the worklet and the JavaScript class with the paint method.

4. To add the module as a paint worklet, we'll use the **addModule** method on the **paintWorklet** object of CSS, which will look something like **CSS.paintWorklet. addModule("workletModule.js");**.

5. With the paint worklet registered and added to our web page, we can use it in CSS wherever we may expect an image. For example, to use a paint worklet as a background image for a paragraph element, we could call the paint worklet with a special CSS **paint** function:

```
<style>
  p {
    background-image: paint(paintWorklet);
  }
</style>
```

Let's run through the creation of a very simple paint worklet to get a better idea of how all of this will tie together.

> **Note**
>
> *Exercise 13.02 – 13.06* and *Activity 13.01* require the use of the web server. If you have not already followed the steps to install and set up your web server, you will need to go back to the instructions in the *Preface* and follow those steps before you begin.
>
> To test these exercises and the activity, you will need to place your files into the XAMPP **htdocs** folder on your computer and access them through your browser using **localhost**. Review the web server instructions if you need help with these steps.

Exercise 13.02: Creating a Red Fill Paint Worklet

In this exercise, we will create our first very simple paint worklet. Our first iteration of a paint worklet will be limited to painting a rectangle, that is, the full size of the canvas, with the fill color red. This exercise will get us used to registering a paint worklet and applying it to CSS. Later, we will expand upon this simple paint worklet to develop some more sophisticated effects.

The steps are as follows:

1. Firstly, we want to create our worklet file. Create a new file and save it as **red-fill-paint.js** under **Chapter13** folder.

2. To be able to use the paint worklet, we will need a web page. In the same directory as our paint worklet file, create a file and save it as **Exercise 13.02.html**. Add the following code for a simple web page:

```html
<!DOCTYPE html>
<html lang="en">
    <head>
        <meta charset="utf-8">
        <title>Exercise 13.02: A Simple Paint Worklet</title>
        <style>
         body {
             padding: 0;
             margin: 0;
             font-family: Arial, Helvetica, sans-serif;
         }
```

```
        </style>
      </head>
      <body>

      </body>
    </html>
```

3. Returning to **red-fill-paint.js**, we need to create a JavaScript class with a **paint** method. The behavior of our paint worklet will be defined in this **paint** method. The paint method will receive a context object that gives us access to drawing methods we can use to paint and a geometry object, which we can use to calculate the width and height of the area the worklet is applied to. To fill the rectangle with red, use the following code:

```
class RedFillPaintWorklet {
  paint(context, geometry) {
    context.fillStyle = "red";
    context.fillRect(0, 0, geometry.width, geometry.height);
  }
}
```

4. For this JavaScript class to be recognized as a paint worklet, we need to register it. We do that with **registerPaint**. We give the worklet a name and pass it the worklet class:

```
registerPaint("red-fill", RedFillPaintWorklet);
```

5. For now, we are done with the paint worklet module, but we need to add it to our web page. Returning to the **Exercise 13.02.html** file, we need to add a script and add the module. This will complete the registration of the paint worklet with our web page:

```
<script>
  CSS.paintWorklet.addModule("red-fill-paint.js");
</script>
```

6. Now, we can use the paint worklet. To do that, we will add a **div** element to the web page with the **red-background** class attribute:

```
<div class="red-background">
  This div should have a red background.
</div>
```

7. We will then use the paint function in our CSS and pass it the **red-fill** name that we have registered our paint worklet with. We'll use the following CSS to style the element, which we can add to the style element in the header of the web page:

```
<style>
 .red-background {
  background: paint(red-fill);
  color: white;
  box-sizing: border-box;
  width: 100%;
  height: 100px;
  padding: 16px;
 }
</style>
```

8. Now save this file and place the **Chapter13** folder within the **htdocs** folder and make sure you have started the web server.

> **Note**
>
> Refer to the instructions in the *Preface* to know how to start the web server.

9. Head to your browser and type **localhost/Chapter13**. Then, hit *Enter*. You will obtain a list of all files present in this folder.

10. Click on the **Exercise 13.02.html** file.

    You will see the result of the above code to be similar to the following screenshot:

This div should have a red background.

Figure 13.13: A red fill paint worklet applied to a div element

We could have simply created the same effect by applying the **background-color: red;** style to the red-background class attribute. However, even with this very simple example, we can start to see some interesting features of paint worklets:

- The worklet is integrated with CSS, meaning we can use it for various properties, such as **background-image**, **background-color**, and **mask-image**, as well as for shorthand properties such as **background**.

- The dimensions of the **worklet** are dynamic and will repaint with the browser render pipeline, which means it is responsive by default.

In the next part of this chapter, we will look at how we can register properties that can be set in CSS and used in our paint worklet to make this background fill behavior more flexible and controllable via CSS styling.

Custom Properties

With CSS custom properties (often called CSS variables), we have a way to store a value for use in several CSS declarations. We can change this value as we like.

The syntax for a CSS custom property is a name prefixed with two dashes (**--**). We can then use the value stored in the property with the CSS **var** function. Here is an example of some CSS custom properties in use:

```
<style>
  :root {
      --dark-text-color: #131313;
      --light-text-color: #cfcfcf;
  }
  p.dark {
      color: var(--dark-text-color);
  }
  p.light {
      color: var(--light-text-color);
  }
</style>
```

The following screenshot shows the result of applying the preceding style rules to one paragraph element with the dark class attribute applied and one paragraph element with the light class attribute applied. The color variables we have created are set as the color style of the paragraph:

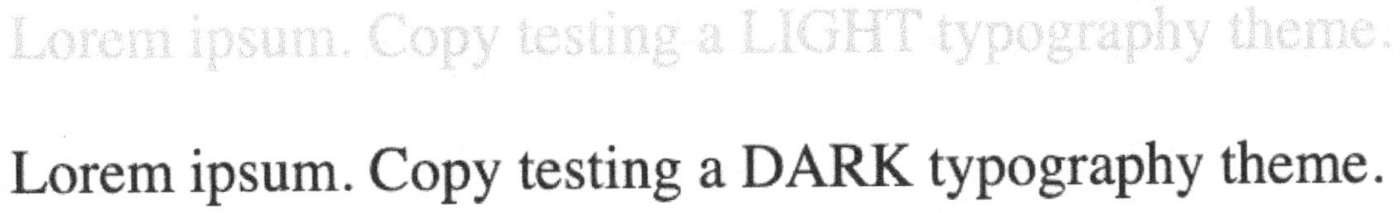

Lorem ipsum. Copy testing a LIGHT typography theme.

Lorem ipsum. Copy testing a DARK typography theme.

Figure 13.14: Custom properties for text color applied to paragraph elements

> **Note**
>
> The `:root` pseudo-class that we used in this example may be new to you. It specifies the root element of a document and is often used when setting global CSS variables.

Custom properties can be used as inputs for a paint worklet, which allows us to reuse the paint worklet with different inputs (such as color), thereby making them a lot more flexible. For example, if we wanted to use more colors than red in our fill color paint worklet, we could use a `--fill-color` custom property to decide which color to fill the HTML element in with.

We can register properties with the paint worklet to allow CSS properties to affect the paint worklet. A static `inputProperties` function can be added to our paint worklet JavaScript class and it will return an array of custom CSS properties the paint worklet expects as input, like so:

```
static get inputProperties() {
  return ["--fill-color"];
}
```

> **Note**
>
> A static function on a JavaScript class or object is a function that is created on the class itself and not created when an instance of the class is created.
>
> It is called through the class, for example, `ColorFillPaintWorklet.inputProperties`.

We can then access the property in our paint method via the properties argument. So, if we want to access the **--fill-color** property in our **paint** method, we can do the following:

```
paint(context, geometry, properties) {
        const color = properties.get("--fill-color");
```

Let's try an exercise to put what we've learned into practice and to see how we can use CSS properties to make a more flexible fill color paint worklet.

Exercise 13.03: The Fill Color Paint Worklet

In this exercise, we will create a new paint worklet that extends the red fill paint worklet to allow any color to be used to fill a rectangle. While still simple and easily achieved with standard CSS, this exercise will let us see how we can use properties with our paint worklets.

The steps are as follows:

1. Make a copy of **red-fill-paint.js** from *Exercise 13.01, Browser Support*, and rename it to **color-fill-paint.js** under **Chapter13** folder. We will change the name of the class to **ColorFillPaintWorklet**.

2. The paint worklet for the fill color paint worklet will be very similar to the red fill paint worklet, although we will take the color value from the **--fill-color** property, thus allowing us to change the color dynamically. To do this, we need to add a static **inputProperties** function that returns an array with the **--fill-color** property:

```
class ColorFillPaintWorklet {
  static get inputProperties() {
    return ["--fill-color"];
  }
```

3. We can then access the properties in the paint worklet through the **properties** argument of the **paint** method. We will get the **--fill-color** property and set the context's **fillStyle** to that value:

```
class ColorFillPaintWorklet {
  static get inputProperties() {
    return ["--fill-color"];
  }
  paint(context, geometry, properties) {
        const color = properties.get("--fill-color");
```

```
        context.fillStyle = color;
        context.fillRect(0, 0, geometry.width, geometry.height);
    }
}
```

4. To complete the work on the paint worklet, we need to register it as **"color-fill"**:

```
registerPaint("color-fill", ColorFillPaintWorklet);
```

5. To make use of the paint worklet, we will make a copy of the **Exercise 13.02. html** file and rename it **Exercise 13.03.html**. We'll change the title of the HTML document to *Exercise 13.03, Fill Color Paint Worklet*:

```
<head>
    <meta charset="utf-8">
    <title>Exercise 13.03: Fill Color Paint Worklet</title>
    <style>
     body {
        padding: 0;
        margin: 0;
        font-family: Arial, Helvetica, sans-serif;
     }
    </style>
</head>
```

6. Also, in the same document, change the paint worklet module to **color-fill-paint.js**:

```
<script>
        CSS.paintWorklet.addModule("color-fill-paint.js");
</script>
</body>
</html>
```

7. Next, we will replace the `.red-background` style with a `.fill-background` style that uses color-fill instead of red-fill. We'll also set up two new class attributes – `.green` and `.orange` – that will set the `--fill-color` property:

```
<style>
    .green {
        --fill-color: #05a505;
    }
    .orange {
        --fill-color: #ff9900;
    }
    .fill-background {
        box-sizing:border-box;
        background: paint(color-fill);
        color: white;
        width: 100%;
        height: 100px;
        padding: 16px;
    }
</style>
```

8. Finally, we'll replace the div element with the `.red-background` class attribute with two div elements, both with the `.fill-background` class attribute applied. However, one will have the `.green` class attribute applied to it, while the other will have the `.orange` class attribute applied to it:

```
<div class="fill-background green">
    This div should have a green background.
</div>
<div class="fill-background orange">
    This div should have an orange background.
</div>
```

9. Save the above file and place it inside the **Chapter13** folder, which should be already present within the **htdocs** folder. Make sure **color-fill-paint.js** is also present in this folder.

10. Start the web server and head to your browser. Type **localhost/Chapter13**. Then, hit *Enter*. You will obtain a list of all files present in this folder.

11. Click on **Exercise 13.03.html** file.

You will see the following screenshot that shows the color fill paint worklet being applied to two different divs with green and orange background fills:

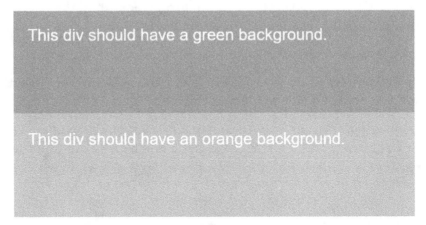

This div should have a green background.

This div should have an orange background.

Figure 13.15: Fill color paint worklet applied with different parameters

Custom properties allow us to create more flexible worklets. We can easily apply different parameters using CSS selectors. An added advantage is that because paint worklets are intrinsically tied to the CSS rendering system, we can take advantage of properties such as CSS **transition** to **animate** a paint worklet.

Next, we will look at how we can use multiple properties and work with user input to update the custom properties. This can make our paint worklet much more interactive and opens up some more possibilities regarding what we can do with a paint worklet.

Input Properties

Using some JavaScript, we can do even more with these custom properties. For example, we can listen to a mouse or touch input and pass the x and y coordinates to a paint worklet. With these values, we can paint an element based on the point where the pointer is moving or the point where the pointer enters and leaves the element.

In the next exercise, we will introduce a few more custom properties and update them based on the movement of the mouse. In so doing, we will create a paint worklet that is a bit more complex but also goes beyond what is easily done with standard CSS properties (such as **background-color**).

Exercise 13.04: Paint Worklet with Mouse Input

In this exercise, we will create a new paint worklet that paints a circle that follows the mouse as it moves over an element that the paint worklet is associated with. Using CSS pseudo-classes, we will then update the custom properties when the user clicks the element in order to paint a clicked state for the element.

The steps are as follows:

1. Create a new file under **Chapter13** folder and name it **pointer-input-paint. js**. We will create a new paint worklet in this file.

2. In pointer-input-paint.js, we want to create a JavaScript class called **PointerInputPaint**. This will encompass the behavior of our paint worklet. We will register the paint worklet as **"pointer"** so that we can use it in our document's CSS:

```
class PointerInputPaint {
    paint(context, geometry, properties) {

    }
}
registerPaint("pointer", PointerInputPaint);
```

3. Next, we can define a set of input properties, which is the set of CSS custom properties we will set via CSS. We want to know the x and y positions of the mouse, so we will store those positions as the **--position-x** and **--position-y** properties. We also want to customize the size and color of our mouse follower and for that, we will store properties for size and two different colors in the following properties: **--size, --primary-fill-color**, and **--secondary-fill-color**:

```
static get inputProperties() {
    return ["--position-x", "--position-y",
        "--primary-fill-color",
        "--secondary-fill-color", "--size"];
}
```

4. In the paint method of our paint worklet, we are going to need to do two things: read the values from our CSS input properties and use the values to draw a gradient onto the element the paint worklet is attached to.

5. First, we need to read the input properties. We do this in the paint function, where the properties argument gives us access to each of the input properties. We can use the get function to get a property by its name. We get each property, that is, the x and y coordinates of the mouse, the primary and secondary colors, and the size property:

```
paint(context, geometry, properties) {
    const x = properties.get("--position-x");
    const y = properties.get("--position-y");
    const primaryColor =
      properties.get("--primary-fill-color");
    const secondaryColor =
      properties.get("--secondary-fill-color");
    const size = properties.get("--size");
```

6. Then, we draw a radial gradient based on the input properties. In the preceding code, we use the x and y coordinates to create a circle of the size of our size input. We then add the primary and secondary colors as color stops to the gradient. We learned about creating gradients on a canvas context in *Chapter 7, Media – Audio, Video, and Canvas*. This technique is the same as it is for a paint worklet because that uses a version of a canvas:

```
var gradient = context.createRadialGradient(
    x, y, 0, x, y, geometry.width * size);
gradient.addColorStop(0.24, primaryColor);
gradient.addColorStop(0.25, secondaryColor);
gradient.addColorStop(1, secondaryColor);
```

The gradient here is positioned at the x and y values, which will be defined by the mouse position.

7. We use this gradient as a fill style and fill the whole rectangle for the element, like we did in the previous exercises:

```
context.fillStyle = gradient;
context.fillRect(0, 0, geometry.width, geometry.height);

}
```

That is the whole of the paint worklet.

8. Next, we will create an HTML document so that we can make use of our paint worklet. Create a new file and name it **Exercise 13.04.html**. Copy the following **markup** into that file. This markup creates a web page with a single button centered on the page. The button is styled to create a large button:

```html
<!DOCTYPE html>
<html lang="en">
    <head>
        <meta charset="utf-8">
        <title>Exercise 13.04: Mouse Input Worklet</title>
        <style>
            body {
                padding: 0;
                margin: 0;
                font-family: Arial, Helvetica, sans-serif;
            }
        </style>
    </head>
    <body>
        <div class="centered-content">
            <button class="button pointer centered-content">
                Click me!
            </button>
        </div>
        <style>
            .centered-content {
                display: flex;
                align-items: center;
                justify-content: center;
            }
            .button {
                outline: none;
                user-select: none;
                -webkit-appearance: none;
                appearance: none;
                margin: 16px;
                width: 375px;
                height: 150px;
                padding: 16px;
                font-size: 24px;
```

```
            }
        </style>
    </body>
</html>
```

As shown in the following screenshot, this simple HTML document creates a single button with some CSS classes and the text **Click me!**. We have added a few styles to replace the default button styles and to center the content:

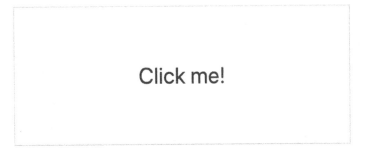

Figure 13.16: A "Click me!" button

9. In a script tag, we'll add our paint worklet to this page so that we can use it in our CSS:

```
<script>
    CSS.paintWorklet.addModule("pointer-input-paint.js");
</script>
```

10. Next, we need to create our CSS custom properties so that we can store the values that our paint worklet will read from. We will create these in a CSS declaration targeted at the **:root** pseudo-class so that they are defined globally for all CSS:

```
<style>
    :root {
        --position-x: 0;
        --position-y: 0;
        --primary-fill-color: #7200ca;
        --secondary-fill-color: #12005e;
        --size: 0.1;
    }
```

11. We can now add the pointer class definition to the CSS. This will use our paint worklet as a background to any element the pointer class attribute is added to. When the button is active (or clicked), we will change the values of two of our CSS properties, that is, the secondary fill color (**--secondary-fill-color**) and the size (**--size**). We do this by setting the properties using the CSS pseudo-classes for **:visited** and **:active** (which we learned about in *Chapter 1, Introduction to HTML and CSS*):

```
.pointer {
    background: paint(pointer);
    color: white;
}
.pointer:visited,
.pointer:active {
    --secondary-fill-color: #12005e;
    --size: 10;
}
</style>
```

The result is a dark purple button, as shown in the following screenshot. There is a small light purple dot in the top left corner of the button. At the moment, this will not move around:

Figure 13.17: The button with the pointer paint worklet applied

When the button is clicked, it will turn a lighter shade of purple, as shown in the following screenshot:

Figure 13.18: The active state of the button with the pointer paint worklet applied

12. To get the mouse follower working correctly, we need a little bit of JavaScript. Let's add a **mousemove** event **listener** to our **button** element. An event listener will trigger when something happens; in this case, when the mouse is moved by the user. When the mouse is moved, we will set the **--position-x** and **--position-y** properties with the new x and y coordinates of the mouse pointer over the element. We'll use the **setProperty** function to set a property on the style object of the document element of the HTML DOM. This is equivalent to setting the CSS properties inline on the HTML element or using the **:root** selector in CSS. We add this to our script tag:

```
const mouseInput = document.querySelector(".
    pointer");
mouseInput.addEventListener("mousemove", () => {
document.documentElement.style.setProperty("--
    position-x", event.offsetX);
document.documentElement.style.setProperty("--
    position-y", event.offsetY);
});
```

13. Now save this file and place the **Chapter13** folder within the **htdocs** folder and make sure you have started the web server.

> **Note**
>
> Refer to the instructions in the *Preface* to know how to start the web server.

14. Head to your browser and type **localhost/Chapter13**. Then, hit *Enter*. You will obtain a list of all files present in this folder.

15. Click on the **Exercise 13.04.html** file.

    You will see the result of the above code to be similar to the following screenshot:

Figure 13.19: The button with pointer following the mouse

In this exercise, we have used the mouse position to update CSS custom properties and used these properties in a paint worklet to draw a mouse follower on an element.

We can begin to see some of the benefits of using a paint worklet here as we haven't had to add any extra elements or pseudo-elements to the HTML to handle the pointer.

In the next section, we will look at another CSS Houdini API, the CSS Properties and Values API, and we will see how we can use this API to register CSS custom properties and how we can have even more control of them in a paint worklet.

CSS Properties and Values API

The CSS Properties and Values API allows us to register custom CSS properties and define their initial value, the syntax they are based on, and whether the value inherits down the HTML DOM tree (as, for example, a font size or family property value would).

To register CSS properties, we need to use JavaScript. The API provides us with a method so that we can register CSS properties with the **registerProperty** method, which is available on the global CSS object.

Let's have a look at an example:

```
CSS.registerProperty({
    name: "--primary-fill-color",
    syntax: "<color>",
    inherits: false,
    initialValue: "white"
});
```

The **registerProperty** method expects an object that defines our CSS custom property. We need to give the property a name, for example, **--primary-fill-color** or **--size**, and define whether it **inherits** – in other words, whether the value is inherited by DOM elements that are children of the element the property is associated with.

We can also define the syntax that the property follows. If we define the syntax that the property follows, we also need to provide an initial value for the property. There are quite a few different syntax types that we can choose from, including **<length>**, **<angle>**, **<color>**, **<number>**, and **<time>**, which define different parts of the CSS syntax.

In the preceding example, we defined a color property that follows the **<color>** syntax and has an initial value set as white. The benefit of defining a property in this way is that we can tell if the value will make sense in a CSS rule and we can use it as a value in a CSS transition where interpolation from one value to another can be calculated.

At the time of writing, this API will only work in Chrome Canary, which presents a further problem that we will discuss in the next section when we talk about progressive enhancement as a technique for developing for the web, where not all users will have the same browser or a browser with the same capabilities.

In the next exercise, we will do just that. We will take the fill color paint worklet from our previous exercise and animate it by changing a custom property with the CSS **transition** and animation properties.

Next, we will look at how we can use CSS animations or transitions to control custom properties. We will use animation to further improve our pointer paint worklet.

Animating Custom Properties

Remember that our paint worklet is integrated into CSS and its rendering pipeline. This means that it will repaint if the style properties change. Because we can provide an input property and the paint worklet will call the paint method if the property changes, we are able to use CSS to animate the custom property. This means that we can easily animate a paint worklet with CSS using the CSS animation and transition properties.

There is one caveat that makes this more difficult, though. The problem is that the type of CSS custom property is not easily recognized. What we need is a way of telling CSS what type of property we are passing it. In other words, we need to be able to tell CSS if we are giving it a percentage value, a number, a color value, a length value (**px**, **em**), or an angle value (**deg**, **turn**, **rad**).

We can use the CSS Properties and Values API to solve this problem. In the next exercise, we will do just that.

Exercise 13.05: Animating a Paint Worklet

In this exercise, we will animate the paint worklet we created in *Exercise 13.04, Paint Worklet with Mouse Input*. We do not need to make any changes to our paint worklet for this exercise, but we will need to register custom properties so that they can be animated using the CSS transition property.

The steps are as follows:

1. Make a copy of the **Exercise 13.04.html** file from our previous exercise. Rename the copied file **Exercise 13.05.html** under **Chapter13** folder and update the title of the web page document:

    ```
    <title>Exercise 13.05: Animated Paint Worklet</title>
    ```

2. We want to add a transition to our pointer class declaration, which will transition the custom properties for **--size, --primary-fill-color**, and **--secondary-fill-color**. This will cause the values for these properties to transition to the values that have been set when the element with the pointer class is clicked:

    ```
    .pointer {
        background: paint(pointer);
        color: white;
        transition: --size 1s, --primary-fill-color 0.5s,
            --secondary-fill-color 0.5s;
    }
    ```

    The transition will not work yet because the custom properties have not had their syntax type defined and the CSS engine does not know how to transition from one value to another without knowing the type of the value.

3. To get the transitions to work, we need to register the properties. We do that with the **registerProperty** method:

    ```
    CSS.registerProperty({
        name: "--primary-fill-color",
        syntax: "<color>",
        inherits: false,
        initialValue: "#7200ca"
    });
    CSS.registerProperty({
        name: "--secondary-fill-color",
        syntax: "<color>",
        inherits: false,
        initialValue: "#12005e"
    });
    ```

```
CSS.registerProperty({
    name: "--size",
    syntax: "<number>",
    inherits: false,
    initialValue: 0.05
});
```

4. Now save this file and place the **Chapter13** folder within the **htdocs** folder and make sure you have started the web server.

> **Note**
>
> Refer to the instructions in the *Preface* to know how to start the web server.

5. Head to your browser and type **localhost/Chapter13**. Then, hit *Enter*. You will obtain a list of all files present in this folder.

6. Click on the **Exercise 13.05.html** file.

    Now that we've registered the properties for the two-color values and the size of the pointer, we can see the animation working when we click the button with the pointer class attribute attached, as shown in the following screenshot:

Figure 13.20: The button with a pointer following the mouse and animated CSS transitions

This exercise will work in Chrome Canary, where the CSS Properties and Values API is currently available. However, it will cause some problems in browsers where that feature is not available.

In the next part of this chapter, we will examine the support for the APIs we have used in browsers. With this information to hand, we can form a strategy for progressive enhancement with which we can deal with browsers that have not yet implemented this feature.

Current Browser Support

We can check the current browser support for the CSS Paint API feature at https://packt.live/2pT7xYi. The following screenshot shows which browsers support the feature at the time of writing. Browser support is still mostly limited to Chrome, Opera, and Android browsers. There is still a lot of red in the screenshot, but it is worth bearing in mind that Chrome makes up a good percentage of browser users:

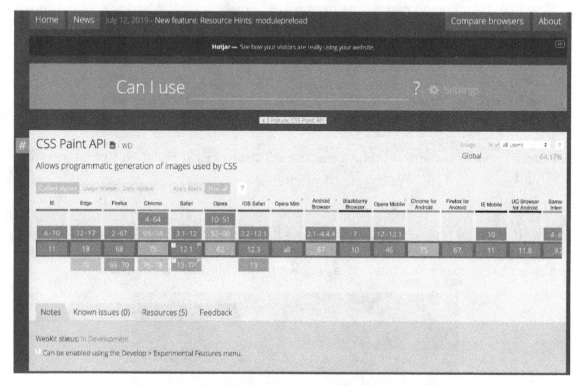

Figure 13.21: caniuse.com browser support data for the CSS Paint API

As well as this information, the Is Houdini Ready Yet site at https://packt.live/2rjZKDj is dedicated to the state of various CSS Houdini APIs. It shows where APIs have already been implemented but also where there is interest in implementing the APIs or where the APIs are currently in development:

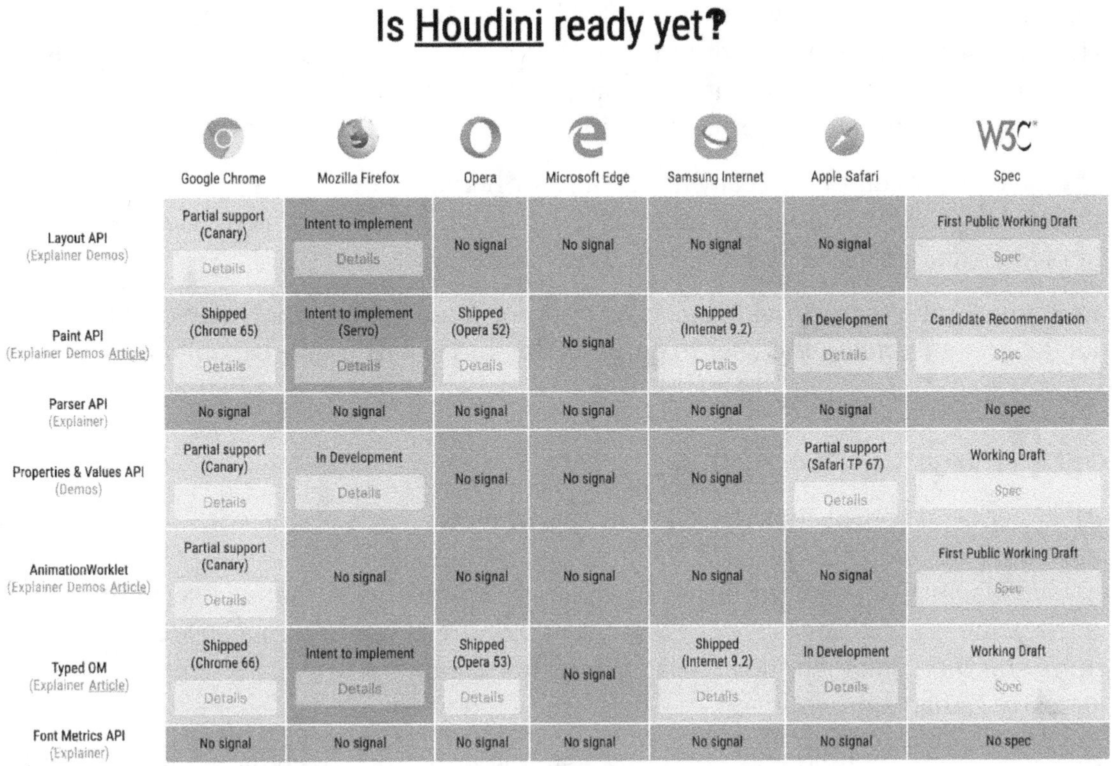

Is Houdini ready yet?

	Google Chrome	Mozilla Firefox	Opera	Microsoft Edge	Samsung Internet	Apple Safari	Spec
Layout API (Explainer Demos)	Partial support (Canary) Details	Intent to implement Details	No signal	No signal	No signal	No signal	First Public Working Draft Spec
Paint API (Explainer Demos Article)	Shipped (Chrome 65) Details	Intent to implement (Servo) Details	Shipped (Opera 52) Details	No signal	Shipped (Internet 9.2) Details	In Development Details	Candidate Recommendation Spec
Parser API (Explainer)	No signal	No signal	No signal	No signal	No signal	No signal	No spec
Properties & Values API (Demos)	Partial support (Canary) Details	In Development Details	No signal	No signal	No signal	Partial support (Safari TP 67) Details	Working Draft Spec
AnimationWorklet (Explainer Demos Article)	Partial support (Canary) Details	No signal	No signal	No signal	No signal	No signal	First Public Working Draft Spec
Typed OM (Explainer Article)	Shipped (Chrome 66) Details	Intent to implement Details	Shipped (Opera 53) Details	No signal	Shipped (Internet 9.2) Details	In Development Details	Working Draft Spec
Font Metrics API (Explainer)	No signal	No signal	No signal	No signal	No signal	No signal	No spec

Figure 13.22: Current state of various CSS Houdini APIs

As we can see from the preceding screenshot, many of the seven APIs that have been developed and specified by the CSS Houdini task force has seen interest from browser developers, and several have seen implementations shipped in major browsers. In particular, the **Paint API** and **Typed OM** have been actively developed and shipped in Chrome, Opera, and Samsung Browser, with Firefox and Safari showing interest in actively developing those features.

We can also see that the Properties and Values API has partial support such as that found in the Chrome Canary browser.

As we can see, the features we are using are not yet available in all the browsers we would want our web pages to work in. It is always a good idea to get as good support as we can, but it is possible to mitigate this issue with the strategy of progressive enhancement, which we will learn about in the next part of this chapter.

Progressive enhancement is really important for helping developers produce web pages that work for as wide a variety of users as possible while allowing us to use the latest technologies and giving us the freedom to experiment.

As a technique, it lets us provide our base content to any user on any device while enriching and enhancing the user experience for those with more capable browsers and devices.

Progressive Enhancement

There are several approaches that we can use when we are working with the web to make sure our websites provide the best experience possible for the greatest number of users. Unlike some closed systems, we cannot trust that all users will have the same capabilities or devices. What we can do is work from a baseline set of features and then enhance the experience for more capable browsers. This is called Progressive Enhancement.

An example of this technique is providing a simple, functional web page with text and links and then adding JavaScript to enhance the experience.

One particular challenge is CSS custom properties, or CSS variables because they are useful when used throughout your CSS declarations and are not available in IE 11. They are difficult to polyfill without having to restrict what they can do. One way of handling CSS variables is with the user of a CSS preprocessor and the build tools that replace the CSS variables at build time to create CSS that works in IE 11. We covered the use of build tools and CSS preprocessors in *Chapter 10, Preprocessors and Tooling*. This could then be served to the restricted browser while more capable browsers make use of CSS variables.

In the case of the CSS Paint API, we can provide a decent experience without a paint worklet, and there are often JavaScript polyfills to reach a similar experience. In the next exercise, we will apply some of these techniques to make our paint worklet example from *Exercise 13.04, Paint Worklet with Mouse Input*, work across more browsers.

Exercise 13.06: Progressive Enhancement

If we take a moment to analyze the code we created in *Exercise 13.05*, *Animating a Paint Worklet*, we will see that it presents a few difficulties as code we may want to use on a live website.

If you open the **Exercise 13.05.html** page in a browser that doesn't support either the CSS Paint API or CSS Properties and Values API, you will experience none of the stylings we expect and also some JavaScript errors.

The following screenshot shows the result of running **Exercise 13.05.html** in the Chrome browser, which does not support the CSS Properties and Values API. The dev tools console panel is open and shows a JavaScript error caused by a lack of support for **CSS.registerProperty**. The button is difficult to use as it appears with none of the expected styles:

Figure 13.23: Unsupported features running in Chrome

In this exercise, we will make our code more robust and provide a decent experience for those browsers that do not support the features we want to use. We will see how we can start from a working version and can then enhance the experience depending on the availability of new features.

Here are the steps:

1. Make a copy of the **Exercise 13.05.html** file from our previous exercise. Rename the copied file **Exercise 13.06.html** under **Chapter13** folder and update the title of the web page document:

```
<title>Exercise 13.06: Progressive Enhancement</title>
```

2. One of the benefits of cascade in CSS is that unrecognized CSS rules will be ignored, and so the last rule to be defined in the CSS, that is, a recognized rule, will be the style that's applied. We can use this technique to progressively enhance CSS. We can set background colors for the pointer class for normal and active states of the button when paint worklets are not recognized:

```css
.pointer {
    background: #12005e;
  background: paint(pointer);
    color: white;
    transition: --size 1s, --primary-fill-color 0.5s,
       --secondary-fill-color 0.5s;
}
.pointer:visited,
.pointer:active {
    background-color: #7200ca;
    --secondary-fill-color: #7200ca;
    --size: 10;
}
```

As a baseline experience for the button, this is enough functionality for browsers without the CSS Paint API. The button still works and there is visual differentiation between the active click state and the normal state. We still have some issues to deal with to stop errors being thrown, which we will handle next.

3. The next step is to prevent some errors being thrown when we register properties and the paint worklet in JavaScript. In this exercise, we can handle any issues well enough with some feature detection. Using an **if** statement, we can check whether a property exists in the browser. We can check that the CSS object is found on the window object, and if so, we run the rest of the code. We do the same to check for the existence of the **registerProperty** method, which is used for registering properties, and the **paintWorklet** property, which is used for adding paintlet modules to our CSS. This protects our code from throwing errors that will prevent further code from running:

```
if (window.CSS) {
    if ("registerProperty" in CSS) {
        CSS.registerProperty({
            name: "--primary-fill-color",
            syntax: "<color>",
            inherits: false,
            initialValue: "#7200ca"
        });
        CSS.registerProperty({
            name: "--secondary-fill-color",
            syntax: "<color>",
            inherits: false,
            initialValue: "#12005e"
        });
        CSS.registerProperty({
            name: "--size",
            syntax: "<number>",
            inherits: false,
            initialValue: 0.05
        });
    }
    if ("paintWorklet" in CSS) {
        CSS.paintWorklet.addModule
          ("pointer-input-paint.js");
    }
}
```

4. We have now prevented any JavaScript errors from being thrown, which could prevent other scripts on the page from functioning. In this simple case, we have done enough to make the button function and look OK and then enhanced it as browser capabilities improved.

5. Now save this file and place the **Chapter13** folder within the **htdocs** folder and make sure you have started the web server.

    > **Note**
    >
    > Refer to the instructions in the *Preface* to know how to start the web server.

6. Head to your browser and type **localhost/Chapter13**. Then, hit *Enter*. You will obtain a list of all files present in this folder.

7. Click on the **Exercise 13.06.html** file.

    You will see the following screenshot that shows the button hover state on a browser that does not support the CSS Paint API or CSS Properties and Values API. The button style is still visible and the button is functional, but we can't see the mouse following the circle. The button is usable but does not have an enhanced experience:

Figure 13.24: Hover state of unsupported features but with improved progressive enhancement

The following screenshot shows the button click state on a browser that does not support the CSS Paint API or CSS Properties and Values API. The click state has no animation but has a color fill and shows that the user is clicking the button:

Figure 13.25: Click state of unsupported features but with improved progressive enhancement

In the following screenshot, we can see the hover state of the button in a browser with full support for the CSS Paint API and the CSS Properties and Values API, along with the mouse follower:

Figure 13.26: Overstate in a browser that supports CSS Houdini features

In the following screenshot, we can see the click state of the button in a browser with full support for the CSS Paint API and the CSS Properties and Values API. The click state is animated:

Figure 13.27: Click state in a browser that supports CSS Houdini features

Having looked at some of the features of the CSS Paint API and CSS Properties and Values API, we have considered a small subset of the CSS Houdini APIs that will provide a lot of low-level controls of the CSS render pipeline and browser layout. We have also considered how to handle experimental browser features in our current web development processes.

We will end this chapter with an activity where we will create our own paint worklet before we summarize our learning.

> **Note**
>
> The following activity requires the use of the web server. If you have not already followed the steps to install and set up your web server, you will need to go back to *Preface* and follow those steps before you begin this activity.
>
> To test this activity, you will need to place your files into the XAMPP **htdocs** folder on your computer and access them through your browser using localhost. Review the web server instructions if you need help with these steps.

Activity 13.01: Button Library

You have been asked to build a library of buttons that can be used as part of a design system for consistent designs across the whole of the site you are working on. To begin with, the client is requesting four button types – default, secondary, ghost, and a special button.

The steps are as follows:

1. Under Chapter13 folder create a new file and name it **Activity 13.01.html**. Copy the following markup into the file:

```
<!DOCTYPE html>
<html lang="en">
    <head>
        <meta charset="utf-8">
        <title>Activity 13.01: Buttons</title>
        <style>
            body {
                padding: 0;
                margin: 0;
                font-family: Arial, Helvetica, sans-serif;
            }
            .button {
                outline: none;
                user-select: none;
                -webkit-appearance: none;
                appearance: none;
                margin: 16px;
                min-width: 128px;
                min-height: 32px;
                padding: 8px 16px;
                font-size: 24px;
                border: 2px solid #fcfcfc;
                box-shadow: 0 0 5px 0 rgba(0, 0, 0, 0.2);
                border-radius: 3px;
                cursor: pointer;
            }
        </style>
    </head>
    <body>
```

```
                <div class="actions">
                    <button class="button button--primary">
                        Primary
                    </button>
                    <button class="button button--secondary">
                        Secondary
                    </button>
                    <button class="button button--ghost">
                        Ghost
                    </button>
                    <button class="button button--special">
                        Special
                    </button>
                </div>
                <script>
                </script>
            </body>
        </html>
```

We have added markup for four button types: **primary**, **secondary**, **ghost**, and **special**.

2. Create styles for the primary button with the text color as white, the font-weight as **bold**, and the background color with a hex value of **#f44336**. When hovered over, the button's background color will change to **#ff9900**.

   The following screenshot shows the expected primary button design:

Figure 13.28: Primary button designs for normal and hover states

3. Create styles for the secondary button with the text color as **white**, the font-weight as **normal**, and the background color with a hex value of **#9e9e9e**. When hovered over, the button's background color will change to **#7c7c7c**.

   The following screenshot shows the expected secondary button design:

Figure 13.29: Secondary button designs for normal and hover states

4. Create styles for the ghost button with the text and border color as **#9e9e9e**, the font-weight as normal, and the background color as white. When hovered over, the button's background color will change to **#efefef**.

The following screenshot shows the expected ghost button design:

Figure 13.30: Ghost button designs for normal and hover states

5. Finally, create a style for the special button. This button will have a linear gradient background with the primary color value **#f44336** and the secondary color value **#ff9900**. The gradient will transition when hovered over since the color values are being swapped. You can use a paint worklet for this button, which you can create in a file called **gradient-paint.js**. The worklet name should be animated-gradient.

The following screenshot shows the expected special button design:

Figure 13.31: Special button designs for normal and hover states

6. To draw the **gradient** of the special button, we will create a gradient paint worklet by creating a JavaScript file called **gradient-paint.js** and copying the following JavaScript code into it:

```
class GradientPaintWorklet {
    static get inputProperties() {
        return ["--primary-fill-color", "--secondary-fill-color"];
    }
    paint(context, geometry, properties) {
        const primaryColor =
          properties.get("--primary-fill-color");
        const secondaryColor =
          properties.get("--secondary-fill-color");
        const gradient =
          context.createLinearGradient(0, 0,
          geometry.width, geometry.height);
```

```
        gradient.addColorStop(0, primaryColor);
        gradient.addColorStop(1.0, secondaryColor);
        context.fillStyle = gradient;
        context.fillRect(0, 0, geometry.width, geometry.height);
    }
  }
  registerPaint("animated-gradient", GradientPaintWorklet);
```

Note

The solution to this activity can be found in page 643.

We have now completed the activity and started creating a design system by including the use of a paint worklet with the CSS Paint API and CSS Properties and Values API.

Next, we will summarize what we've learned in this chapter.

Summary

In this chapter, we have looked at the future of the web and how we can keep up with changes that are being made to the web platform as web developers.

We have considered some techniques and resources for keeping up with cutting-edge browser features.

We have also looked at several experimental features that have started getting some traction in official and upcoming web browsers, such as the CSS Paint API and the CSS Properties and Values API. We have learned how to create a CSS paint worklet and how we can use CSS properties as inputs.

Then, we looked at how we can get a good understanding of browser support for features and use techniques such as progressive enhancement to work with new features while providing a good experience for less capable browsers.

Ultimately, the future of the web is down to web developers who love the platform they work on and want to create great websites for their users. The techniques we've learned about in this chapter and throughout the previous chapters can help you on that journey.

Appendix

About

This section is included to assist the students to perform the activities present in the book. It includes detailed steps that are to be performed by the students to complete and achieve the objectives of the book.

Chapter 1: Introduction to HTML and CSS

Activity 1.01: Video Store Page Template

Solution

1. To start, we open the **Chapter01** folder in VSCode (**File** > **Open Folder...**) and we will create a new plain text file by clicking **File** > **New File**. Then, save it in HTML format, by clicking **File** > **Save As...** and enter the file name: **Activity 1.01. html**. Start with the doctype and the **html**, **head**, and **body** elements:

```
<!DOCTYPE html>
<html lang="en">
    <head>
    </head>
    <body>
    </body>
</html>
```

2. Add the title in the **head** element to describe the site and add a placeholder for the part of the title specific to the page:

```
<!DOCTYPE html>
<html lang="en">
    <head>
        <title>Films on Demand - <!-- Title for page goes
            here -->
        </title>
    </head>
    <body>
    </body>
</html>
```

3. Next, add the metadata to provide the document character encoding, site description, and the viewport:

```
<!DOCTYPE html>
<html lang="en">
    <head>
        <meta charset="utf-8">
        <title>Films on Demand - <!-- Title for page goes
            here -->
        </title>
```

```
        <meta name="description" content="Buy films from our
          great selection. Watch movies on demand.">
        <meta name="viewport" content="width=device-width,
          initial-scale=1">
    </head>
    <body>
    </body>
</html>
```

4. Add an **h1** element in the **body** element to provide a page heading. Add the text Lorem ipsum to this element:

```
<!DOCTYPE html>
<html lang="en">
    <head>
        <meta charset="utf-8">
        <title>Films on Demand - <!-- Title for page goes
          here -->
        </title>
        <meta name="description" content="Buy films from our
          great selection. Watch movies on demand.">
        <meta name="viewport" content="width=device-width,
          initial-scale=1">
    </head>
    <body>
        <h1>Lorem ipsum</h1>
    </body>
</html>
```

5. Finally, add a **p** element beneath the **h1** element to provide content. Add the paragraph of lorem ipsum provided above to the **p** element:

```
<!DOCTYPE html>
<html lang="en">
    <head>
        <meta charset="utf-8">
        <title>Films on Demand - <!-- Title for page goes
          here -->
        </title>
        <meta name="description" content="Buy films from our
          great selection. Watch movies on demand.">
        <meta name="viewport" content="width=device-width,
          initial-scale=1">
    </head>
    <body>
        <h1>Lorem ipsum</h1>
```

```
        <p>
        Lorem ipsum dolor sit amet, consectetur adipiscing elit.
        Nullam quis scelerisque mauris. Curabitur aliquam ligula
        in erat placerat finibus. Mauris leo neque, malesuada
        et augue at, consectetur rhoncus libero. Suspendisse
        vitae dictum dolor. Vestibulum hendrerit iaculis ipsum,
        ac ornare ligula. Vestibulum efficitur mattis urna vitae
        ultrices. Nunc condimentum blandit tellus ut mattis.
        Morbi eget gravida leo. Mauris ornare lorem a mattis
        ultricies. Nullam convallis tincidunt nunc, eget rhoncus
        nulla tincidunt sed. Nulla consequat tellus lectus, in
        porta nulla facilisis eu. Donec bibendum nisi felis, sit
        amet cursus nisl suscipit ut. Pellentesque bibendum id
        libero at cursus. Donec ac viverra tellus. Proin sed
        dolor quis justo convallis auctor sit amet nec orci.
        Orci varius natoque penatibus et magnis dis parturient
        montes, nascetur ridiculus mus.
        </p>
    </body>
</html>
```

6. If you now right-click on the filename in VSCode on the left-hand side of the screen and select **Open In Default Browser**, you will see the following web page in your browser:

Lorem ipsum

Lorem ipsum dolor sit amet, consectetur adipiscing elit. Nullam quis scelerisque mauris. Curabitur aliquam ligula in erat placerat finibus. Mauris leo neque, malesuada et augue at, consectetur rhoncus libero. Suspendisse vitae dictum dolor. Vestibulum hendrerit iaculis ipsum, ac ornare ligula. Vestibulum efficitur mattis urna vitae ultrices. Nunc condimentum blandit tellus ut mattis. Morbi eget gravida leo. Mauris ornare lorem a mattis ultricies. Nullam convallis tincidunt nunc, eget rhoncus nulla tincidunt sed. Nulla consequat tellus lectus, in porta nulla facilisis eu. Donec bibendum nisi felis, sit amet cursus nisl suscipit ut. Pellentesque bibendum id libero at cursus. Donec ac viverra tellus. Proin sed dolor quis justo convallis auctor sit amet nec orci. Orci varius natoque penatibus et magnis dis parturient montes, nascetur ridiculus mus.

Figure 1.38: The result

Activity 1.02: Styling the Video Store Template Page

Solution

1. To start, we open the **Chapter01** folder in VSCode (**File** > **Open Folder**...) and we will open the previously saved **Activity 1.01.html** from the previous activity. Create a new file named **Activity 1.02.html** and save all codes of **Activity 1.01.html** in it. Now, we will add a link to the **head** element of **Activity 1.02. html** pointing to a local copy of **styles/normalize.css**:

```
<link href="styles/normalize.css" rel="stylesheet">
```

2. Next, beneath the link, we add a style element to the head of **Activity 1.02.html**:

```
<style>

</style>
```

3. In the style element, we want to set the page to use **border-box** as its **box-sizing** model.

```
<style>
    html {
            box-sizing: border-box;
    }
    *, *:before, *:after {
        box-sizing: inherit;
    }
</style>
```

4. Then, we are going to add the following styles to the **body** element of the style element to set the font size and family to **16px** and **Arial**, respectively:

```
<style>
    html {
            box-sizing: border-box;
    }
    *, *:before, *:after {
        box-sizing: inherit;
    }
    body {
            font-family: Arial, Helvetica, sans-serif;
            font-size: 16px;
    }
</style>
```

5. Next, we are going to add a **div** element to the body element. We will give it an **ID** of **pageWrapper** and a class of full-page. We will wrap this element around the **h1** element and **p** element that is the content of the **body** element:

```
<div id="pageWrapper" class="full-page">
  <h1>Lorem ipsum</h1>
  <p>
  Lorem ipsum dolor sit amet, consectetur adipiscing
  elit. ullam quis scelerisque mauris. Curabitur aliquam
  ligula in erat placerat finibus. Mauris leo neque,
  malesuada et augue at, consectetur rhoncus libero.
  Suspendisse vitae dictum dolor. Vestibulum hendrerit
  iaculis ipsum, ac ornare ligula. Vestibulum efficitur
  mattis urna vitae ultrices. Nunc condimentum blandit
  tellus ut mattis. Morbi eget gravida leo. Mauris ornare
  lorem a mattis ultricies. Nullam convallis tincidunt
  nunc, eget rhoncus nulla tincidunt sed. Nulla consequat
  tellus lectus, in porta nulla facilisis eu. Donec
  bibendum nisi felis, sit amet cursus nisl suscipit
  ut. Pellentesque bibendum id libero at cursus. Donec ac
  viverra tellus. Proin sed dolor quis justo convallis
  auctor sit amet nec orci. Orci varius natoque penatibus
  et magnis dis parturient montes, nascetur ridiculus mus.
  </p>
</div>
```

6. We will add styles to set the background color and a **padding** of **16px** for the page and to set the minimum height to **100vh**, the full height of the viewport:

```
<style>
    html {
        box-sizing: border-box;
    }
    *, *:before, *:after {
        box-sizing: inherit;
    }
    body {
        font-family: Arial, Helvetica, sans-serif;
        font-size: 16px;
    }
    #pageWrapper {
        background-color: #eeeae4;
        padding: 16px;
    }
    }
```

```
        .full-page {
            min-height: 100vh;
        }
    </style>
```

7. Finally, we will add change the **margin** and **padding** of the **h1** element. We can add this to the bottom of the style element:

```
        .full-page {
            min-height: 100vh;
        }
        h1 {
            margin: 0;
            padding: 0 0 16px 0;
        }
    </style>
```

8. The resulting code is:

```
<!DOCTYPE html>
<html lang="en">
    <head>
        <meta charset="utf-8">
        <title>Films on Demand - <!-- Title for page goes
        here --></title>
        <meta name="description" content="Buy films from our
        great selection. Watch movies on demand.">
        <meta name="viewport" content="width=device-width,
        initial-scale=1">
          <link href="styles/normalize.css" rel="stylesheet">
          <style>
          html {
              box-sizing: border-box;
          }
          *, *:before, *:after {
              box-sizing: inherit;
          }
          body {
              font-family: Arial, Helvetica, sans-serif;
              font-size: 16px;
          }
        #pageWrapper {
              background-color: #eeeae4;
```

```
            padding: 16px;
        }
        .full-page {
            min-height: 100vh;
        }
        h1 {
            margin: 0;
            padding: 0 0 16px 0;
        }
        </style>
    </head>
    <body>
        <div id="pageWrapper" class="full-page">
            <h1>Lorem ipsum</h1>
            <p>
            Lorem ipsum dolor sit amet, consectetur adipiscing
              elit. Nullam quis scelerisque mauris. Curabitur
              aliquam ligula n erat placerat finibus. Mauris leo
              neque, malesuada et augue at, consectetur rhoncus
              libero. Suspendisse vitae dictum dolor. Vestibulum
              hendrerit iaculis ipsum, ac ornare ligula.
              Vestibulum efficitur mattis urna vitae
              ultrices. Nunc condimentum blandit tellus ut
              mattis. Morbi eget gravida leo. Mauris ornare
              lorem a mattis ultricies. Nullam convallis
              tincidunt nunc, eget rhoncus nulla tincidunt sed.
              Nulla consequat tellus lectus, in porta nulla
              facilisis eu. Donec bibendum nisi felis, sit amet
              cursus nisl suscipit ut. Pellentesque bibendum id
              libero at cursus. Donec ac viverra tellus. Proin
              sed dolor quis justo convallis auctor sit amet nec
              orci. Orci varius natoque penatibus et magnis dis
              parturient montes, nascetur ridiculus mus.
            </p>
        </div>
    </body>
</html>
```

9. If you now right-click on the filename in VSCode on the left-hand side of the screen and select **Open In Default Browser**, you will see the result that looks like the following figure:

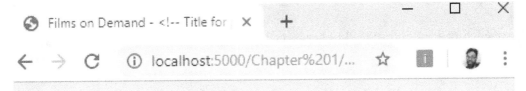

Figure 1.39: The resulting web template

Chapter 2: Structure and Layout

Activity 2.01: Video Store Home Page

Solution

1. Start with the following HTML skeleton:

```
<!DOCTYPE html>
<html lang = "en">
    <head>
        <title>Video store home page</title>
        <style>
          header,
          nav,
          section,
          footer {
             background: #659494;
             border-radius: 5px;
             color: white;
             font-family: arial, san-serif;
             font-size: 30px;
             text-align: center;
             padding: 30px;
             margin-bottom: 20px;
          }
          header:before,
          nav:before,
          section:before,
          footer:before {
             content: '<';
          }
          header:after,
          nav:after,
          section:after,
          footer:after {
             content: '>';
          }
        </style>
    </head>
```

```
    <body>
<!-- your code will go here -->
    </body>
</html>
```

2. Now, let's add the **header** tag:

```
<body>
  <header>header</header>
</body>
```

3. Then, the **navigation**:

```
<body>
  <header>header</header>
  <nav>nav</nav>
</body>
```

4. Now, add the main page content section:

```
<body>
  <header>header</header>
  <nav>nav</nav>
  <section>section</section>
</body>
```

5. And finally, let's add the **footer** tag:

```
<body>
  <header>header</header>
  <nav>nav</nav>
  <section>section</section>
  <footer>footer</footer>
</body>
```

6. If you now right-click on the filename in VSCode on the left-hand side of the screen and select **Open In Default Browser**, you will see the output in a browser.

Activity 2.02: Online Clothes Store Home Page

Solution

1. Start with the following HTML skeleton that contains all of the CSS we need:

```html
<!DOCTYPE html>
<html lang = "en">
<head>
  <title>Video store product page</title>
  <style>
    header,
    nav,
    section,
    footer {
      background: #659494;
      border-radius: 5px;
      color: white;
      font-family: arial, san-serif;
      font-size: 30px;
      text-align: center;
    }
    header:before,
    nav:before,
    footer:before {
      content: '<';
    }
    header:after,
    nav:after,
    footer:after {
      content: '>';
    }
    header,
    nav,
    section,
    footer {
      border: 1px solid gray;
      padding: 20px;
      margin-bottom: 25px;
    }
    .products {
      display: grid;
```

```
        grid-template-columns: auto auto auto auto;
      }
      .products div {
        border: 2px solid white;
        padding: 30px;
        margin: 10px;
      }
    </style>
  </head>
  <body>
    <!-- your code goes here -->
  </body>
</html>
```

2. Now, let's add the **header** element between the opening and closing **body** tags:

```
<body>
  <header>header</header>
</body>
```

3. Then, the navigation element, which will be placed after the **header** element:

```
<body>
  <header>header</header>
  <nav>nav</nav>
</body>
```

4. Then, a **section** element for an image carousel, which comes after the **nav** element:

```
<body>
  <header>header</header>
  <nav>nav</nav>
  <section>Image carousel</section>
</body>
```

5. Then, a **section** element for the product items. Observe how we add a **class** name to allow for different styling between the two **section** elements:

```
<body>
  <header>header</header>
  <nav>nav</nav>
  <section>Image carousel</section>
  <section class="products">
```

```
      <div>product 1</div>
      <div>product 2</div>
      <div>product 3</div>
      <div>product 4</div>
   </section>
</body>
```

6. Finally, we add the **footer** element, which will go after the **section** element used for the product items:

```
<body>
   <header>header</header>
   <nav>nav</nav>
   <section>Image carousel</section>
   <section class="products">
      <div>product 1</div>
      <div>product 2</div>
      <div>product 3</div>
      <div>product 4</div>
   </section>
   <footer>footer</footer>
</body>
```

7. If you now right-click on the filename in VSCode on the left-hand side of the screen and select **Open In Default Browser**, you will see the output in a browser.

Chapter 3: Text and Typography

Activity 3.01: Converting a Newspaper Article to a Web Page

Solution

1. Start with the following HTML code, which will form the starting point:

```html
<!DOCTYPE html>
<html lang = "en">
<head>
  <title>Newspaper article</title>
</head>
<body>
  <!-- ...your markup goes here -->
</body>
</html>
```

2. Now, let's add the HTML, starting with the article heading and then followed by all text-based elements:

```html
<!DOCTYPE html>
<html lang = "en">
<head>
  <title>Newspaper article</title>
</head>
<body>
  <article>
    <h1>News article heading</h1>
    <p>
    <strong>Introduction text</strong>
    </p>
    <p>
    Lorem ipsum dolor sit amet, consectetur adipiscing elit.
      Curabitur consequat egestas mauris, non auctor justo
      sagittis sit amet. Donec mattis ut magna non rutrum. Morbi
      dolor risus, venenatis non semper id, maximus ac lectus.
      Phasellus pulvinar felis nibh, eu imperdiet metus finibus
      vel.
    </p>
    <ul>
      <li>Lorem ipsum dolor</li>
      <li>Donec mattis ut</li>
      <li>Phasellus pulvinar</li>
    </ul>
```

```
        <p>
        Lorem ipsum dolor sit amet, consectetur adipiscing elit.
           Curabitur consequat egestas mauris, non auctor justo
           sagittis sit amet. Donec mattis ut magna non rutrum. Morbi
           dolor risus, venenatis non semper id, maximus ac lectus.
           Phasellus pulvinar felis nibh, eu imperdiet metus finibus
           vel.
        </p>
     </article>
   </body>
   </html>
```

3. Now we will add some styling for the news article. Notice that we add a font family and customize the size of each text-based element:

```
<!DOCTYPE html>
<html lang = "en">
<head>
   <title>Newspaper article</title>
   <style>
     body {
        font-family: "Times New Roman", Times, serif;
     }
     h1 {
        font-size: 50px;
     }
     p,
     ul {
        color: #666;
        font-size: 16px;
     }
   </style>
</head>
<body>
```

```
<article>
  <h1>News article heading</h1>
  <p>
  <strong>Introduction text</strong>
  </p>
  <p>
  Lorem ipsum dolor sit amet, consectetur adipiscing elit.
    Curabitur consequat egestas mauris, non auctor justo
    sagittis sit amet. onec mattis ut magna non rutrum. Morbi
    dolor risus, venenatis non semper id, maximus ac lectus.
    Phasellus pulvinar felis nibh, eu imperdiet metus finibus
    vel.</p>
  <ul>
    <li>Lorem ipsum dolor</li>
    <li>Donec mattis ut</li>
    <li>Phasellus pulvinar</li>
  </ul>
  <p>
  Lorem ipsum dolor sit amet, consectetur adipiscing elit.
    Curabitur consequat egestas mauris, non auctor justo
    sagittis sit amet. Donec mattis ut magna non rutrum. Morbi
    dolor risus, venenatis non semper id, maximus ac lectus.
    Phasellus pulvinar felis nibh, eu imperdiet metus finibus
    vel.</p>
  </article>
</body>

</html>
```

4. If you now right-click on the filename in VSCode on the left-hand side of the screen and select **Open In Default Browser**, you will see the output in a browser.

Chapter 4: Forms

Activity 4.01: Building an Online Property Portal Website Form

Solution

1. Start with the following skeleton HTML and create a new file in VSCode called **`Activity 4.01.html`**:

```
<!DOCTYPE html>
<html lang = "en">
<head>
  <title>Activity 4.01</title>
</head>
<body>
  <!-- your markup goes here -->
</body>
</html>
```

2. Now, let's add the HTML for our property search form. We will start by adding the HTML for the search radius, price range, and bedroom fields. The HTML for the **radius** field will consist of a **label** and select box; likewise for the price range and bedrooms fields:

```
<form action = "url_to_send_form_data" method = "post">
    <fieldset>
        <h2>Property for sale in London</h2>
        <div class="row">
            <div>
                <div class="item">
                    <label for="radius">Search radius:</label>
                    <div class="select-wrapper">
                        <select id="radius" required>
                            <option value="">This area only
                            </option>
                            <option value="1">1 mile</option>
                            <option value="5">5 miles</option>
                            <option value="10">10 miles
                                </option>
                        </select>
                    </div>
                </div>
                <div class="item">
                    <label for="price">Price range:</label>
```

```
                <div class="select-wrapper">
                    <select id="price" required>
                        <option value="">Any</option>
                        <option value="100">Up to 100k
                        </option>
                        <option value="250">Up to 250k
                        </option>
                        <option value="500">Up to 500k
                        </option>
                    </select>
                </div>
            </div>
            <div class="item">
                <label for="beds">Bedrooms:</label>
                <div class="select-wrapper">
                    <select id="beds" required>
                        <option value="">Any</option>
                        <option value="1">1 Bed</option>
                        <option value="2">2 Beds</option>
                        <option value="3">3 Beds</option>
                    </select>
                </div>
            </div>
        </div>
```

3. Now we will continue writing out the HTML for the remaining form fields. For the property type and added to site fields, we will use a **label** and a select box. For the "**include sold properties**" option, we will use a **label** with a checkbox. Finally, we will add a submit button to allow the user to submit the form:

```
                <div class="item">
                    <input id="sold" class="checkbox"
                      type="checkbox" name="sold" />
                    <label for="sold">Include sold properties
                    </label>
                </div>
            </div>
        </div>
        <button type="submit">Find properties</button>
    </fieldset>
  </form>
 </body>
</html>
```

4. Now that we have our HTML in place, we can add the CSS to improve the look and feel of the form. We will start by adding a font to all text on the web page and setting some spacing properties for the form elements:

```
<style>
  body {
    font-family: arial, sans-serif;
  }
  fieldset {
    border: 0;
    padding: 0;
    width: 900px;
  }
  fieldset > div {
    margin-bottom: 30px;
  }
</style>
```

5. Next, we will add some CSS for the individual form elements, including inputs, selects, and labels:

```
label {
  display: block;
  margin-bottom: 10px;
}
input,
select {
  border: 1px solid gray;
  padding: 10px;
  width: 200px;
}
select {
  background: transparent;
  border-radius: 0;
  box-shadow: none;
  color: #666;
  -webkit-appearance: none;
  width: 100%;
}
.select-wrapper {
  position: relative;
  width: 100%;
```

```css
    }
    .select-wrapper:after {
      content: '<>';
      color: #666;
      font-size: 14px;
      top: 8px;
      right: 0;
      transform: rotate(90deg);
      position: absolute;
      z-index: -1;
    }
    button {
      background: green;
      border: 0;
      color: white;
      width: 224px;
      padding: 10px;
      text-transform: uppercase;
      float: right;
      margin-right: 30px;
    }
```

6. Finally, we will add some styles that handle the layout of the form elements and also the styles for form validation:

```css
    .row,
    .item {
      display: flex;
    }
    .item {
      margin-bottom: 15px;
      margin-right: 30px;
    }
    .row > div {
      flex-basis: 50%;
    }
    label {
      margin-right: 15px;
      line-height: 35px;
      width: 150px;
    }
```

```
.checkbox {
  margin-left: 122px;
  margin-right: 10px;
  width: auto;
}
.checkbox+label {
  width: auto;
  line-height: 20px;
}
select:valid {
  border: 2px solid green;
}
select:invalid {
  border: 2px solid red;
}
```

7. If you now right-click on the filename in VSCode, on the left-hand side of the screen, and select **Open In Default Browser**, you will see the form in your browser.

    You should now have a form that looks like the following figure:

Figure 4.28: Property portal website form

8. Finally, we will check that the validation works for our required fields by submitting the form with empty fields:

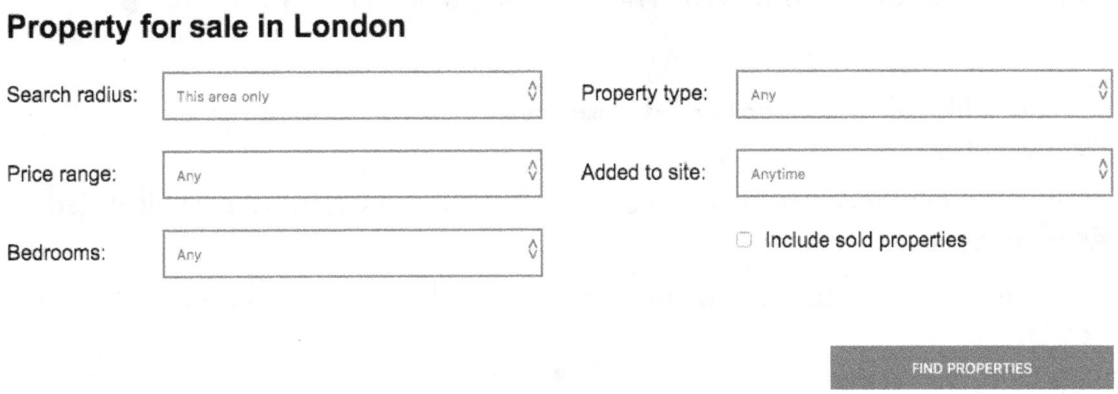

Property for sale in London

Search radius:	This area only	Property type:	Any
Price range:	Any	Added to site:	Anytime
Bedrooms:	Any		☐ Include sold properties

FIND PROPERTIES

Figure 4.29: A form highlighting fields that are required to be filled in to submit the form

Chapter 5: Themes, Colors, and Polish

Activity 5.01: Creating Your Own Theme Using a New Color Palette

Solution

1. Create a file called **new-colors-theme.css** in the assets folder of your sample project.

2. Copy the contents of the **baseline.html** file and paste it into a new file called **Activity 5.01.html**.

3. Add the style sheet to your document with a new link element (see highlighted code):

```
<head>
  <meta http-equiv="Content-Type" content="text/html;
    charset=UTF-8">
  <meta name="viewport" content="width=device-width,
    initial-scale=1">
  <title>HTML5 Boilerplate 7.2.0 and main.css 2.0.0
    released - HTML + CSS + JavaScript</title>
  <link rel="stylesheet" id="hcj2-0-style-css"
    href="./assets/style.css" type="text/css" media="all">
  <link rel="stylesheet" id="dark-theme"
    href="./assets/new-colors-theme.css"
    type="text/css" media="all">

</head>
```

4. Now, open **new-colors-theme.css** and set the text color for **body**, **button**, **input**, **select**, and **textarea** to **#335**:

```
body,
button,
input,
select,
textarea {
    color: #335;
}
```

5. Next, change the background of the **body** element to a light blue **#9db3f4**:

```
body {
    background: #9db3f4;
}
```

6. Next, we change the color of the links. The default link color is set to be the darkest brand blue, **#0e3ece**, while **visited** links are set to be **#bf0ece**, which is a purple of the same value as the darkest brand blue. The **active**, **focus**, and **hover** links are then set to be another of the brand colors, **#2a56db**:

```
a {
    color: #0e3ece;
}
a:visited {
    color: #bf0ece;
}
a:hover,
a:focus,
a:active {
    color: #2a56db;
}
```

7. Next, **.site-content** is updated with a new background color and two new borders on the right and left. The background color is pure white, and the 10-pixel border is set to be the brand blue:

```
.site-content {
    background: #ffffff;
    border-right: 10px solid #4c72e5;
border-left: 10px solid #4c72e5;
}
```

8. Now, update the text in the site header. Change the **h1**, the link inside the **h1**, and the main navigation menu items to **#fffff** and add a new rule for **.site-description** so that it's the lightest brand yellow, **#ffef98**:

```
header.site-header h1 {
    color: #ffffff;
}
header.site-header h1 a {
    color: #ffffff;
}
header.site-header nav .menu,
header.site-header nav li a {
    color: #ffffff;
}
.site-description {
    color: #ffef98;
}
```

9. Now, change the background of the site header to brand blue, **#4c72e5**, and remove the **box-shadow** from the element:

```
.blog .site-header,
.single-post .site-header{
    background: #4c72e5;
    box-shadow: none;
}
```

10. Now, make some changes to the text in the **article** element. Change the **h1** to brand blue, **#4c72e5**, and then add a new rule to make the **h2** elements brand orange, **#FF7a00**:

```
article h1 {
    color: #4c72e5;
}
article h2 {
    color: #FF7a00;
}
article .content {
    color: #999999;
}
article picture img {
    border-top: 2px solid #6F66F6;
    border-bottom: 4px solid #6F66F6;
}
```

11. Remove the **box-shadow** from the **.content-area**:

```
.content-area{
    box-shadow: none;
}
```

12. Finally, change the background of the **footer** to brand blue and change the text to pure white:

```css
footer.site-footer {
    color: #ffffff;
    background: #4c72e5;
}

footer.site-footer a {
    color: #ffffff;
}
```

13. Save **new-colors-theme.css** if you now right-click on the filename in VSCode on the left-hand side of the screen and select **Open In Default Browser**, you'll see the following screenshot in your browser:

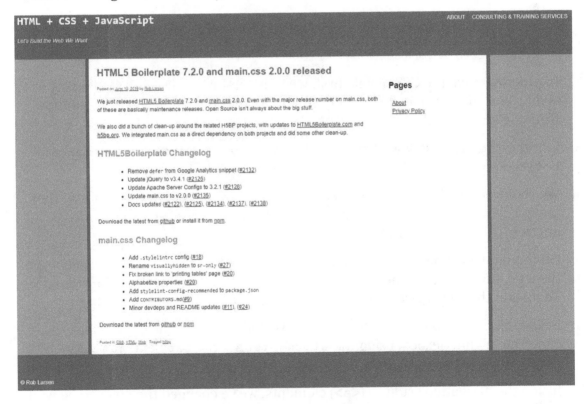

Figure 5.20: Resulting web page

Chapter 6: Responsive Web Design/Media Queries

Activity 6.01: Refactoring the Video Store Product Cards into a Responsive Web Page

Solution

1. Save a new HTML file called **Activity 6.01.html** using the existing code from *Chapter 3*, *Text and Typography Exercise 3.06*, *Putting It All Together*. Insert the viewport meta tag into the **<head>** of the HTML document. This would be inserted at the top of the **<head>** section, ideally, as follows:

```
<!DOCTYPE html>
<html lang = "en">
<head>
   <meta name="viewport" content="width=device-width,
      initial-scale=1" />
   <title>Video store home page</title>
```

2. In order to make the CSS mobile first, we must first remove the desktop-only styling and add new mobile-friendly styles. Remove the fixed width of **1,200px** from the **<body>** tag.

3. Edit the **<header>** styling to remove the **display**, **align-items**, and **margin-bottom** properties, as these are not required in mobile-first styling.

4. We removed the display grid properties from the product cards element, as they are not required in the mobile-first version. We'll use them later in the media queries for higher resolutions.

5. On the **<nav>** element, we added a **20px** top margin to separate it vertically from the logo, then removed the left margin as this isn't required for mobile-first styles. To update the styling, we added a black background to the navigation bar.

6. Inside the navigation list (**<ul>**) element, we added horizontal and vertical centering within flexbox, using the **align-items: center;** and **justify-content: center;** properties.

7. Inside the navigation list links (**<a>**) elements, we've changed the color to be white, so that it's visible on the new black navigation background.

8. Next, we updated the navigation list link's hover state (**a:hover**) to have a font color of red, and also gave the breadcrumb element a light gray background.

9. We updated the width and padding on the **intro** element and removed the **display** property from the product card to make it mobile-first suitable. The updated code after removing the desktop-only styling and adding the mobile-friendly styles can be found in https://packt.live/2NFdEc3.

10. We've created the following new CSS additions to provide a different style for screens of **768px** or higher. This will be appended to the end of the mobile-first CSS we created in *step* 3. The CSS is as follows:

The **header** element is restored back to the original flexbox model and centered. The navigation background is removed and the margin updated for larger screens. For the navigation links, the colors have been reset back to the former colors (before *step* 3), without the navigation background anymore. We've also updated the **width** in the intro section.

For the product cards section, we've restored the former grid display, and for the product cards, we've restored the display inline block style to ensure that the product cards are laid out in grid format for display on larger screens.

The code for *step* 4 looks as follows:

```
<style type="text/css">
    @media (min-width: 768px) {
        header {
            align-items: center;
            display: flex;
        }
        nav {
            background: transparent;
            margin-top: 0;
            margin-left: 30px;
        }
        nav a {
            color: black;
        }
        nav a:hover {
            text-decoration: underline;
            color: black;
        }
        .intro {
            width: 50%;
        }
```

```
                    .product-cards {
                         display: grid;
                    grid-template-columns: auto auto auto auto;
                    }
                    .product-card {
                         display: inline-block;
                    }
               }
          </style>
```

11. The additional CSS with the print media query could be as follows. We've hidden the **<nav>** element when in printing mode. You can test this by doing a print preview from your web browser:

```
<style type="text/css">
   @media only print {
          nav {
               display: none;
          }
     }
</style>
```

12. If you now right-click on the filename in VSCode on the left-hand side of the screen and select **Open In Default Browser**, you will see the end result that would look something like the following figure:

Videos **Mobile**

Lorem ipsum dolor sit amet, consectetur adipiscing eit. In bibendm non purus quis vestbulum. Pelientesque ultricies quam lacus, ut tristique sapien tristique et.

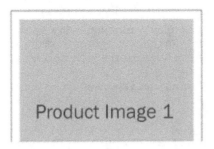

Home / Used DVDs / Less than £10

Videos than £10

Lorem ipsum dolor sit amet, consectetur adipiscing eit. In bibendm non purus quis vestbulum. Pelientesque ultricies quam lacus, ut tristique sapien tristique et.

Desktop

Figure 6.25: Screenshot on mobile and desktop screen sizes

Chapter 7: Media - Audio, Video, and Canvas

Activity 7.01: Media Page

Solution

1. Create an HTML file and name it **Activity 7.01.html**. Save the following code in the file. This is a copy of the template provided. We have changed the title to *Films on Demand – Video Promo*:

```html
<!DOCTYPE html>
<html lang="en">
    <head>
        <meta charset="utf-8">
        <title>Films on Demand - Video Promo</title>
        <meta name="description" content="Buy films from our
            great selection. Watch movies on demand.">
        <meta name="viewport" content="width=device-width,
            initial-scale=1">
        <link href="styles/normalize.css" rel="stylesheet">
        <style>
            body {
                font-family: Arial, Helvetica, sans-serif;
                font-size: 16px;
            }
            #pageWrapper {
                background-color: #eeeae4;
            }
            .full-page {
                min-height: 100vh;
            }
        </style>
    </head>
    <body>
        <div id="pageWrapper" class="full-page">
         <!-- content goes here -->
        </div>
    </body>
</html>
```

2. Next, we need to replace the **<!-- content goes here -->** comment with a **video** element. The **video** element has an ID value of **video-1**, which the JavaScript will use to find the **video** element. We need to add a poster image (**media/poster.png**) and the WebM and MP4 source for the video. We also create a fallback message for users with no video support in their browser:

```html
<div id="pageWrapper" class="full-page">
    <div id="media-player" class="media-player">
        <video id="media-1" poster="media/poster.png"
          preload="metadata">
          <source src="media/html_and_css.webm"
            type="video/webm"/>
          <source src="media/html_and_css.mp4"
            type="video/mpeg"/>
          It looks like your browser does not support
            <code>video</code>.
        </video>
    </div>
</div>
```

3. The next step is to create the markup for the video player. We have added this after the **video** element as a **div** with a **class** attribute with the value controls. The **controls** container hosts a **play** and **pause** button and a **div** for the current time and total duration of the **video** element. After the **controls** markup, we also add the **media-controls.js** JavaScript file, which will add event listeners and behavior for the controls of the media player:

```html
<div id="media-player" class="media-player">
    <video id="media-1" poster="media/poster.png"
      preload="metadata">
      <source src="media/html_and_css.webm"
        type="video/webm" />
      <source src="media/html_and_css.mp4"
        type="video/mpeg" />
      It looks like your browser does not support
        <code>video</code>.
    </video>
    <div class="controls">
    <button id="play-btn" class="button">Play</button>
    <button id="pause-btn" class="button">Pause</button>
    <div class="status">
        <span id="played">00:00</span> / <span
          id="duration">00:00</span>
    </div>
    </div>
    <script src="scripts/media-controls.js"></script>
</div>
```

4. Next, we will add styles to the media player and controls. We give the **.media-player** container a width of 768 pixels and the **video** element a width of 100% so that it will fill the area of the container while maintaining its aspect ratio. We also set general color and font styles for the controls:

```
<style>
    .media-player {
        position: relative;
        margin: 0 auto;
        padding-top: 24px;
        width: 768px;
    }
    video {
        width: 100%;
        margin: 0;
        padding: 0;
    }
    .media-player,
    .controls,
    .controls .button {
        color: white;
        font-family: Arial, Helvetica, sans-serif;
        font-size: 16px;
        font-weight: 200;
    }
```

5. Next, we lay the controls out so that they're left- and right-aligned with space between them by using flexbox CSS for the layout. The **controls** bar is black but has slight transparency so that you can see the underlying video. It is positioned absolutely at the bottom of the **video** element:

```
    .controls {
        align-items: center;
        background-color: rgba(0, 0, 0, 0.75);
        bottom: 5px;
        display: flex;
        justify-content: space-between;
        margin: 0;
        padding: 10px 0;
        position: absolute;
        width: 100%;
    }
```

6. We remove the default button appearance and style both the **play** and **pause** controls buttons with a bright purple to stand out on the dark **controls** bar:

```
.controls .button {
    -webkit-appearance: none;
    appearance: none;
    background-color: #4717F6;
    border: 1px solid #E7DFDD;
    border-radius: 3px;
    cursor: pointer;
    outline: none;
    padding: 4px 16px;
    text-transform: uppercase;
    min-width: 128px;
    margin-left: 8px;
}
.status {
    margin-right: 16px;
}
```

7. Finally, using a **data** attribute to express the state of the media player, we hide the **pause** button or **play** button depending on whether the video is playing or paused, respectively:

```
[data-state="playing"] #play-btn,
[data-state="paused"] #pause-btn {
    display: none;
}
</style>
```

8. If you now right-click on the filename in VSCode on the left-hand side of the screen and select **Open In Default Browser**, you will see the output in a browser.

Chapter 8: Animation

Activity 8.01: Animating Our Video Store Home Page

Solution

Complete this activity by following these steps:

1. First, take a copy of **Activity 6.01.html** from your **Chapter06** folder and save this in your **Chapter08** folder as **Activity 8.01.html**. Now, we're going to edit the navigation links so that they have a CSS transition on the active hovered navigation link item. We will do this by moving it slightly upwards and then smoothing the transition of its hover color to red. Both transitions will be completed over **500** milliseconds. This will be triggered upon hovering the mouse over the link. Note that we only want to apply this to our desktop screen sizes since this could impair the usability of the mobile layout for users otherwise. We'll achieve this now.

   In the CSS **<style>** tag, within the **@media (min-width: 768px)** media query selector, while editing the **nav a** selector, we're going to add the following new properties:

   ```
   position: relative;
   top: 0;
   transition: top .5s, color .5s;
   ```

   We're then going to edit the **nav a:hover** selector with the following properties. Note that we are changing the text's color:

   ```
   top: -10px;
   text-decoration: none;
   color: red;
   ```

   Now, you can observe the navigation transition effect in your web browser. The result is shown in the following screenshot:

Videos to rent   Videos to buy   Used DVDs   Offers

Home / Used DVDs / Less than £10

Videos less than £10

Lorem ipsum dolor sit amet, consectetur adipiscing elit. In

Figure 8.36: Navigation hover

Next, we're going to use keyframes to create a hover effect on the product cards as the mouse hovers over each of them. On the animation, we're going to change the background color of the card, add a thicker border to the card (while reducing the padding simultaneously so as not to increase the overall width of the box), and change the color of the text so that it works with the new background color.

We will apply this style to mobile and desktop screen sizes, so it doesn't need to be desktop-specific this time.'

2. Add the keyframes animation text in the CSS **<style>** tag, just before the **.product-card** selector statement. Add the following code:

```
@keyframes productCards {
    from {
        background-color: white;
        color: black;
        border-width: 1px;
        padding: 15px;
    }
    to {
        background-color: black;
        color: white;
        border-width: 5px;
        padding: 11px;
    }
}
```

Next, we're editing the **.product-card:hover** selector and adding it to the animation property code, as follows:

```
animation: productCards 2s;
```

Finally, we'll add the **.product-card:hover a** selector statement and make the color change for the link when we hover over the product card. Add the following code:

```
.product-card:hover a {
    color: white;
}
```

3. If you now right-click on the filename in VSCode on the left-hand side of the screen and select **Open In Default Browser**, you will see the following screenshot that shows the expected result:

Figure 8.37: Appearance of the product cards when we hover over them on a desktop screen

To recap, the following screenshot shows the expected output, with the animations completed:

Figure: 8.38: The output with the hover states and completed animations shown

Chapter 9: Accessibility

Activity 9.01: Making a Page Accessible

Solution

1. To start, create a file called **Activity 9.01.html** and copy the code from **activity_1_inaccessible.html** into it.

2. Open that page in the browser and run the Axe accessibility checker in the developer tools.

   The following screenshot shows the results you should see for this page. There are five distinct issues, with multiple occurrences of some of the issues (form label issues):

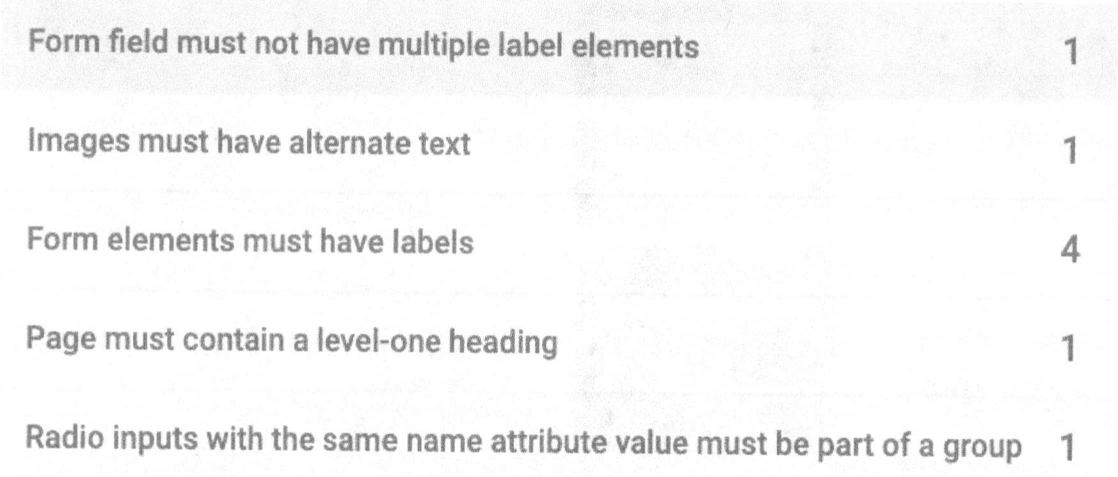

Form field must not have multiple label elements	1
Images must have alternate text	1
Form elements must have labels	4
Page must contain a level-one heading	1
Radio inputs with the same name attribute value must be part of a group	1

   Figure 9.22: Issues with the page reported by Axe

3. The first issue to solve is changing the second label in the Rating the product section. It and the first label are associated with the same form field. Changing it to associate the label with the form field with the "**rate-2**" ID makes sense:

```
<input type="radio" name="rating" value="1" id=
"rate-1">
       <label for="rate-1">1</label>
       <input type="radio" name="rating" value="2" id=
"rate-2">
   <label for="rate-2">2</label>
```

   If we run the Axe tool again, we will see that the first issue has been solved and that the **Form elements must have labels** issues have decremented by one.

4. The next issue is to fix the lack of alt text on the product image. For now, we will add an alt attribute with a "**Product**" value:

```
<img src="images/product.png" class="product-image"
   alt="Product">
```

5. Next, we will fix the third issue, which relates to label elements not being associated with form fields. We need to add the **for** attribute to the first two **label** elements to fix them:

```
                          <label for="first-name">First name:
   </label>
                          <input id="first-name" type="text" />
                          <br />
                          <label for="last-name">Last name:</label>
                          <input id="last-name" type="text"  />
```

We also need to change the fourth label of the Rating the product section, which has an incorrect value:

```
                          <input type="radio" name="rating"
   value="4" id="rate-4">
                          <label for="rate-4">4</label>
```

6. The penultimate issue is that the page does not have a top-level heading. To fix this, we can change the first heading with the text **Product Store** so that it's an **h1** element. This will improve the structure of the page:

```
          <h1>Product Store</h1>
```

7. Finally, we need to resolve an issue where the five radio inputs for rating the product need to be grouped properly. To do this, we can use a legend element in the second fieldset instead of a p element:

```
          <legend>Rating the product:</legend>
```

8. If we run the Axe tool again, we should see that all the issues have been fixed. We should see the following message from Axe:

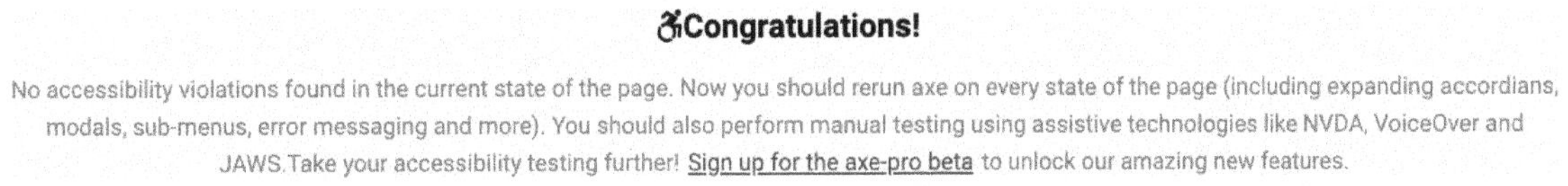

⚙Congratulations!

No accessibility violations found in the current state of the page. Now you should rerun axe on every state of the page (including expanding accordians, modals, sub-menus, error messaging and more). You should also perform manual testing using assistive technologies like NVDA, VoiceOver and JAWS.Take your accessibility testing further! Sign up for the axe-pro beta to unlock our amazing new features.

Figure 9.23: All issues fixed in Axe

Chapter 10: Preprocessors and Tooling

Activity 10.01: Converting the Video Store Home Page into SCSS

Solution

1. Get a copy of your code from *Chapter 6, Responsive Web Design and Media Queries,
    Activity 6.01, Refactoring the Video Store Product Cards into a Responsive Web Page,*
    and save it as a new file within your **Chapter10** project folder called **Activity
    10.01.html**. Remove the CSS code from within the **<style>** tags and save it as
    activity1.scss within your **SCSS** subfolder in the project. Add a link to the
    activity1.css file in your document from the CSS subfolder of your project.

    Activity 10.01.html can be found in https://packt.live/36YzkXZ.

    activity1.scss code can be found in https://packt.live/32FW5fY.

2. Update the SCSS to use **nesting** and **color** variables, keeping the media
    queries within the relevant tags, instead of at the end of file. Here's the updated
    activity1.scss file:

```scss
$primary-color: black;
$secondary-color: white;
$hover-color: red;
$bg-color: lightgray;
$saving-color: green;
$original-color: gray;
* {
    margin: 0;
    padding: 0;
}
body {
    font-family: sans-serif;
    margin: 0 auto;
}
header {
    text-align: center;
    @media (min-width: 768px) {
        align-items: center;
            display: flex;
    }
}
.product-cards {
  margin-bottom: 30px;
}
```

```scss
nav {
    margin-top: 20px;
    background: $primary-color;
    @media (min-width: 768px) {
        background: transparent;
        margin-top: 0;
        margin-left: 30px;
    }
    @media only print {
        display: none;
    }
    ul {
        display: flex;
        list-style: none;
        align-items: center;
        justify-content: center;
    }
    a {
        color: $secondary-color;
        font-weight: bold;
        display: block;
        padding: 15px;
        text-decoration: none;
        @media (min-width: 768px) {
            color: $primary-color;
        }
        &:hover {
            color: $hover-color;
            @media (min-width: 768px) {
                text-decoration: underline;
                color: $primary-color;
            }
        }
    }
}
.breadcrumb {
    display: flex;
    list-style: none;
    background: $bg-color;
    li {
        padding: 10px;
        &:after {
```

```scss
                        content: "/";
                    margin-left: 20px;
            }
            &:last-child:after {
              content: "";
            }
        }
        a {
            color: $primary-color;
              text-decoration: none;
              &:hover {
                  text-decoration: underline;
              }
        }
    }
.intro {
    margin: 30px 0;
    padding: 0 10px;
    @media (min-width: 768px) {
        width: 50%;
    }
    h1 {
        margin-bottom: 15px;
    }
    p {
        line-height: 1.5;
    }
}
.product-cards {
    @media (min-width: 768px) {
        display: grid;
        grid-template-columns: auto auto auto auto;
    }
}
.product-card {
    border: 1px solid $primary-color;
    padding: 15px;
    margin: 10px;
    @media (min-width: 768px) {
        display: inline-block;
    }
```

```scss
        a {
            color: $primary-color;
            text-decoration: none;
        }
        img {
            width: 100%;
        }
        h2 {
            margin: 30px 0 15px;
        }
        p {
            line-height: 1.5;
        }
    }
    .original-price {
      color: $original-color;
      text-transform: uppercase;
    }
    .current-price span {
      font-weight: bold;
      text-decoration: underline;
    }
    .saving {
      color: $saving-color;
    }
```

3. Edit your **package.json** in your project to have the SCSS script output in compressed format.

    package.json file:

```json
    {
        "name": "chapter10",
        "version": "1.0.0",
        "description": "HTML5 & CSS3 Workshop Chapter10 Exercises",
        "main": "index.js",
        "scripts": {
            "test": "echo \"Error: no test specified\" && exit 1",
            "scss": "node-sass --watch scss -o css --output-style compressed"
        },
```

```
    "author": "Matt Park",
    "license": "ISC",
    "dependencies": {
      "node-sass": "^4.12.0"
    }
  }
}
```

4. Using Terminal, compile your SCSS code into minified CSS code.

   Navigate to the project directory: **cd ~/Desktop/Chapter10**.

   Run the following command to run the compile script: **npm run scss**.

   Then, save the **activity1.scss** file to generate the **activity1.css** file.

5. The end result will be compressed (minified) CSS code, as shown in the succeeding snippet. The browser's visual output should be the same as in *Chapter 6, Responsive Web Design and Media Queries, Activity 6.01, Refactoring the Video Store Product Cards into a Responsive Web Page*, if done correctly. Check *Figure 10.14* for reference:

```
*{margin:0;padding:0}body{font-family:sans-serif;margin:0 auto}
header{text-align:center}@media (min-width: 768px){header{align-
items:center;display:flex}}.product-cards{margin-bottom:30px}nav{margin-
top:20px;background:#000}@media (min-width: 768px){nav{background:transpar
ent;margin-top:0;margin-left:30px}}@media only print{nav{display:none}}nav
ul{display:flex;list-style:none;align-items:center;justify-content:center}nav
a{color:#fff;font-weight:bold;display:block;padding:15px;text-decoration:none}@
media (min-width: 768px){nav a{color:#000}}nav a:hover{color:red}@media
(min-width: 768px){nav a:hover{text-decoration:underline;color:#000}}.
breadcrumb{display:flex;list-style:none;background:#d3d3d3}.breadcrumb
li{padding:10px}.breadcrumb li:after{content:"/";margin-left:20px}.breadcrumb
li:last-child:after{content:""}.breadcrumb a{color:#000;text-decoration:none}.
breadcrumb a:hover{text-decoration:underline}.intro{margin:30px 0;padding:0
10px}@media (min-width: 768px){.intro{width:50%}}.intro h1{margin-
bottom:15px}.intro p{line-height:1.5}@media (min-width: 768px){.product-
cards{display:grid;grid-template-columns:auto auto auto auto}}.product-
card{border:1px solid #000;padding:15px;margin:10px}@media (min-width:
768px){.product-card{display:inline-block}}.product-card a{color:#000;text-
decoration:none}.product-card img{width:100%}.product-card h2{margin:30px
0 15px}.product-card p{line-height:1.5}.original-price{color:gray;text-
transform:uppercase}.current-price span{font-weight:bold;text-
decoration:underline}.saving{color:green}
```

Chapter 11: Maintainable CSS

Activity 11.01: Making Our Video Store Web Page Maintainable

Solution

1. First, we'll take a copy of the *Chapter 10, Preprocessors and Tooling, Activity 10.01, Converting a Video Store Home Page into SCSS*, HTML file and save it in our **Chapter11** project folder under **Activity 11.01.html**. Next, we need to take a copy of the *Chapter 10, Preprocessors and Tooling, Activity 10.01, Converting a Video Store Home Page into SCSS*, SCSS file and save this in our **Chapter11** project folder, under the **scss/activity1.scss** path.

2. Edit the **Activity 11.01.html** file so that it uses a BEM semantic markup. This involves changing the existing CSS classes to follow the BEM structure; for example, **header** would become **.header** and **header img** would become **.header__logo**.

   Once the HTML file has been updated, the markup will look like this:

```html
<!DOCTYPE html>
<html lang = "en">
<head>
  <meta name="viewport" content="width=device-width,
    initial-scale=1" />
  <title>Video store home page</title>
  <link href="css/activity1.css" rel="stylesheet" />
</head>
<body>
  <header class="header">
    <img class="header__logo" src=
  "https://dummyimage.com/200x100/000/fff&text=Logo" alt="" />
    <nav class="nav">
      <ul class="nav__list">
        <li class="nav__item"><a class="nav__link" href="">Videos
          to rent</a></li>
        <li class="nav__item"><a class="nav__link" href="">Videos
          to buy</a></li>
        <li class="nav__item"><a class="nav__link" href="">Used
          DVDs</a></li>
        <li class="nav__item"><a class="nav__link" href="">
          Offers</a></li>
      </ul>
    </nav>
  </header>
  <section class="breadcrumb">
    <ol class="breadcrumb__list">
```

```
        <li class="breadcrumb__item"><a class="breadcrumb__link"
          href="">Home</a></li>
        <li class="breadcrumb__item"><a class="breadcrumb__link"
          href="">Used DVDs</a></li>
        <li class="breadcrumb__item">Less than £10</li>
      </ol>
  </section>
  <section class="intro">
    <h1 class="intro__title">Videos less than £10</h1>
    <p class="intro__paragraph">Lorem ipsum dolor sit amet,
      consectetur adipiscing elit. In bibendum non purus quis
      vestibulum. Pellentesque ultricies quam lacus, ut tristique
      sapien tristique et.</p>
  </section>
  <section class="products">
    <div class="products__card">
      <img class="products__image" src="https://dummyimage.
        com/300x300/7EC0EE/000&text=Product+Image+1" alt="" />
      <h2 class="products__title"><a class="products__link"
        href="">Video title 1</a></h2>
      <p class="products__price products__price--original">RRP:
        £18.99</p>
      <p class="products__price products__price--current">Price
        you pay <span>£9.99</span></p>
      <p class="products__saving">Your saving £9</p>
    </div>
    <div class="products__card">
      <img class="products__image" src="https://dummyimage.
        com/300x300/7EC0EE/000&text=Product+Image+2" alt="" />
      <h2 class="products__title"><a class="products__link"
        ref="">Video title 2</a></h2>
      <p class="products__price products__price--original">RRP:
        £18.99</p>
      <p class="products__price products__price--current">Price
        you pay <span>£9.99</span></p>
      <p class="products__saving">Your saving £9</p>
    </div>
    <div class="products__card">
      <img class="products__image" src="https://dummyimage.
        com/300x300/7EC0EE/000&text=Product+Image+3" alt="" />
      <h2 class="products__title"><a class="products__link"
        href="">Video title 3</a></h2>
      <p class="products__price products__price--original">RRP:
        £18.99</p>
      <p class="products__price products__price--current">Price
        you pay <span>£9.99</span></p>
      <p class="products__saving">Your saving £9</p>
    </div>
    <div class="products__card">
      <img class="products__image" src="https://dummyimage.
```

```
        com/300x300/7EC0EE/000&text=Product+Image+4" alt="" />
      <h2 class="products__title"><a class="products__link"
        href="">Video title 4</a></h2>
      <p class="products__price products__price--original">RRP:
        £18.99</p>
      <p class="products__price products__price--current">Price
        you pay <span>£9.99</span></p>
      <p class="products__saving">Your saving £9</p>
    </div>
  </section>
</body>
</html>
```

3. Then following the updates, we made to the **Activity 11.01.html** file, we'll
    need to edit the **activity1.scss** file to follow the same updated BEM semantic
    markup. In the following code, you've seen that we've now got a block for **.header**,
    .nav, **.breadcrumb**, **.intro** and **.products**, and in each of these new blocks, we
    have the elements directly related to these blocks, which are nested within.

    In the **.products__price** element, you can see that's we've added two modifiers
    for **.products__price--original** and **.products__price--current**. These
    modify the **.products_price** element to extend them with the right styles for the
    original and current prices.

    Once the updates have been done, the updated code will look similar to this:

```scss
$primary-color: black;
$secondary-color: white;
$hover-color: red;
$bg-color: lightgray;
$saving-color: green;
$original-color: gray;
* {
    margin: 0;
    padding: 0;
}
body {
    font-family: sans-serif;
    margin: 0 auto;
}
.header {
    text-align: center;
    @media (min-width: 768px) {
        align-items: center;
        display: flex;
    }
}
```

```scss
    }
.nav {
    margin-top: 20px;
    background: $primary-color;
    @media (min-width: 768px) {
        background: transparent;
        margin-top: 0;
        margin-left: 30px;
    }
    @media only print {
        display: none;
    }
    &__list {
        display: flex;
        list-style: none;
        align-items: center;
        justify-content: center;
    }
    &__link {
        color: $secondary-color;
        font-weight: bold;
        display: block;
        padding: 15px;
        text-decoration: none;
        @media (min-width: 768px) {
            color: $primary-color;
        }
        &:hover {
            color: $hover-color;
            @media (min-width: 768px) {
                text-decoration: underline;
                color: $primary-color;
            }
        }
    }
}
.breadcrumb {
    &__list {
        display: flex;
        list-style: none;
        background: $bg-color;
    }
```

```
    &__item {
        padding: 10px;
        &:after {
            content: "/";
            margin-left: 20px;
        }
        &:last-child:after {
            content: "";
        }
    }
    &__link {
        color: $primary-color;
        text-decoration: none;
        &:hover {
            text-decoration: underline;
        }
    }
}
.intro {
    margin: 30px 0;
    padding: 0 10px;
    @media (min-width: 768px) {
        width: 50%;
    }
    &__title {
        margin-bottom: 15px;
    }
    &__paragraph {
        line-height: 1.5;
    }
}
.products {
    margin-bottom: 30px;
    @media (min-width: 768px) {
        display: grid;
        grid-template-columns: auto auto auto auto;
    }
    &__card {
        border: 1px solid $primary-color;
        padding: 15px;
        margin: 10px;
        @media (min-width: 768px) {
```

```scss
        display: inline-block;
      }
    }
    &__link {
      color: $primary-color;
      text-decoration: none;
    }
    &__image {
      width: 100%;
    }
    &__title {
      margin: 30px 0 15px;
    }
    &__price,
    &__saving {
      line-height: 1.5;
    }
    &__price {
      &--original {
        color: $original-color;
        text-transform: uppercase;
      }
      &--current span {
        font-weight: bold;
        text-decoration: underline;
      }
    }
    &__saving {
      color: $saving-color;
    }
  }
```

4. Now that we've written the SCSS, we're going to split the **activity1.scss** file's SCSS code into suitable subfolders and files, and then import these into the **activity1.scss** file.

   Listed in the following code blocks are all the files you can separate **activity1. scss** into to provide a more maintainable structure in which this project could grow:

scss/activity1/abstracts/_variables.scss:

```scss
$primary-color: black;
$secondary-color: white;
$hover-color: red;
$bg-color: lightgray;
$saving-color: green;
$original-color: gray;
```

scss/activity1/base/_reset.scss:

```scss
* {
    margin: 0;
    padding: 0;
}
```

scss/activity1/base/_body.scss:

```scss
body {
    font-family: sans-serif;
    margin: 0 auto;
}
```

scss/activity1/components/_breadcrumb.scss:

```scss
.breadcrumb {
    &__list {
        display: flex;
        list-style: none;
        background: $bg-color;
    }
    &__item {
        padding: 10px;
        &:after {
            content: "/";
            margin-left: 20px;
        }
        &:last-child:after {
            content: "";
        }
    }
    &__link {
        color: $primary-color;
        text-decoration: none;
```

```scss
        &:hover {
            text-decoration: underline;
        }
    }
}
```

scss/activity1/components/_intro.scss:

```scss
.intro {
    margin: 30px 0;
    padding: 0 10px;
    @media (min-width: 768px) {
        width: 50%;
    }
    &__title {
        margin-bottom: 15px;
    }
    &__paragraph {
        line-height: 1.5;
    }
}
```

scss/activity1/layout/_header.scss:

```scss
.header {
    text-align: center;
    @media (min-width: 768px) {
        align-items: center;
        display: flex;
    }
}
```

scss/activity1/layout/_nav.scss:

```scss
.nav {
    margin-top: 20px;
    background: $primary-color;
    @media (min-width: 768px) {
        background: transparent;
        margin-top: 0;
        margin-left: 30px;
    }
    @media only print {
        display: none;
    }
}
```

```scss
    &__list {
        display: flex;
        list-style: none;
        align-items: center;
        justify-content: center;
    }
    &__link {
        color: $secondary-color;
        font-weight: bold;
        display: block;
        padding: 15px;
        text-decoration: none;
        @media (min-width: 768px) {
            color: $primary-color;
        }
        &:hover {
            color: $hover-color;
            @media (min-width: 768px) {
                text-decoration: underline;
                color: $primary-color;
            }
        }
    }
}
```

scss/activity1/layout/_products.scss:

```scss
.products {
    margin-bottom: 30px;
    @media (min-width: 768px) {
        display: grid;
        grid-template-columns: auto auto auto auto;
    }
    &__card {
        border: 1px solid $primary-color;
        padding: 15px;
        margin: 10px;
        @media (min-width: 768px) {
            display: inline-block;
        }
    }
    &__link {
        color: $primary-color;
```

```scss
        text-decoration: none;
    }
    &__image {
        width: 100%;
    }
    &__title {
        margin: 30px 0 15px;
    }
    &__price,
    &__saving {
        line-height: 1.5;
    }
    &__price {
        &--original {
            color: $original-color;
            text-transform: uppercase;
        }
        &--current span {
            font-weight: bold;
            text-decoration: underline;
        }
    }
    &__saving {
        color: $saving-color;
    }
}
```

scss/activity1.scss:

```scss
@import 'activity1/abstracts/_variables';
@import 'activity1/base/_reset';
@import 'activity1/base/_body';
@import 'activity1/components/_breadcrumb';
@import 'activity1/components/_intro';
@import 'activity1/layout/_header';
@import 'activity1/layout/_nav';
@import 'activity1/layout/_products';
```

As you can see, in **scss/activity1.scss**, we have imported all the files we created from splitting up the file originally. It's important that the variables are in the first file that's imported as this will be used by the files after it.

5. Now that we've written the SCSS, it's time to compile it in your Terminal to generate the **css/activity1.css** file.

   First, navigate to the project directory, **cd ~/Desktop/Chapter11**.

   Next, run the following command to run the compile script: **npm run scss**.

   Finally, save the **scss/activity1.scss** file to generate the **css/activity1.css** file.

6. The last step is to test the web page in your browser to ensure it's loading as expected. Check this against *Figure 12.10* by opening **Activity 11.01.html** in your web browser.

Chapter 12: Web Components

Activity 12.01: Creating a Profile

Solution

The following is the solution to this activity:

1. We will start with the template that we used to create the *Films on Demand page*, which we created in *Activity 1.01, Video Store Page Template, Chapter 1, Introduction to HTML and CSS*. Copy and paste the code into a new file called **Activity 12.01.html**. We can set the title of the document as Profile:

    ```
    <title>Films on Demand - Profile</title>
    ```

2. For the content of the element with the ID attribute **pageWrapper**, we will add a custom element, which we will call **fod-profile**.

    ```
    <div id="pageWrapper" class="full-page">
        <fod-profile>
        </fod-profile>
    </div>
    ```

3. At the end of the HTML document, we can add a template element that will hold the HTML structure for our **fod-profile** component. It will have slots for the avatar, name, and email. These slots contain content that we can replace for each instance of the component:

    ```
    <template id="fod-profile-template">
        <section class="profile-container">
            <div class="profile-body">
                <div class="avatar"><slot name="avatar"><img
                  src="media/placeholder.png"></slot>
                </div>
                <div class="info">
                    <p class="name"><span class="attribute">Name:
                      </span> <slot name="name">John
                      Doe</slot></p>

                    <p class="email"><span class="attribute">
                      E-mail:</span>
    <slot name="email">john.doe@gmail.com
                      </slot></p>
                </div>
            </div>
        </section>
    </template>
    ```

4. We then style the custom element by providing the following styles for the template:

```html
<style>
    .profile-container {
        display: flex;
        padding: 16px;
    }
    .profile-body {
        display: flex;
        margin: 16px;
        background-color: #fff;
        border-radius: 3px;
        box-shadow: 0 4px 12px rgba(0, 0, 0, 0.6);
    }
    .avatar {
        line-height: 0;
        margin: 8px;
    }
    .info {
        padding: 8px 16px 8px 8px;
    }
    .attribute {
        font-weight: bold;
    }
</style>
```

5. Now that we've created our template, we need to use it in the **fod-profile** custom element, which we can do with the script for defining our **fod-profile** custom element:

```
<script>
    customElements.define("fod-profile", class FODProfile
      extends
      HTMLElement {
        constructor() {
            super();
            this.attachShadow({ mode: "open" });
        }
        connectedCallback() {
            const tmpl = document.getElementById
              ("fod-profile-template");
            const node = document.importNode
              (tmpl.content, true);
            this.shadowRoot.appendChild(node);
        }
    });
</script>
```

6. If you now right-click on the filename in VSCode on the left-hand side of the screen and select **Open In Default Browser**, you will see the output in a browser.

Chapter 13: Future of HTML and CSS

Activity 13.01: Button Library

Solution

1. After creating the **Activity 13.01.html** file with the markup, the first task is to add styles for the primary button. We do this by adding a class selector (**.button--primary**) to our style rules and adding the appropriate styles (background and text color) for the default and hover states (with a **:hover** pseudo-class):

```
.button--primary {
    background: #f44336;
    color: white;
    font-weight: bold;
}
.button--primary:hover {
    background: #ff9900;
}
```

2. Next, we do the same for the secondary button. We add styling for the default and **hover** states. The only difference is the colors and styling that's applied:

```
.button--secondary {
    background: #9e9e9e;
    color: white;
    font-weight: normal;
}
.button--secondary:hover {
    background: #7c7c7c;
}
```

3. We repeat this again for the ghost button. The main difference for this button is that we have a white background and have set a darker border color to define the outline of the button:

```
.button--ghost {
    border-color: #9e9e9e;
    background: white;
    color: #9e9e9e;
    font-weight: normal;
}
.button--ghost:hover {
    background: #efefef;
}
```

4. Next, we add the styles for the special button. This button requires some different techniques to provide an animated gradient. We will add a :root selector to initialize the custom properties for --primary-fill-color and --secondary-fill-color. These will be the colors of the gradient and will change during the animation:

```
:root {
    --primary-fill-color: #f44336;
    --secondary-fill-color: #ff9900;
}
```

5. We will then add our style rules for the special button. First, we'll add the default state, which will include a transition for both of the custom properties we defined in the previous step:

```
.button--special {
    background: #f44336;
    color: white;
    font-weight: bold;
    transition: --primary-fill-color 0.5s,
      --secondary-fill-color 0.7s;
}
```

6. The hover state for the special button will set the background color and also set the color values for our two custom properties (**--primary-fill-color** and **--secondary-fill-color**). When we hover over the button, these values, which will be used in a paint worklet to draw the gradient, will be changed and the change in state will be animated:

```
.button--special:hover {
    background-color: #ff9900;
    --secondary-fill-color: #f44336;
    --primary-fill-color: #ff9900;
}
```

7. For the color properties to be animated, we need to register them using JavaScript:

```
if ("registerProperty" in CSS) {
    CSS.registerProperty({
        name: "--primary-fill-color",
        syntax: "<color>",
        inherits: false,
        initialValue: "#f44336"
    });
    CSS.registerProperty({
        name: "--secondary-fill-color",
        syntax: "<color>",
        inherits: false,
        initialValue: "#ff9900"
    });
}
```

8. We also need to create a paint worklet to draw the gradient. We will add this to the web page using the JavaScript **CSS.paintWorklet.addModule** method:

```
if ("paintWorklet" in CSS) {
    CSS.paintWorklet.addModule
        ("gradient-paint.js");
}
```

9. We also need to apply the paint worklet to our special button:

```
.button--special {
    background: #f44336 paint(animated-gradient);
```

10. The complete **GradientPaintWorklet** in **gradient-paint.js** is as follows:

```
class GradientPaintWorklet {
    static get inputProperties() {
        return ["--primary-fill-color", "--secondary-fill-color"];
    }
    paint(context, geometry, properties) {
        const primaryColor = properties.get
          ("--primary-fill-color");
        const secondaryColor =
          properties.get("--secondary-fill-color");
        const gradient = context.createLinearGradient
          (0, 0, geometry.
          width, geometry.height);
        gradient.addColorStop(0, primaryColor);
        gradient.addColorStop(1.0, secondaryColor);

        context.fillStyle = gradient;
        context.fillRect(0, 0, geometry.width, geometry.height);
    }
}
registerPaint("animated-gradient", GradientPaintWorklet);
```

11. If you now right-click on the filename in VSCode on the left-hand side of the screen and select **Open In Default Browser**, you will see the following screenshot that shows the expected final result:

Figure 13.32: The four buttons when styled

Index

About

All major keywords used in this book are captured alphabetically in this section. Each one is accompanied by the page number of where they appear.